T0156013

UNITEXT for Physics

UNITEXT for Physics series, formerly UNITEXT Collana di Fisica e Astronomia, publishes textbooks and monographs in Physics and Astronomy, mainly in English language, characterized of a didactic style and comprehensiveness. The books published in UNITEXT for Physics series are addressed to graduate and advanced graduate students, but also to scientists and researchers as important resources for their education, knowledge and teaching.

More information about this series at http://www.springer.com/series/13351

Alessandro Teta

A Mathematical Primer
on Quantum Mechanics

 Springer

Alessandro Teta
Dipartimento di Matematica Guido
    Castelnuovo
Università degli Studi di Roma "La
    Sapienza"
Rome
Italy

ISSN 2198-7882   ISSN 2198-7890   (electronic)
UNITEXT for Physics
ISBN 978-3-030-08566-7   ISBN 978-3-319-77893-8   (eBook)
https://doi.org/10.1007/978-3-319-77893-8

Printed on acid-free paper

This Springer imprint is published by the registered company Springer International Publishing AG
part of Springer Nature
The registered company address is: Gewerbestrasse 11, 6330 Cham, Switzerland

To Ade

Preface

This book is an elaboration of lecture notes written during the preparation of courses for mathematics students given for some years at the Universities of L'Aquila and "La Sapienza" of Roma. The aim of the book is to provide an elementary introduction to nonrelativistic Quantum Mechanics for a single particle from the point of view of Mathematical Physics. It is mainly addressed to students in Mathematics at master's degree level, but it could also be useful for students in Physics interested in a deeper understanding of the mathematical aspects of the theory.

No previous physical knowledge of Quantum Mechanics is required and the mathematical language necessary to develop theory and applications is explicitly introduced. The prerequisites are basic notions of Classical Physics, elementary theory of Hilbert spaces, Fourier transform, and elements of real and complex analysis.

The first four chapters of the book are aimed to provide some physical and mathematical background. In Chap. 1, the basic notions of Hamiltonian Mechanics and Electromagnetism are recalled, with an emphasis on some aspects relevant for understanding the origin and the development of Quantum Mechanics. In Chap. 2, the historical evolution of the ideas in Physics from Planck's hypothesis to Bohr's model of the hydrogen atom is briefly described. In Chap. 3, we introduce the Schrödinger equation, discuss Born's statistical interpretation, and outline a first sketch of the new theory. It is also stressed the need of more advanced notions from the theory of linear operators in Hilbert spaces to obtain a rigorous and consistent description of the theory. Such notions are introduced in Chap. 4, where we describe the main concepts, i.e., self-adjoint and unitary operators, resolvent and spectrum, spectral theorem (stated without proof), different classifications of the spectrum, all supported by examples and exercises.

In Chap. 5, an axiomatic and rigorous formulation of nonrelativistic Quantum Mechanics is given. We avoid generality and neglect some technical difficulties. The aim is to provide the theoretical instruments to approach elementary problems of Quantum Mechanics for a single particle. We also briefly mention some interpretational problems of the theory connected with the measurement process.

The applications of the formalism are described in Chaps. 6–9. The emphasis is on a detailed analysis of simple models as free particle, harmonic oscillator, point interaction, and hydrogen atom. For each model, we construct the self-adjoint Hamiltonian, characterize the spectrum, and discuss the main dynamical properties. These models are of crucial importance to understand the qualitative behavior of a quantum particle in some relevant physical situations. We believe that a deep understanding of their physical and mathematical properties is an unavoidable prerequisite to deal with more difficult problems in Quantum Mechanics.

Some more advanced topics are also treated starting from the concrete analysis of the models. In Chap. 9, elements of spectral analysis for Schrödinger operators are introduced to study the spectrum of the Hamiltonian of the hydrogen atom. In Appendix A, the semiclassical evolution of a Gaussian state is described using the results obtained for the harmonic oscillator in Chap. 7. In Appendix B, the basic concepts of scattering theory are discussed and then used to revisit the simple one-dimensional scattering problem analyzed in Chap. 8 for a point interaction.

Many exercises are proposed in the text and their solution is crucial for a good understanding of the subject.

After studying this book, the reader should have gained the essential background to approach the analysis of advanced mathematical problems in Quantum Mechanics. Among the many excellent textbooks on the subject appeared in the literature in recent times, we mention

Strocchi, F.: An Introduction to the Mathematical Structure of Quantum Mechanics. World Scientific (2008).
Faddeev, L. D., Yakubovskii, O. A.: Lectures on Quantum Mechanics for Mathematics Students. AMS (2009).
Teschl, G.: Mathematical Methods in Quantum Mechanics. AMS (2009).
Dimock, J.: Quantum Mechanics and Quantum Field Theory. Cambridge University Press (2011).
Gustafson, S. J., Sigal, I. M.: Mathematical Concepts of Quantum Mechanics. Springer (2011).
Hall, B. C.: Quantum Theory for Mathematicians. Springer (2013).
Dell'Antonio, G.: Lectures on the Mathematics of Quantum Mechanics. Atlantis Press (2015).

This book could not have been written without the precious collaboration of many friends, colleagues, and students. They have helped me to correct many errors and to considerably improve the presentation. I warmly thank all of them: G. Dell'Antonio, R. Figari, A. Posilicano, A. Sacchetti, D. Noja, P. Buttà, R. Adami, D. Finco, R. Carlone, M. Correggi, G. Panati, C. Cacciapuoti, S. Cenatiempo, L. Tentarelli, D. Dimonte, E. Giacomelli, G. Basti, and M. Olivieri.

Rome, Italy																			Alessandro Teta
February 2018

Contents

Chapter 1
Elements of Hamiltonian Mechanics and Electromagnetism

1.1 Hamilton's Equations

The Hamiltonian form of Newton's law of dynamics is particularly relevant for the formulation of Quantum Mechanics. Here, we derive Hamilton's equations for a mechanical system starting from the Lagrangian description. Let us consider a mechanical system with n degrees of freedom described by the generalized coordinates $q = (q_1, \ldots, q_n) \in D$, where D is a domain of $\mathbb{R}^n$ called configuration space of the system. Let $\mathscr{L}(q, \eta, t)$ be the Lagrangian, where $\eta = (\eta_1, \ldots, \eta_n) \in \mathbb{R}^n$ and $t \in \mathbb{R}$. We assume that $\mathscr{L}$ is regular, i.e., twice differentiable and such that

$$\det \left(\frac{\partial^2 \mathscr{L}}{\partial \eta_h \partial \eta_k}(q, \eta, t) \right) \neq 0 \tag{1.1}$$

for $(q, \eta, t) \in D \times \mathbb{R}^n \times \mathbb{R}$. The motion of the system $t \to q(t) = (q_1(t), \ldots, q_n(t))$ is the solution of Lagrange's equations

$$\frac{d}{dt} \frac{\partial \mathscr{L}}{\partial \eta_k}(q(t), \dot{q}(t), t) = \frac{\partial \mathscr{L}}{\partial q_k}(q(t), \dot{q}(t), t), \qquad k = 1, \ldots, n. \tag{1.2}$$

Note that in (1.2) the derivatives of the Lagrangian are evaluated along the motion, i.e., for $q = q(t)$ and $\eta = \dot{q}(t)$. Lagrange's equations are n second-order differential equations and the condition (1.1) guarantees that they can be written in normal form. Therefore, they have a unique solution satisfying the initial conditions $(q(0), \dot{q}(0))$.

Example 1.1 Newton's equations for a system of N point particles in $\mathbb{R}^d, d = 1, 2, 3$, are

$$m_i \ddot{\underline{x}}_i = -\nabla_{\underline{x}_i} U(\underline{x}_1, \ldots, \underline{x}_N), \qquad i = 1, \ldots, N, \tag{1.3}$$

where m_i denotes the mass and $\underline{x}_i \in \mathbb{R}^d$ the position of the i-th particle and U is the potential energy. Let $\underline{v}_i \in \mathbb{R}^d$ be the velocity of the i-th particle. Consider an arbitrary change of coordinates

$$\underline{x}_i = \underline{x}_i(q), \qquad i = 1, \ldots, N, \tag{1.4}$$

© Springer International Publishing AG, part of Springer Nature 2018
A. Teta, *A Mathematical Primer on Quantum Mechanics*,
UNITEXT for Physics, https://doi.org/10.1007/978-3-319-77893-8_1

where $q = (q_1, \ldots, q_n)$, $n = Nd$, with inverse denoted by $q = f(\underline{x}_1, \ldots, \underline{x}_N)$. Define $\eta = (\eta_1, \ldots, \eta_n)$ by

$$\underline{v}_i := \sum_{k=1}^n \frac{\partial \underline{x}_i}{\partial q_k}(q)\eta_k, \qquad i = 1, \ldots, N, \tag{1.5}$$

and the Lagrangian

$$\mathscr{L}(q, \eta) := \frac{1}{2}\sum_{i=1}^N m_i |\underline{v}_i(q, \eta)|^2 - U(\underline{x}_1(q), \ldots, \underline{x}_N(q)) = \frac{1}{2}\sum_{h,k=1}^n a_{hk}(q)\eta_h\eta_k - V(q),$$

$$\tag{1.6}$$

where

$$V(q) := U(\underline{x}_1(q), \ldots, \underline{x}_N(q)), \qquad a_{hk}(q) := \left(\sum_{i=1}^N m_i \frac{\partial \underline{x}_i}{\partial q_h}(q) \cdot \frac{\partial \underline{x}_i}{\partial q_k}(q)\right)^2. \tag{1.7}$$

One can verify that
(i) $a_{hk}(q)$ is a symmetric, positive definite matrix, so that condition (1.1) holds.
(ii) If $t \to \underline{x}_i(t)$, $i = 1, \ldots, N$, is a solution of Newton's equations (1.3) then $t \to q(t) := f(\underline{x}_1(t), \ldots, \underline{x}_N(t))$ is a solution of Lagrange's equations (1.2) with Lagrangian (1.6).

Vice versa, if $t \to q(t) = (q_1(t), \ldots, q_n(t))$ is a solution of Lagrange's equations (1.2) with Lagrangian (1.6) then $t \to \underline{x}_i(t) := \underline{x}_i(q(t))$, $i = 1, \ldots, N$, is a solution of Newton's equations (1.3).

Observe that Eq. (1.2) can be equivalently written as a system of $2n$ equations of first order in the unknowns (q, η)

$$\dot{q}_k(t) = \eta_k(t), \qquad \frac{d}{dt}\frac{\partial \mathscr{L}}{\partial \eta_k}(q(t), \eta(t), t) = \frac{\partial \mathscr{L}}{\partial q_k}(q(t), \eta(t), t). \tag{1.8}$$

Hamilton's equations of motion are a new system of first-order equations obtained from (1.8) by replacing η with the (conjugate) momenta $p = (p_1 \ldots, p_n)$, where

$$p_k := \frac{\partial \mathscr{L}}{\partial \eta_k}(q, \eta, t), \tag{1.9}$$

such that the vector field is written in a peculiar symmetric form and the variables (q, p) play essentially the same role.

Due to condition (1.1), we first solve (1.9) for the η_k as functions of q, p, t

$$\eta = \alpha(q, p, t), \qquad \alpha(q, p, t) = (\alpha_1(q, p, t), \ldots, \alpha_n(q, p, t)). \tag{1.10}$$

Using (1.9) and (1.10) in (1.8), we obtain the following $2n$ equations of first order in the variables q, p

$$\dot{q}_k(t) = \alpha_k(q(t), p(t), t), \qquad \dot{p}_k(t) = \frac{\partial \mathscr{L}}{\partial q_k}(q(t), \alpha(q(t), p(t), t), t). \quad (1.11)$$

Let us now define the Hamiltonian function of the system as the Legendre transform of the Lagrangian, i.e.,

$$H(q, p, t) := \sum_{h=1}^{n} \alpha_h(q, p, t) p_h - \mathscr{L}(q, \alpha(q, p, t), t) \qquad (1.12)$$

and compute the derivatives

$$\frac{\partial H}{\partial q_k} = \sum_{h=1}^{n} \frac{\partial \alpha_h}{\partial q_k} p_h - \frac{\partial \mathscr{L}}{\partial q_k} - \sum_{h=1}^{n} \frac{\partial \mathscr{L}}{\partial \eta_h} \frac{\partial \alpha_h}{\partial q_k} = -\frac{\partial \mathscr{L}}{\partial q_k}, \qquad (1.13)$$

$$\frac{\partial H}{\partial p_k} = \sum_{h=1}^{n} \frac{\partial \alpha_h}{\partial p_k} p_h + \alpha_k - \sum_{h=1}^{n} \frac{\partial \mathscr{L}}{\partial \eta_h} \frac{\partial \alpha_h}{\partial p_k} = \alpha_k. \qquad (1.14)$$

Using (1.13) and (1.14) in (1.11) we finally obtain Hamilton's equations

$$\dot{q}_k(t) = \frac{\partial H}{\partial p_k}(q(t), p(t), t), \quad \dot{p}_k(t) = -\frac{\partial H}{\partial q_k}(q(t), p(t), t), \quad k = 1, \ldots, n. \qquad (1.15)$$

Equation (1.15) are $2n$ equations of first order in the variables $(q, p) \in D \times \mathbb{R}^n := \Gamma$, where Γ is called phase space of the system. The solution $t \to (q(t), p(t))$ is therefore a curve in the phase space Γ. Thus, we have proved the following proposition.

Proposition 1.1 *Let us assume that $\mathscr{L}(q, \eta, t)$ is a regular Lagrangian and $t \to q(t)$ is the solution of the corresponding Lagrange's equations with initial conditions $(q_0, \dot{q}_0)$. Then $t \to (q(t), p(t))$, with $p_i(t) = (\partial \mathscr{L}/\partial \eta_i)(q(t), \dot{q}(t), t)$, $i = 1, \ldots, n$, is the solution of Hamilton's equations (1.15) with Hamiltonian (1.12) and initial conditions (q_0, p_0), where $p_{0i} = (\partial \mathscr{L}/\partial \eta_i)(q(0), \dot{q}(0), 0)$.*

It is not difficult to prove that the converse is also true. Therefore, one obtains that Hamilton's equations are equivalent to Lagrange's equations.

A systems of equations of the form (1.15) is called Hamiltonian system. In the following, we shall recall some of their relevant properties (for further details we refer to [1, 3–5, 7–9, 11, 13]).

If we denote $z = (q, p)$, Hamilton's equations can also be written in the more compact form

$$\dot{z} = F_H(z, t), \qquad F_H := \left(\frac{\partial H}{\partial p_1}, \ldots, \frac{\partial H}{\partial p_n}, -\frac{\partial H}{\partial q_1}, \ldots, -\frac{\partial H}{\partial q_n} \right), \qquad (1.16)$$

where F_H is called Hamiltonian vector field. We assume that $H(q, p, t)$ is sufficiently regular in order to guarantee existence and uniqueness of the solution with given initial conditions.

Moreover, we recall that the generalized energy for a Lagrangian system is

$$\mathscr{E}(q, \eta, t) = \sum_{i=1}^{n} \frac{\partial \mathscr{L}}{\partial \eta_i}(q, \eta, t)\eta_i - \mathscr{L}(q, \eta, t) \tag{1.17}$$

and therefore we have

$$H(q, p, t) = \mathscr{E}(q, \alpha(q, p), t), \tag{1.18}$$

i.e., the Hamiltonian is the generalized energy in the variables (q, p) and it is a constant of motion in the autonomous case.

Hamilton's equations can be derived from a variational principle, with the action functional defined on a set of paths in phase space. More precisely, we denote $z = (q, p) \in \Gamma$ and $w = (\dot{q}, \dot{p}) \in \mathbb{R}^{2n}$, we fix $T > 0$ and the points $z^{(1)}, z^{(2)} \in \Gamma$. Then, we define the set of paths

$$M = \left\{ \gamma \mid \gamma : [0, T] \to \Gamma, \ \gamma \in C^2, \ \gamma(0) = z^{(1)}, \ \gamma(T) = z^{(2)} \right\}, \tag{1.19}$$

the Lagrangian

$$L(z, w, t) = \sum_{k=1}^{n} z_{n+k} w_k - H(z, t) = \sum_{k=1}^{n} p_k \dot{q}_k - H(p, q, t) \tag{1.20}$$

and the action functional $S : M \to \mathbb{R}$, where

$$S(\gamma) = \int_0^T dt \, L(\gamma(t), \dot{\gamma}(t), t) = \int_0^T dt \left[\sum_{k=1}^{n} p_k(t)\dot{q}_k(t) - H(p(t), q(t), t) \right]. \tag{1.21}$$

We have

Proposition 1.2 *Let $\gamma \in M$. Then γ is a critical point of S if and only if $t \to \gamma(t) = (q(t), p(t))$ is a solution of Hamilton's equations with Hamiltonian H.*

Proof We know that γ is a critical point of S if and only if $t \to \gamma(t)$ is a solution of the problem

$$\frac{d}{dt} \frac{\partial L}{\partial w_k}(\gamma(t), \dot{\gamma}(t), t) = \frac{\partial L}{\partial z_k}(\gamma(t), \dot{\gamma}(t), t), \qquad k = 1, \ldots, 2n, \tag{1.22}$$

$$\gamma(0) = z^{(1)}, \qquad \gamma(T) = z^{(2)}, \tag{1.23}$$

where (1.22) are called Euler–Lagrange equations associated to S. Equation (1.22) can be rewritten as

$$\frac{d}{dt}\frac{\partial L}{\partial \dot{q}_k} = \frac{\partial L}{\partial q_k} \qquad \text{if} \quad k = 1, \ldots, n, \qquad (1.24)$$

$$\frac{d}{dt}\frac{\partial L}{\partial \dot{p}_k} = \frac{\partial L}{\partial p_k} \qquad \text{if} \quad k = n+1, \ldots, 2n. \qquad (1.25)$$

From the definition of L, we have

$$\frac{\partial L}{\partial \dot{q}_k} = p_k, \qquad \frac{\partial L}{\partial q_k} = -\frac{\partial H}{\partial q_k}, \qquad \frac{\partial L}{\partial \dot{p}_k} = 0, \qquad \frac{\partial L}{\partial p_k} = \dot{q}_k - \frac{\partial H}{\partial p_k}. \qquad (1.26)$$

Using these equations in (1.24) and (1.25) we obtain Hamilton's equations. $\quad\square$

Remark 1.1 For any fixed $z^{(1)}$, there is only one solution of Hamilton's equations $t \to \gamma(t)$ such that $\gamma(0) = z^{(1)}$. Therefore, if we fix $z^{(2)} \neq \gamma(T)$, there is no solution connecting the points $z^{(1)}$ and $z^{(2)}$. This means that a critical point of S does not exist for an arbitrary choice of $z^{(1)}$ and $z^{(2)}$, but only for $z^{(2)} = \gamma(T)$, where $t \to \gamma(t)$ is the unique solution of Hamilton's equations with initial condition $\gamma(0) = z^{(1)}$.

Let us derive the explicit form of the Hamiltonian for a system of point particles subject to conservative forces. In this case, the Lagrangian is given by (1.6) and the momenta are

$$p_i = \frac{\partial \mathscr{L}}{\partial \eta_i}(q, \eta) = \sum_{k=1}^{n} a_{ik}(q)\eta_k \qquad (1.27)$$

and

$$\eta_l = \sum_{j=1}^{n} a_{lj}^{-1}(q)\, p_j, \qquad (1.28)$$

where $a_{lj}^{-1}(q)$ is the inverse of $a_{lj}(q)$. Then, using (1.12), we find

$$H(q, p) = \frac{1}{2} \sum_{l,j=1}^{n} a_{lj}^{-1}(q)\, p_l p_j + V(q). \qquad (1.29)$$

Example 1.2 For a point particle of mass m in Cartesian coordinates, we have

$$H(x_1, x_2, x_3, p_1, p_2, p_3) = \frac{p_1^2}{2m} + \frac{p_2^2}{2m} + \frac{p_3^2}{2m} + V(x_1, x_2, x_3), \qquad (1.30)$$

while in spherical coordinates

$$H(r, \theta, \phi, p_r, p_\theta, p_\phi) = \frac{p_r^2}{2m} + \frac{p_\theta^2}{2mr^2} + \frac{p_\phi^2}{2mr^2 \sin^2\theta} + V(r, \theta, \phi). \qquad (1.31)$$

Let us consider a charged point particle of mass m and charge e subject to given electric $E(q, t)$ and magnetic $H(q, t)$[1] fields. Newton's law is

$$m \ddot{q} = F_L(q, \dot{q}, t), \qquad F_L(q, \dot{q}, t) = eE(q, t) + \frac{e}{c} \dot{q} \wedge H(q, t), \quad (1.32)$$

where F_L denotes the Lorentz force and c is the speed of light in vacuum. By a direct computation, one verifies that (1.32) are the Lagrange equations with Lagrangian given by

$$\mathcal{L}_{em}(q, \eta, t) = \frac{1}{2} m |\eta|^2 + \frac{e}{c} \eta \cdot A(q, t) - e V(q, t), \qquad (1.33)$$

where $A(q, t)$ and $V(q, t)$ are the vector and scalar potentials respectively, with

$$E = -\nabla V - \frac{\partial A}{\partial t}, \qquad H = \nabla \wedge A. \qquad (1.34)$$

By (1.33), we find the momentum

$$p = m \eta + \frac{e}{c} A(q, t) \qquad (1.35)$$

and therefore the Hamiltonian reads

$$H_{em}(p, q, t) = \frac{1}{2m} \left(p - \frac{e}{c} A(q, t)\right)^2 + e V(q, t). \qquad (1.36)$$

1.2 Constants of Motion and Poisson Brackets

We recall the notion of constant of motion in the Hamiltonian formalism.

Definition 1.1 The function $G(q, p, t)$ is called a constant of motion for the Hamiltonian system (1.15) if

$$\frac{d}{dt} G(q(t), p(t), t) = 0 \qquad (1.37)$$

for any solution $t \rightarrow (q(t), p(t))$ of (1.15).

Note that (1.37) means that the function G, evaluated along the solutions, is a constant, i.e., $G(q(t), p(t)) = G(q(0), p(0))$ for any t.

Condition (1.37) can be written in an equivalent form using the Poisson bracket.

---

[1]Unfortunately, the symbols denoting the Hamiltonian function and the magnetic field coincide. The distinction between the two will be clear from the context.

Definition 1.2 Given two differentiable functions $F(q, p, t)$, $G(q, p, t)$, the Poisson bracket of F and G is the function

$$\{F, G\} = \sum_{k=1}^{n} \left(\frac{\partial F}{\partial q_k} \frac{\partial G}{\partial p_k} - \frac{\partial F}{\partial p_k} \frac{\partial G}{\partial q_k} \right). \tag{1.38}$$

The following proposition holds.

Proposition 1.3 *The function $G(q, p, t)$ is a constant of motion for the system* (1.15) *if and only if*

$$\{G, H\} + \frac{\partial G}{\partial t} = 0. \tag{1.39}$$

In particular, if G does not depend on time then Eq. (1.39) *reduces to*

$$\{G, H\} = 0. \tag{1.40}$$

Proof The claim follows from

$$\frac{d}{dt} G(q(t), p(t), t) = \sum_{k=1}^{n} \left(\frac{\partial G}{\partial q_k} \dot{q}_k + \frac{\partial G}{\partial p_k} \dot{p}_k \right) + \frac{\partial G}{\partial t}$$

$$= \sum_{k=1}^{n} \left(\frac{\partial G}{\partial q_k} \frac{\partial H}{\partial p_k} - \frac{\partial G}{\partial p_k} \frac{\partial H}{\partial q_k} \right) + \frac{\partial G}{\partial t} = \{G, H\} + \frac{\partial G}{\partial t}. \tag{1.41}$$

$\square$

Exercise 1.1 Verify that the following properties hold.
Linearity:

$$\{\lambda F_1 + \mu F_2, G\} = \lambda\{F_1, G\} + \mu\{F_2, G\} \qquad \lambda, \mu \in \mathbb{R}. \tag{1.42}$$

Antisymmetry:
$$\{F, G\} = -\{G, F\}. \tag{1.43}$$

Jacobi identity:

$$\{F, \{G, L\}\} + \{L, \{F, G\}\} + \{G, \{L, F\}\} = 0. \tag{1.44}$$

Leibniz rule:
$$\{F, GL\} = \{F, G\}L + G\{F, L\}. \tag{1.45}$$

Fundamental brackets:

$$\{q_i, p_j\} = \delta_{ij} \quad \{q_i, q_j\} = \{p_i, p_j\} = 0. \tag{1.46}$$

Using the above properties, we prove the following proposition.

Proposition 1.4 *Consider a Hamiltonian system with a time-independent Hamiltonian $H(q, p)$. Then,*

(i) *H is a constant of motion.*
(ii) *If F, G are constants of motion, so is $\{F, G\}$.*
(iii) *Hamilton's equations can be written*

$$\dot{q}_i = \{q_i, H\} \quad , \quad \dot{p}_i = \{p_i, H\}. \tag{1.47}$$

Proof Point (i) is a consequence of antisymmetry. If F, G are constants of motion then $\{F, H\} = 0 = \{G, H\}$. Using Jacobi identity $\{F, \{G, H\}\} + \{H, \{F, G\}\} + \{G, \{H, F\}\} = 0$, we have $\{H, \{F, G\}\} = 0$ and therefore point (ii) is proved. For point (iii) we observe that

$$\{q_i, H\} = \sum_{k=1}^{n} \left(\frac{\partial q_i}{\partial q_k} \frac{\partial H}{\partial p_k} - \frac{\partial q_i}{\partial p_k} \frac{\partial H}{\partial q_k} \right) = \frac{\partial H}{\partial p_i} \tag{1.48}$$

and analogously $\{p_i, H\} = -\partial H/\partial q_i$. □

Remark 1.2 Poisson brackets play an important role in the Hamiltonian formalism, in particular in the study of canonical transformations, of symmetries and, more generally, of the algebraic structure of Classical Mechanics (see e.g. [11]).

1.3 Canonical Transformations

A canonical transformation is a local change of coordinates (i.e., a differentiable and invertible map) in phase space preserving the form of Hamilton's equations, possibly with a new Hamiltonian. In the following, we recall the precise definition and some important properties.

Definition 1.3 A change of coordinates in phase space

$$\begin{cases} Q = Q(q, p, t) \\ P = P(q, p, t) \end{cases} \text{ with } \text{ inverse } \begin{cases} q = q(Q, P, t) \\ p = p(Q, P, t) \end{cases} \tag{1.49}$$

is canonical if the following holds: for any Hamiltonian $H(q, p, t)$ there exists a function $K(Q, P, t)$ such that if $t \to (q(t), p(t))$ is any solution of (1.15) then $t \to (Q(t), P(t)) := (Q(q(t), p(t), t), P(q(t), p(t), t))$ is a solution of

$$\dot{Q}_i = \frac{\partial K}{\partial P_i}, \qquad \dot{P}_i = -\frac{\partial K}{\partial Q_i}, \qquad i = 1, \dots, n. \tag{1.50}$$

The transformation is called completely canonical if it is independent of time and $K(Q, P, t) = H(q(Q, P), p(Q, P), t)$.

Exercise 1.2 Let $Q_i = Q_i(q_1, \dots, q_n)$, $i = 1, \dots, n$, be an arbitrary change of Lagrangian coordinates and let $q_i = q_i(Q_1, \dots, Q_n)$ be the inverse. Verify that

$$Q_i = Q_i(q_1, \dots, q_n), \qquad P_i = \sum_{j=1}^{n} \frac{\partial q_j}{\partial Q_i}\bigg|_{Q=Q(q)} p_j \tag{1.51}$$

is completely canonical.

Exercise 1.3 Verify that $Q = -p$, $P = q$ is completely canonical.

Exercise 1.4 Verify that $Q = p$, $P = q$ is canonical and the new Hamiltonian is $K = -H$.

Exercise 1.5 Verify that

$$P = \frac{p^2 + q^2}{2}, \qquad Q = \tan^{-1}\frac{q}{p} \tag{1.52}$$

is completely canonical.

Exercise 1.6 Verify that $P = pq^2$, $Q = qp^2$ is not canonical.
   (Hint: consider the Hamiltonian system with Hamiltonian $H = p^2$).

Completely canonical transformations can be characterized in terms of Poisson brackets.

Proposition 1.5 *A change of coordinates in phase space* $Q_i = Q_i(q, p)$, $P_i = P_i(q, p)$, $i = 1, \dots, n$, *is completely canonical if and only if*

$$\{Q_i, P_j\} = \delta_{ij}, \qquad \{Q_i, Q_j\} = \{P_i, P_j\} = 0. \tag{1.53}$$

Proof Consider the case $n = 1$. Let us define

$$H'(Q, P) = H(q(Q, P), p(Q, P)) \tag{1.54}$$

and then $H(q, p) = H'(Q(q, p), P(q, p))$. We have, along a solution,

$$
\begin{aligned}
\dot{Q} &= \frac{\partial Q}{\partial q}\dot{q} + \frac{\partial Q}{\partial p}\dot{p} = \frac{\partial Q}{\partial q}\frac{\partial H}{\partial p} - \frac{\partial Q}{\partial p}\frac{\partial H}{\partial q} \\
&= \frac{\partial Q}{\partial q}\left(\frac{\partial H'}{\partial Q}\frac{\partial Q}{\partial p} + \frac{\partial H'}{\partial P}\frac{\partial P}{\partial p}\right) - \frac{\partial Q}{\partial p}\left(\frac{\partial H'}{\partial Q}\frac{\partial Q}{\partial q} + \frac{\partial H'}{\partial P}\frac{\partial P}{\partial q}\right) \\
&= \frac{\partial H'}{\partial Q}\left(\frac{\partial Q}{\partial q}\frac{\partial Q}{\partial p} - \frac{\partial Q}{\partial p}\frac{\partial Q}{\partial q}\right) + \frac{\partial H'}{\partial P}\left(\frac{\partial Q}{\partial q}\frac{\partial P}{\partial p} - \frac{\partial Q}{\partial p}\frac{\partial P}{\partial q}\right) \\
&= \frac{\partial H'}{\partial Q}\{Q, Q\} + \frac{\partial H'}{\partial P}\{Q, P\}.
\end{aligned}
\tag{1.55}
$$

Analogously

$$
\dot{P} = -\frac{\partial H'}{\partial Q}\{Q, P\} + \frac{\partial H'}{\partial P}\{P, P\}.
\tag{1.56}
$$

Taking into account of (1.53), (1.55) and (1.56), the proposition is proved for $n = 1$. For $n > 1$ the proof is similar and it is left as an exercise. $\square$

Exercise 1.7 Determine $\beta, \gamma \in \mathbb{R}$ such that

$$
\begin{aligned}
q_1 &= \frac{1}{\sqrt{2}}(Q_1 + Q_2), & p_1 &= \frac{1}{\sqrt{2}}(P_1 + \gamma P_2), \\
q_2 &= \frac{1}{\sqrt{2}}(Q_1 - Q_2), & p_2 &= \frac{1}{\sqrt{2}}(P_1 + \beta P_2)
\end{aligned}
\tag{1.57}
$$

is completely canonical.

Another useful characterization of a canonical transformation can be given using differential forms. To explain this fact, we introduce the extended phase space

$$
\tilde{\Gamma} = \Gamma \times \mathbb{R}
\tag{1.58}
$$

and denote a point in this space by $\tilde{z} = (q, p, t)$, where $(q, p) \in \Gamma$ and $t \in \mathbb{R}$. A curve in $\tilde{\Gamma}$ is a smooth map $\lambda \to \tilde{z}(\lambda) = (q(\lambda), p(\lambda), t(\lambda))$. Notice that the system of equations in $\tilde{\Gamma}$

$$
\frac{d\tilde{z}}{d\lambda} = \tilde{F}_H(\tilde{z}), \qquad \tilde{F}_H(\tilde{z}) := \left(\frac{\partial H}{\partial p}, -\frac{\partial H}{\partial q}, 1\right)
\tag{1.59}
$$

is equivalent to Hamilton's equations in Γ, since $t(\lambda) = \lambda + c$, with c constant.

Let us define the differential form in $\tilde{\Gamma}$

$$\pi = \sum_{i=1}^{n} p_i dq_i - H dt \,. \tag{1.60}$$

We have

Proposition 1.6 *A change of coordinates in phase space* $Q = Q(q, p, t)$, $P = P(q, p, t)$ *is canonical if and only if for any Hamiltonian* $H(q, p, t)$ *there exist* $c \neq 0$ *and two differentiable functions* $G(q, p, t)$ *and* $K(Q, P, t)$ *such that*

$$\sum_{i=1}^{n} p_i dq_i - H dt = c \left(\sum_{i=1}^{n} P_i dQ_i - K dt \right) + dG \,. \tag{1.61}$$

Proof We only prove that if (1.61) holds then the transformation is canonical and K is the new Hamiltonian (for the complete proof see e.g. [5, 9]).

Let $\hat{\gamma}(t) = (\hat{q}(t), \hat{p}(t))$ be a solution of Hamilton's equations with Hamiltonian H. Then, by Proposition 1.2, $\hat{\gamma}$ is a critical point of the functional

$$S(\gamma) = \int_{0}^{T} dt\, (p(t)\dot{q}(t) - H(p(t), q(t), t)) = \int_{\gamma} \pi, \tag{1.62}$$

where $\gamma(t) = (q(t), p(t))$ is a path in M (see (1.19)). Using (1.61), we have

$$S(\gamma) = c \int_{0}^{T} dt\, \left(P(t)\dot{Q}(t) - K(P(t), Q(t), t) \right) + G(z^{(2)}, T) - G(z^{(1)}, 0)$$
$$:= c\, S'(\gamma') + G(z^{(2)}, T) - G(z^{(1)}, 0), \tag{1.63}$$

where the functional S' is defined on

$$M' = \left\{ \gamma' \mid \gamma'(t) = (Q(\gamma(t), t), P(\gamma(t), t)) := (Q(t), P(t)),\ t \in [0, T],\ \gamma \in M \right\}. \tag{1.64}$$

Let us also denote $\hat{\gamma}'(t) = (Q(\hat{\gamma}(t), t), P(\hat{\gamma}(t), t))$. From (1.63) we have that if $\hat{\gamma}$ is a critical point of S in M then $\hat{\gamma}'$ is a critical point of S' in M' and therefore $\hat{\gamma}'(t)$ is a solution of Hamilton's equations with Hamiltonian K. $\qquad\square$

Exercise 1.8 Verify that the transformation

$$Q = pf(t), \qquad P = \frac{q}{f(t)} \tag{1.65}$$

with f differentiable and different from zero, is canonical.

The next proposition shows how to construct canonical transformations.

Proposition 1.7 *Let $F(x, y, t)$ be a twice differentiable function such that*

$$det\left(\frac{\partial^2 F}{\partial x_i \partial y_j}(x, y, t)\right) \neq 0. \tag{1.66}$$

For $i = 1, \ldots, n$, denote $x_i = q_i$, $y_i = P_i$ and define

$$p_i = \frac{\partial F}{\partial q_i}(q, P, t), \tag{1.67}$$

$$Q_i = \frac{\partial F}{\partial P_i}(q, P, t). \tag{1.68}$$

Then, Eqs. (1.67) and (1.68) define a canonical transformation $Q = Q(q, p, t)$, $P = P(q, p, t)$ and the Hamiltonian in the new coordinates (Q, P) is

$$K(Q, P, t) = H(q(Q, P, t), p(Q, P, t), t) + \frac{\partial F}{\partial t}(q(Q, P, t), P, t). \tag{1.69}$$

Proof Due to condition (1.66), from (1.67), we can solve for $P = P(q, p, t)$ and then, using this last equation in (1.68), we find $Q = Q(q, p, t)$. Analogously, we can obtain the inverse $p = p(Q, P, t)$, $q = q(Q, P, t)$. Therefore, (1.67) and (1.68) define a change of coordinates in phase space.

Let us verify that the transformation is canonical. Let $H(q, p, t)$ be a Hamiltonian and denote $\hat{H}(P, q, t) = H(p(q, P, t), q, t)$, where $p(q, P, t)$ is given by (1.67). Then

$$\sum_{i=1}^{n} p_i dq_i - \hat{H}dt = \sum_{i=1}^{n} \frac{\partial F}{\partial q_i}dq_i - \hat{H}dt = dF - \sum_{i=1}^{n} \frac{\partial F}{\partial P_i}dP_i - \frac{\partial F}{\partial t}dt - \hat{H}dt$$

$$= dF - \sum_{i=1}^{n} Q_i dP_i - \left(\hat{H} + \frac{\partial F}{\partial t}\right)dt$$

$$= \sum_{i=1}^{n} P_i dQ_i - \left(\hat{H} + \frac{\partial F}{\partial t}\right)dt + d\left(F - \sum_{i=1}^{n} P_i Q_i\right). \tag{1.70}$$

Therefore, by Proposition 1.6, the transformation is canonical and the new Hamiltonian is (1.69).   □

The construction of the canonical transformation starting from (1.67) and (1.68) is called the method of second type and $F(q, P, t)$ is called the generating function of second type. With different choices of old and new coordinates as arguments of the function F, one obtains other types of generating functions.

Exercise 1.9 Given $F(x, y, t)$ satisfying (1.66) and denoted $x_i = q_i$, $y_i = Q_i$, verify that the equations

$$p_i = \frac{\partial F}{\partial q_i}(q, Q, t), \qquad P_i = -\frac{\partial F}{\partial Q_i}(q, Q, t) \tag{1.71}$$

for $i = 1, \ldots, n$, define a canonical transformation (method of first type).

Exercise 1.10 Construct the canonical transformations generated by

$$F(q, P) = \sum_{i=1}^{n} q_i P_i, \qquad F(q, P) = \sum_{i=1}^{n} f_i(q_1, \ldots, q_n) P_i, \qquad F(q, Q) = \sum_{i=1}^{n} q_i Q_i. \tag{1.72}$$

Exercise 1.11 Find a generating function $F(q, P)$ of the completely canonical transformation

$$q = \sqrt{P}\, e^{Q/2}, \qquad p = 2\sqrt{P}\, e^{-Q/2}. \tag{1.73}$$

Exercise 1.12 Using the generating function

$$F(x, y, P_x, P_y) = x P_x + y P_y - \frac{P_x P_y}{a}, \tag{1.74}$$

where a is an arbitrary positive parameter, solve the equation of motion for a point charge moving in the xy plane and subject to a uniform magnetic field described by the vector potential $A = (0, Bx, 0)$, $B > 0$.

1.4 Hamilton–Jacobi Equation

Hamilton–Jacobi equation is at the basis of the so-called Hamilton–Jacobi method for the solution of Hamilton's equations. It is relevant in Mechanics, e.g., for the analysis of integrable systems and their perturbations [1, 7], and also in many other contexts. In particular, we shall encounter the equation in the study of the analogy between Mechanics and Optics (Sects. 1.6 and 1.8), the short wavelength limit of Wave Optics (Sect. 1.10), the Sommerfeld approach to the Old Quantum Theory (Sect. 2.6), the WKB method for the classical limit of Quantum Mechanics (Sect. A.4).

The idea of the method is to find a suitable generating function of a canonical transformation $Q = Q(q, p, t)$, $P = P(q, p, t)$) such that in the new coordinates (Q, P) the new Hamiltonian K is identically zero, and therefore the solution of the corresponding Hamilton's equations is trivial. Taking into account of (1.69), the problem is reduced to find an F solution of

$$H(q, \nabla F, t) + \frac{\partial F}{\partial t} = 0. \tag{1.75}$$

Equation (1.75) is called Hamilton–Jacobi equation. Notice that, for any given $H(q, p, t)$, (1.75) is a partial differential equation of first order for the unknown function $F(q, t)$.

Definition 1.4 A complete integral of (1.75) is a twice differentiable solution $F(q, \alpha, t)$, depending on n real parameters $\alpha = (\alpha_1, \ldots, \alpha_n)$ and such that

$$\det \left(\frac{\partial^2 F}{\partial q_i \partial \alpha_j}(q, \alpha, t) \right) \neq 0. \tag{1.76}$$

Given a complete integral, one can construct the solution of Hamilton's equations.

Theorem 1.1 *(Jacobi) Let $F(q, \alpha, t)$ be a complete integral of the Hamilton–Jacobi equation (1.75). Then the equations*

$$\beta_i = \frac{\partial F}{\partial \alpha_i}(q, \alpha, t), \tag{1.77}$$

$$p_i = \frac{\partial F}{\partial q_i}(q, \alpha, t), \tag{1.78}$$

for $i = 1, \ldots, n$, define $2n$ functions $q = q(\alpha, \beta, t)$, $p = p(\alpha, \beta, t)$ which are solutions of Hamilton's equations with Hamiltonian $H(q, p, t)$ for any value of the constants α, β. Moreover, assigned the initial conditions (q_0, p_0), there exist (and are unique) the constants $\bar{\alpha}, \bar{\beta}$ such that $q(\bar{\alpha}, \bar{\beta}, t), p(\bar{\alpha}, \bar{\beta}, t)$ are the solutions of Hamilton's equations with initial conditions (q_0, p_0).

Proof Given the complete integral $F(q, \alpha, t)$, we identify the constants α with the new momenta so that $F(q, \alpha, t)$ is a generating function of second type and (1.77), (1.78) define the corresponding canonical transformation $q = q(\alpha, \beta, t)$, $p = p(\alpha, \beta, t)$. Since F solves Eq. (1.75), the new Hamiltonian is identically zero and the solution of Hamilton's equations in the new coordinates (α, β) is trivial

$$\alpha(t) = \alpha(0), \qquad \beta(t) = \beta(0), \tag{1.79}$$

with $\alpha(0)$ e $\beta(0)$ arbitrary constants. Then, the solution of Hamilton's equations in the original coordinates (q, p) is $q(t) = q(\alpha(0), \beta(0), t), p(t) = p(\alpha(0), \beta(0), t)$.

Let (q_0, p_0) be assigned initial conditions. Due to (1.76), we can solve (1.77) and (1.78) for $\beta = \beta(q, p, t), \alpha = \alpha(q, p, t)$. Defining $\bar{\beta} := \beta(q_0, p_0, 0)$ and $\bar{\alpha} := \alpha(q_0, p_0, 0)$, the solution of Hamilton's equations with initial conditions (q_0, p_0) is $q(\bar{\alpha}, \bar{\beta}, t), p(\bar{\alpha}, \bar{\beta}, t)$ and this concludes the proof. $\square$

Exercise 1.13 Using Hamilton–Jacobi method, solve the equations of motion for a system with Hamiltonian $H(q, p, t) = t^2(p^2 + 6q)$ and initial conditions $q(0) = 0$, $p(0) = 1$.

(Hint: look for a solution in the form $F(q, t) = f(q) + g(t)$.)

When the Hamiltonian is time independent, a variation of the Hamilton–Jacobi method can be used. Let us consider the time independent Hamilton–Jacobi equation

$$H(q, \nabla W) = \alpha_n, \tag{1.80}$$

where $H(q, p)$ is a given time-independent Hamiltonian and $\alpha_n := E$ is the energy of the system. In analogy with the time-dependent case, we look for a complete integral of Eq. (1.80), i.e., a twice differentiable solution $W(q, \alpha)$ depending on n real parameters $\alpha = (\alpha_1, \ldots, \alpha_n)$ and such that

$$\det \left(\frac{\partial^2 W}{\partial q_i \partial \alpha_j}(q, \alpha) \right) \neq 0. \tag{1.81}$$

Given such a solution, we write

$$p_i = \frac{\partial W}{\partial q_i}(q, \alpha), \qquad \beta_i = \frac{\partial W}{\partial \alpha_i}(q, \alpha), \qquad i = 1, \ldots, n \tag{1.82}$$

and construct the completely canonical transformation generated by $W(q, \alpha)$

$$q = q(\alpha, \beta), \qquad p = p(\alpha, \beta). \tag{1.83}$$

By (1.80), the Hamiltonian in the coordinates (α, β) is

$$H(q(\alpha, \beta), p(\alpha, \beta)) = \alpha_n \tag{1.84}$$

so that the solution of Hamilton's equations is

$$\alpha_i(t) = \alpha_i(0), \qquad \beta_i(t) = \delta_{in} t + \beta_i(0), \qquad i = 1, \ldots, n. \tag{1.85}$$

The solution in the coordinates (q, p) is obtained replacing (1.85) in (1.83)

$$\begin{aligned} q(t) &= q(\alpha(0), \beta_1(0), \ldots, \beta_{n-1}(0), t + \beta_n(0)), \\ p(t) &= p(\alpha(0), \beta_1(0), \ldots, \beta_{n-1}(0), t + \beta_n(0)) \end{aligned} \tag{1.86}$$

and the constants $\alpha(0), \beta(0)$ are determined imposing the initial conditions.

We note that in the case of a time-independent Hamiltonian, a complete integral F of (1.75) is given by $F = W - \alpha_n t$, where W is a complete integral of (1.80).

Exercise 1.14 Using Hamilton–Jacobi method, solve the equations of motion for a system with Hamiltonian $H(q, p) = p \tan q$ and initial conditions $q(0) = \frac{\pi}{4}$, $p(0) = 1$.

1.5 Separation of Variables

By the method of separation of variables, one can find the solution of Hamilton–Jacobi equation for a certain class of Hamiltonians. Let us consider the simpler case of a time-independent Hamiltonian $H(q, p)$.

Definition 1.5 The Hamiltonian $H(q, p)$ is called separable if we can find a complete integral of Eq. (1.80) in the form

$$W(q, \alpha) = \sum_{k=1}^{n} W_k(q_k, \alpha), \tag{1.87}$$

where each W_k is the solution of a first-order ordinary differential equation.

In the following, we shall consider the particular but relevant case where the equation for W_k has the form

$$\left(\frac{dW_k}{dq_k}\right)^2 + U_k(q_k, \alpha) = c_k(\alpha), \qquad k = 1, \dots, n, \tag{1.88}$$

for some functions $U_k(q_k, \alpha)$ and $c_k(\alpha)$. The solution of (1.88) is then explicitly found by quadratures

$$W_k^{\pm}(q_k, \alpha) = \pm \int_{q_k^0}^{q_k} dq' \sqrt{c_k(\alpha) - U_k(q', \alpha)}, \qquad k = 1, \dots, n, \tag{1.89}$$

where $q_k^0, k = 1, \dots, n$, are arbitrary constants.

Remark 1.3
(i) The separability of a Hamiltonian depends on the system of coordinates. A list of separable Hamiltonians in various coordinates can be found in [13], Chap. 7.
(ii) We note that, from (1.88), the constants α must satisfy the condition

$$\inf_q U_k(q, \alpha) < c_k(\alpha). \tag{1.90}$$

(iii) Taking into account that $p_k = \partial W / \partial q_k$, in (1.89) we choose the sign $+$ to generate a completely canonical transformation in a neighborhood of a point (q, p) with $p_k > 0$ and the sign $-$ otherwise.

Example 1.3 Consider the Hamiltonian of the planar Kepler problem

$$H = \frac{p_r^2}{2m} + \frac{p_\phi^2}{2mr^2} - \frac{k}{r}, \qquad k > 0, \tag{1.91}$$

where $V(r) = -k/r$ is the Newtonian potential. The corresponding time-independent Hamilton–Jacobi equation is

$$\frac{1}{2m}\left(\frac{\partial W}{\partial r}\right)^2 + \frac{1}{2mr^2}\left(\frac{\partial W}{\partial \phi}\right)^2 - \frac{k}{r} = \alpha_2 . \tag{1.92}$$

We look for a solution in the form $W(r, \phi) = W_1(\phi) + W_2(r)$. Then

$$\frac{1}{2m}\left(\frac{dW_2}{dr}\right)^2 + \frac{1}{2mr^2}\left(\frac{dW_1}{d\phi}\right)^2 - \frac{k}{r} = \alpha_2 , \tag{1.93}$$

which can be rewritten as

$$\frac{1}{2m}\left(\frac{dW_1}{d\phi}\right)^2 = -\frac{r^2}{2m}\left(\frac{dW_2}{dr}\right)^2 + kr + \alpha_2 r^2 . \tag{1.94}$$

Since the r.h.s. is a function of r and the l.h.s. is a function of ϕ the above equation holds if and only if both r.h.s and l.h.s. are equal to the same constant α_1. Therefore, we obtain two first-order differential equations of the form (1.88)

$$\frac{1}{2m}\left(\frac{dW_1}{d\phi}\right)^2 = \alpha_1 , \tag{1.95}$$

$$\frac{1}{2m}\left(\frac{dW_2}{dr}\right)^2 + \frac{\alpha_1}{r^2} - \frac{k}{r} = \alpha_2 , \tag{1.96}$$

which can be easily solved by quadratures. Moreover, the separation constants must satisfy

$$\alpha_1 > 0, \qquad \alpha_2 > \inf_r \left(\frac{\alpha_1}{r^2} - \frac{k}{r}\right) = -\frac{k^2}{4\alpha_1} . \tag{1.97}$$

Exercise 1.15 Find a complete integral of the Hamilton–Jacobi equation associated to the Hamiltonian

$$H(x, \theta, p_x, p_\theta) = \frac{p_x^2}{2(1+x^2)} + \frac{p_\theta^2}{2}\frac{1+x^2}{1+\sin^2\theta} + \frac{1+x^2}{2}(1+\sin^2\theta), \tag{1.98}$$

where $x \in \mathbb{R}, \theta \in [-\frac{\pi}{2}, \frac{\pi}{2})$.

Under suitable conditions, a useful set of canonical coordinates, named action-angle variables, can be introduced. They are relevant in the analysis of integrable systems and their perturbations [1, 7]. In our context, action-angle variables are interesting since they were used in the formulation of the Bohr–Sommerfeld quantization rule in the Old Quantum Theory (Sect. 2.6).

We describe the construction in the special case of the separable Hamiltonians considered above. We first note that, since $p_k = \partial W/\partial q_k$, Eq. (1.88) can be rewritten as

$$p_k^2 + U_k(q_k, \alpha) = c_k(\alpha), \qquad k = 1, \ldots, n . \tag{1.99}$$

For each k, this equation can be interpreted as the conservation of energy for a one-dimensional motion of a point particle of mass $1/2$ subject to the potential $U_k(q_k, \alpha)$, where $c_k(\alpha)$ denotes the value of the energy. Furthermore, Eq. (1.99) defines a curve $\gamma_k(\alpha)$ in the plane (q_k, p_k).

Action-angle variables can be introduced if $\gamma_k(\alpha), k = 1, \ldots, n$ are periodic orbits of the one-dimensional motion described by (1.99). We distinguish two cases.

(1) Libration.
In this case $\gamma_k(\alpha)$ is a closed and regular curve, symmetric with respect to the q_k-axis. Therefore, there exist two simple zeroes q_k^0, q_k^1 of the function $c_k(\alpha) - U_k(q_k, \alpha)$ and

$$\gamma_k(\alpha) = \gamma_k^+(\alpha) \cup \gamma_k^-(\alpha), \tag{1.100}$$

$$\gamma_k^\pm(\alpha) = \left\{ (q_k, p_k) \mid q_k^0 \leq q_k \leq q_k^1, \quad p_k = \pm\sqrt{c_k(\alpha) - U_k(q_k, \alpha)} \right\}. \tag{1.101}$$

(2) Rotation.
This case occurs when there exists $\lambda > 0$ such that q_k and $q_k + m\lambda, m \in \mathbb{Z}$, correspond to the same configuration of the one-dimensional motion described by (1.99) and the curve $\gamma_k(\alpha)$ is the graph of a regular function $p_k = p_k(q_k, \alpha)$, periodic with period λ in the variable q_k.

Example 1.4 The typical case where periodic motions of libration and rotation are both present is the simple pendulum of mass m and length l, whose Hamiltonian is

$$H = \frac{p_\theta^2}{2ml^2} - mgl\cos\theta. \tag{1.102}$$

Denoting by α the energy, we have libration if $-mgl < \alpha < mgl$ and rotation if $\alpha > mgl$. Note that the values $\alpha = \pm mgl$ must be excluded. For $\alpha = -mgl$, the only possible motion is the stable equilibrium position and for $\alpha = mgl$, the possible motions are the unstable equilibrium position and the two motions asymptotic to the unstable equilibrium position.

In the next definition, we introduce new constants of motion in presence of libration or rotation.

Definition 1.6 Assume that (1.99) describes a periodic motion. In the case of libration, the action variable J_k is defined by

$$J_k(\alpha) = \frac{1}{\pi} \int_{q_k^0}^{q_k^1} dq_k \sqrt{c_k(\alpha) - U_k(q_k, \alpha)}, \tag{1.103}$$

while in the case of rotation it is defined by

$$J_k(\alpha) = \frac{1}{2\pi} \int_{\bar{q}_k}^{\bar{q}_k + \lambda} dq_k\, p_k(q_k, \alpha), \tag{1.104}$$

where $(\bar{q}_k, \bar{q}_k + \lambda)$ is any interval of periodicity of the function $p_k(q_k, \alpha)$.

Note that $2\pi J_k(\alpha)$ is the area inside the closed orbit in the case of libration and the area under one cycle of the orbit in the case of rotation.

In the next proposition, we characterize the action-angle variables.

Proposition 1.8 *Assume that Eq. (1.99) describe periodic motions of libration or rotation for any $k = 1, \ldots, n$. Then,*
(i) The map $\alpha \rightarrow J(\alpha)$ is invertible.
(ii) Given the generating function of second type

$$\hat{W}(q, J) = \sum_{i=1}^{n} \hat{W}_i(q_i, J) = \sum_{i=1}^{n} W_i(q_i, \alpha_1(J), \ldots, \alpha_n(J)), \qquad (1.105)$$

the equations

$$\phi_k = \frac{\partial \hat{W}}{\partial J_k}(q, J), \qquad p_k = \frac{\partial \hat{W}}{\partial q_k}(q, J), \qquad k = 1, \ldots, n, \qquad (1.106)$$

define a completely canonical transformation $q = q(\phi, J)$, $p = p(\phi, J)$ such that the new momenta J_k are constants of motion and the new positions ϕ_k are angles, i.e., they satisfy

$$\oint_{\gamma_l} d\phi_k = 2\pi \delta_{lk}. \qquad (1.107)$$

(iii) The Hamiltonian in the new variables is $H(q(\phi, J), p(\phi, J)) = \alpha_n(J)$. Then, the solution of Hamilton's equations

$$\dot{J}_k = 0, \qquad \dot{\phi}_k = \frac{\partial \alpha_n}{\partial J_k}(J) \qquad (1.108)$$

is

$$J_k(t) = J_k^0, \qquad \phi_k(t) = \omega_k t + \phi_k^0, \qquad (1.109)$$

where J_k^0, ϕ_k^0 are arbitrary constants and

$$\omega_k := \frac{\partial \alpha_n}{\partial J_k}(J^0) \qquad (1.110)$$

are called the frequencies of the system.

Proof Concerning (i), it is sufficient to show that

$$\det \left(\frac{\partial J_k}{\partial \alpha_l}(\alpha) \right) \neq 0. \qquad (1.111)$$

We shall limit to the case of libration and, for notational simplicity, we only consider the case $n = 2$ (the extension to the general case is not difficult). We have

$$
\begin{aligned}
\det\!\left(\frac{\partial J_k}{\partial \alpha_l}(\alpha)\right) &= \det\!\left(\frac{1}{\pi}\frac{\partial}{\partial \alpha_l}\int_{q_k^0}^{q_k^1} dq_k\,\frac{\partial W_k^+}{\partial q_k}(q_k,\alpha)\right) = \det\!\left(\frac{1}{\pi}\int_{q_k^0}^{q_k^1} dq_k\,\frac{\partial^2 W_k^+}{\partial \alpha_l \partial q_k}(q_k,\alpha)\right) \\
&= \frac{1}{\pi^2}\int_{q_1^0}^{q_1^1} dq_1 \int_{q_2^0}^{q_2^1} dq_2\,\Bigg(\frac{\partial^2 W_1^+}{\partial \alpha_1 \partial q_1}(q_1,\alpha)\,\frac{\partial^2 W_2^+}{\partial \alpha_2 \partial q_2}(q_2,\alpha) \\
&\qquad\qquad\qquad\qquad\qquad -\frac{\partial^2 W_1^+}{\partial \alpha_2 \partial q_1}(q_1,\alpha)\,\frac{\partial^2 W_2^+}{\partial \alpha_1 \partial q_2}(q_2,\alpha)\Bigg) \\
&= \frac{1}{\pi^2}\int_{q_1^0}^{q_1^1} dq_1 \int_{q_2^0}^{q_2^1} dq_2\,\det\!\left(\frac{\partial^2 W}{\partial q_k \partial \alpha_l}(q,\alpha)\right),
\end{aligned}
\tag{1.112}
$$

where the last integral is different from zero by definition of complete integral. This proves (1.111) and then $J(\alpha)$ is invertible.

Let us consider (ii) Using the functions $\alpha_k = \alpha_k(J)$, we define $\hat{W}(q,J)$ as in (1.105). It is easy to verify that

$$
\det\left(\frac{\partial^2 \hat{W}}{\partial q_k \partial J_l}(q,J)\right) \neq 0
\tag{1.113}
$$

and therefore $\hat{W}(q,J)$ is a generating function of second type and (1.106) define a completely canonical transformation $(q,p) \to (\phi,J)$.

In order to prove (1.107), we first show that $\hat{W}(q,J)$, as a function, is only locally defined. Note that

$$
\hat{W}(q,J) = \sum_{i=1}^{n} \hat{W}_i(q,J) = \sum_{i=1}^{n} \int_{q_i^0}^{q_i} dq'\, p_i(q',\alpha(J)),
\tag{1.114}
$$

where the integration is along the closed and regular curve $\hat{\gamma}_i(J) = \gamma_i(\alpha(J))$. The variation of $\hat{W}(q,J)$ when the variables J and q_i, with $i \neq l$, are fixed and the variable q_l goes through one complete cycle $\hat{\gamma}_l(J)$ is

$$
\Delta_l \hat{W} = \oint_{\hat{\gamma}_l(J)} dq'\, p_l(q',\alpha(J)) = 2\pi J_l.
\tag{1.115}
$$

Therefore $\hat{W}(q,J)$ is multivalued. Let us now consider the variation of ϕ_k when the variables J and q_i, with $i \neq l$, are fixed and the variable q_l goes through one complete cycle $\hat{\gamma}_l(J)$

$$\Delta_l \phi_k = \oint_{\hat{\gamma}_l(J)} d\phi_k = \oint_{\hat{\gamma}_l(J)} d\left(\frac{\partial \hat{W}}{\partial J_k}(q, J)\right) = \oint_{\hat{\gamma}_l(J)} d\left(\frac{\partial \hat{W}_l}{\partial J_k}(q_l, J)\right)$$

$$= \oint_{\hat{\gamma}_l(J)} \frac{\partial^2 \hat{W}_l}{\partial q_l \partial J_k}(q_l, J) dq_l = \oint_{\hat{\gamma}_l(J)} \frac{\partial}{\partial J_k} p_l(q_l, \alpha(J)) dq_l$$

$$= \frac{\partial}{\partial J_k} \oint_{\hat{\gamma}_l(J)} p_l(q_l, \alpha(J)) dq_l = 2\pi \frac{\partial J_l}{\partial J_k} = 2\pi \, \delta_{lk} \,. \tag{1.116}$$

Point (iii) is easy to verify and then the proof is complete.   □

Remark 1.4 The solution of Hamilton's equations in the original variables (q, p) is written in the form

$$q = q(\omega_1 t + \phi_1^0, \ldots, \omega_n t + \phi_n^0, J_1^0, \ldots, J_n^0), \quad p = p(\omega_1 t + \phi_1^0, \ldots, \omega_n t + \phi_n^0, J_1^0, \ldots, J_n^0) \tag{1.117}$$

i.e., it is a multiperiodic function of the time with frequencies $\omega_1, \ldots, \omega_n$.

Remark 1.5 An important generalization of the previous proposition is the theorem of Arnold–Liouville (see, e.g., [1, 7]).

Exercise 1.16 Consider a harmonic oscillator, i.e., a point particle with mass m in dimension one subject to the potential $V(x) = \frac{1}{2}m\omega^2 x^2$, where $\omega > 0$ denotes the frequency of the oscillator. Verify that the action-angle variables for the harmonic oscillator are

$$J = \frac{1}{\omega}\left(\frac{p^2}{2m} + \frac{1}{2}m\omega^2 q^2\right), \qquad \phi = \tan^{-1}\left(m\omega \frac{q}{p}\right). \tag{1.118}$$

Exercise 1.17 Verify that the action variables for the planar Kepler problem (see Example 1.3) are

$$J_1 = \sqrt{2m\alpha_1}, \qquad J_2 = -\sqrt{2m\alpha_1} + k\sqrt{\frac{m}{2|\alpha_2|}} \tag{1.119}$$

and the Hamiltonian written in the variables J_1, J_2 is

$$\alpha_2(J_1, J_2) = -\frac{m k^2}{2(J_1 + J_2)^2}. \tag{1.120}$$

1.6 Wave Front Method

In this section, we discuss another approach for solving Hamilton's equations starting from a suitable solution of Hamilton–Jacobi equation. Such a method is particularly useful to show the formal analogy between mechanics and geometrical optics.

We describe the method in the simple case of a point particle of mass m moving under the action of a force with potential energy $V(x)$. Using Cartesian coordinates, the Hamiltonian takes the form (1.30). Let us also fix the initial condition x_0, p_0, with $p_0 \neq 0$, and denote E the corresponding energy. The time-independent Hamilton–Jacobi equation can be written as

$$|\nabla W|^2 = \hat{n}^2(x), \qquad \hat{n}(x) := \sqrt{2m(E - V(x))}. \tag{1.121}$$

Note that, by energy conservation, we have

$$\hat{n}^2(x_0) = |p_0|^2. \tag{1.122}$$

Let us suppose that $W(x, \alpha_1, \alpha_2)$ is a solution of (1.121) depending on two real parameters α_1, α_2 in such a way that the matrix

$$\left(\frac{\partial^2 W}{\partial x_i \partial \alpha_j}(x, \alpha_1, \alpha_2) \right) \tag{1.123}$$

with $i = 1, 2, 3$, $j = 1, 2$, has rank two. With a slight abuse, we shall call complete integral also this type of solution. Given a complete integral, we consider the surface in $\mathbb{R}^3$

$$\Sigma_{\alpha_1, \alpha_2, C} = \left\{ x \in \mathbb{R}^3 \mid W(x, \alpha_1, \alpha_2) = C \right\}, \tag{1.124}$$

where C is an arbitrary constant. We determine α_1, α_2, C requiring that x_0 belongs to the surface and that the unit normal in x_0 is parallel to p_0. Taking into account of (1.121) and (1.122), this means

$$W(x_0, \alpha_1, \alpha_2) = C, \tag{1.125}$$

$$\frac{\partial W}{\partial x_i}(x_0, \alpha_1, \alpha_2) = p_{0i} \tag{1.126}$$

with $i = 1, 2, 3$. By the assumption on the matrix (1.123), from (1.126) we can find α_1^0, α_2^0 as functions of x_0, p_0. Then from (1.125) we find $C_0 = W(x_0, \alpha_1^0, \alpha_2^0)$. Let us define the surface

$$\Sigma_0 = \left\{ x \in \mathbb{R}^3 \mid W(x, \alpha_1^0, \alpha_2^0) = C_0 \right\}. \tag{1.127}$$

By construction, $x_0 \in \Sigma_0$ and the normal in x_0 is parallel to p_0. Moreover, we have $|\nabla W(x_0, \alpha_1^0, \alpha_2^0)| = |p_0|$.

Starting from Σ_0, we define the following family of surfaces parametrized by $t \in I$, where I is a (possibly small) interval containing zero

$$\Sigma_t = \left\{ x \in \mathbb{R}^3 \mid W(x, \alpha_1^0, \alpha_2^0) = C_0 + Et \right\}. \tag{1.128}$$

With a terminology taken from optics and explained in Sect. 1.8, these surfaces are called wave fronts or surfaces of constant phase.

Exercise 1.18 Verify that a complete integral of (1.121) in the case of a free particle is $W = \alpha \cdot x$, where $\alpha = \left(\alpha_1, \alpha_2, \sqrt{2mE - \alpha_1^2 - \alpha_2^2}\right)$, and Σ_t is the family of planes $\left\{x \in \mathbb{R}^3 \mid p_0 \cdot x = p_0 \cdot x_0 + \frac{|p_0|^2}{2m}t\right\}$.

Using the family of surfaces Σ_t, we can find the motion of the point particle. More precisely, let us consider the curve $t \to x(t)$, $t \in I$, starting from x_0 and orthogonal to Σ_t, obtained as the solution of the problem

$$\frac{dx(t)}{dt} = \frac{1}{m}\nabla W(x(t), \alpha_1^0, \alpha_2^0), \qquad x(0) = x_0. \tag{1.129}$$

The following proposition shows that such a curve coincides with the solution of the Hamilton's equations for the point particle.

Proposition 1.9 *Given x_0, p_0, with $p_0 \neq 0$, let $W(x, \alpha_1^0, \alpha_2^0)$ be the complete integral of (1.121) satisfying condition (1.126). Then, the solution of problem (1.129) is the motion of the point particle with initial conditions x_0, p_0.*

Proof Let $t \to x(t)$ be the solution of (1.129). We first note that

$$\dot{x}(0) = \frac{1}{m}\nabla W(x_0, \alpha_1^0, \alpha_2^0) = \frac{p_0}{m} \tag{1.130}$$

and therefore $(x(0), \dot{x}(0)) = (x_0, p_0/m)$. Moreover, dropping for brevity the dependence on α_1^0, α_2^0, we have for $k = 1, 2, 3$

$$m\frac{d^2x_k(t)}{dt^2} = \frac{d}{dt}\frac{\partial W}{\partial x_k}(x(t)) = \sum_j \frac{\partial^2 W}{\partial x_j \partial x_k}(x(t))\frac{dx_j(t)}{dt}$$

$$= \frac{1}{m}\sum_j \frac{\partial^2 W}{\partial x_j \partial x_k}(x(t))\frac{\partial W}{\partial x_j}(x(t))$$

$$= \frac{1}{2m}\frac{\partial}{\partial x_k}\sum_j \left(\frac{\partial W}{\partial x_j}\right)^2(x(t)) = \frac{1}{2m}\frac{\partial}{\partial x_k}\hat{n}^2(x(t))$$

$$= -\frac{\partial V}{\partial x_k}(x(t)) \tag{1.131}$$

and this completes the proof. □

We conclude this section defining the phase velocity associated to the family of surfaces (1.128), i.e., the velocity $\hat{v}_f(x_t)$ of a point of $x_t \in \Sigma_t$ computed along the normal to Σ_t. More precisely, we fix $x_t \in \Sigma_t$ and consider the point $x_{t+\Delta t} \in \Sigma_{t+\Delta t}$, for Δt sufficiently small, obtained as the intersection of the line orthogonal to Σ_t and passing through x_t with the surface $\Sigma_{t+\Delta t}$. Then, by definition

$$\hat{v}_f(x_t) := \lim_{\Delta t \to 0} \frac{|x_t - x_{t+\Delta t}|}{\Delta t} = |\dot{x}_t| . \tag{1.132}$$

In order to find an explicit expression for $\hat{v}_f(x_t)$, we observe that $W(x_t) = C_0 + Et$, so that

$$\nabla W(x_t) \cdot \dot{x}_t = E . \tag{1.133}$$

Taking into account that the vectors $\nabla W(x_t)$ and $\dot{x}_t$ are parallel, we find

$$\hat{v}_f(x_t) = \frac{|E|}{|\nabla W(x_t)|} = \frac{|E|}{\sqrt{2m(E - V(x_t))}} . \tag{1.134}$$

Note that the phase velocity is different from the velocity of the point particle.

Exercise 1.19 Using the wave front method, find the motion of a point particle in a plane subject to a constant force.

1.7 Liouville's Equation

In this section, we study Liouville's equation and recall its role in the evolution laws of observables and (mixed) states in Classical Mechanics. Liouville's equation will also be relevant for the short wavelength limit of Wave Optics (Sect. 1.10) and for the classical limit of Quantum Mechanics (Sect. A.4).

Given a vector field $z \to F(z, t) = (F_1(z, t), \ldots, F_n(z, t))$ in $\mathbb{R}^n$, with F smooth, and $x \in \mathbb{R}^n$ an arbitrary initial condition, we consider the Cauchy problem

$$\dot{z} = F(z, t), \qquad z(t_0) = x . \tag{1.135}$$

We denote by ϕ^{t,t_0} the flow generated by the vector field F, so that $\phi^{t,t_0}(x) = (\phi_1^{t,t_0}(x), \ldots, \phi_n^{t,t_0}(x))$ is the solution of problem (1.135) at time t (in the autonomous case we simply write $\phi^t(x)$). We first study the time evolution of the matrix

$$D\phi^{t,t_0}(x)_{ij} := \frac{\partial \phi_i^{t,t_0}}{\partial x_j}(x) \tag{1.136}$$

and of its determinant

$$J^{t,t_0}(x) := \det \left(D\phi^{t,t_0}(x) \right) . \tag{1.137}$$

Note that

$$D\phi^{t,t_0}(x)\Big|_{t=t_0} = \mathbb{I}, \qquad J^{t,t_0}(x)\Big|_{t=t_0} = 1, \qquad (1.138)$$

where $\mathbb{I}$ denotes the identity matrix. We have

Proposition 1.10

$$(i) \qquad \frac{d}{dt}D\phi^{t,t_0}(x) = DF(\phi^{t,t_0}(x),t)\,D\phi^{t,t_0}(x), \qquad (1.139)$$

$$(ii) \qquad \frac{d}{dt}J^{t,t_0}(x) = \nabla \cdot F(\phi^{t,t_0}(x),t)\,J^{t,t_0}(x) \qquad (1.140)$$

where DF denotes the matrix $\partial F_i/\partial z_j$ and $\nabla \cdot F$ is the divergence of F.

Proof Let us write problem (1.135) in integral form

$$\phi_i^{t,t_0}(x) = x_i + \int_{t_0}^t ds\, F_i(\phi^{s,t_0}(x),s) \qquad (1.141)$$

and compute the derivative with respect to x_j

$$\frac{\partial \phi_i^{t,t_0}}{\partial x_j}(x) = \delta_{ij} + \int_{t_0}^t ds \sum_k \frac{\partial F_i}{\partial z_k}(\phi^{s,t_0}(x),s)\,\frac{\partial \phi_k^{s,t_0}}{\partial x_j}(x). \qquad (1.142)$$

Computing now the derivative with respect to time, we obtain (i). In order to prove (ii), we observe that from (1.142) we have

$$\frac{\partial \phi_i^{t+h,t_0}}{\partial x_j}(x) = \frac{\partial \phi_i^{t,t_0}}{\partial x_j}(x) + \int_t^{t+h} ds \sum_k \frac{\partial F_i}{\partial z_k}(\phi^{s,t_0}(x),s)\,\frac{\partial \phi_k^{s,t_0}}{\partial x_j}(x) \qquad (1.143)$$

and therefore we can write

$$D\phi^{t+h,t_0}(x) = D\phi^{t,t_0}(x) + h\,DF(\phi^{t,t_0}(x),t)\,D\phi^{t,t_0}(x) + O(h^2). \qquad (1.144)$$

Multiplying (1.144) by the matrix $(D\phi^{t,t_0}(x))^{-1}$, we find

$$D\phi^{t+h,t_0}(x)(D\phi^{t,t_0}(x))^{-1} = \mathbb{I} + h\,DF(\phi^{t,t_0}(x),t) + O(h^2). \qquad (1.145)$$

From the above equation, we have (see e.g. Lemma 12.6 in [5])

$$J^{t+h,t_0}(x)\big(J^{t,t_0}(x)\big)^{-1} = 1 + h\,\nabla \cdot F(\phi^{t,t_0}(x),t) + O(h^2). \qquad (1.146)$$

Equation (1.146) implies

$$\frac{J^{t+h,t_0}(x) - J^{t,t_0}(x)}{h} = \nabla \cdot F(\phi^{t,t_0}(x), t) J^{t,t_0}(x) + O(h) \tag{1.147}$$

and taking the limit $h \to 0$ we obtain (ii).   □

Remark 1.6 In the case of a Hamiltonian system described by the Hamiltonian $H(q, p)$, the vector field F_H satisfies $\nabla \cdot F_H = 0$. Therefore, the solution of Eq. (1.140) is $J_H^t(x) = 1$. In particular, this means that the volume in phase space is preserved under the Hamiltonian flow ϕ_H^t.

Moreover, one can prove that the map $x \to \phi_H^t(x)$ is completely canonical for any t. Verify that in dimension $n = 1$ the result simply follows from $J_H^t(x) = 1$.

Remark 1.7 The solution of Eq. (1.139) is the linearized flow around $\phi^{t,t_0}(x)$. It describes how the trajectory $\phi^{t,t_0}(x)$ changes when the initial datum is varied. As an example, we explicitly write the equation in the case of a point particle of mass m, subject to a force with potential energy $V(q)$ and initial conditions (q_0, p_0).

Equation (1.135) reduces to

$$\dot{q} = \frac{p}{m}, \qquad \dot{p} = -V'(q). \tag{1.148}$$

Denoting by $\phi_H^t(q_0, p_0) = (q(t), p(t))$ the solution corresponding to the initial conditions (q_0, p_0), we have

$$\begin{aligned}
\frac{d}{dt}\frac{\partial q(t)}{\partial q_0} &= \frac{1}{m}\frac{\partial p(t)}{\partial q_0}, & \frac{d}{dt}\frac{\partial p(t)}{\partial q_0} &= -V''(q(t))\frac{\partial q(t)}{\partial q_0}, \\
\frac{d}{dt}\frac{\partial q(t)}{\partial p_0} &= \frac{1}{m}\frac{\partial p(t)}{\partial p_0}, & \frac{d}{dt}\frac{\partial p(t)}{\partial p_0} &= -V''(q(t))\frac{\partial q(t)}{\partial p_0}.
\end{aligned} \tag{1.149}$$

Let us now consider the Cauchy problem for Liouville's continuity equation

$$\frac{\partial u}{\partial t}(x, t) + \nabla \cdot (u(x, t)F(x, t)) = 0, \qquad u(x, 0) = u_0(x), \tag{1.150}$$

where F is a given vector field and u_0 is the initial datum. The solution of problem (1.150) can be explicitly written in terms of the solution $\phi^{t,t_0}(x)$ of problem (1.135).

Proposition 1.11 *The unique solution of problem* (1.150) *is*

$$u(x, t) = \frac{u_0(\phi^{t_0,t}(x))}{J^{t,t_0}(\phi^{t_0,t}(x))}. \tag{1.151}$$

Proof We rewrite the equation for $u(x, t)$ in the form

$$\frac{\partial u}{\partial t}(x, t) + \nabla_x u(x, t) \cdot F(x, t) + u(x, t) \nabla \cdot F(x, t) = 0 \tag{1.152}$$

and we evaluate (1.152) in $x = \phi^{t,t_0}(y)$. We have

$$
\begin{aligned}
0 &= \frac{\partial u}{\partial t}(\phi^{t,t_0}(y), t) + \nabla_x u(\phi^{t,t_0}(y), t) \cdot F(\phi^{t,t_0}(y), t) + u(\phi^{t,t_0}(y), t) \nabla \cdot F(\phi^{t,t_0}(y), t) \\
&= \frac{\partial u}{\partial t}(\phi^{t,t_0}(y), t) + \nabla_x u(\phi^{t,t_0}(y), t) \cdot \frac{d}{dt}\phi^{t,t_0}(y) + u(\phi^{t,t_0}(y), t) \frac{1}{J^{t,t_0}(y)} \frac{d}{dt} J^{t,t_0}(y) \\
&= \frac{1}{J^{t,t_0}(y)} \left(J^{t,t_0}(y) \frac{d}{dt} u(\phi^{t,t_0}(y), t) + u(\phi^{t,t_0}(y), t) \frac{d}{dt} J^{t,t_0}(y) \right) \\
&= \frac{1}{J^{t,t_0}(y)} \frac{d}{dt} \left(J^{t,t_0}(y)\, u(\phi^{t,t_0}(y), t) \right) .
\end{aligned}
\tag{1.153}
$$

Then $u(x, t)$ is solution of problem (1.150) if and only if

$$
J^{t,t_0}(y)\, u(\phi^{t,t_0}(y), t) = u_0(y)
\tag{1.154}
$$

It is now sufficient to evaluate the above equation in $y = \phi^{t_0,t}(x)$ to conclude the proof.   □

Remark 1.8 In the case $\nabla \cdot F = 0$, Eq. (1.150) is called Liouville's transport equation. In particular, for a Hamiltonian vector field, the equation reads

$$
\frac{\partial u}{\partial t} + \{u, H\} = 0
\tag{1.155}
$$

and the solution is $u(x, t) = u_0(\phi_H^{-t}(x))$.

We conclude the section recalling the notions of state and observable in Classical Mechanics and their relations with Liouville's equation (see also [6, 15]). Let us consider a Hamiltonian system described by the Hamiltonian $H(q, p)$. A (pure) state of the system is, by definition, a point $x = (q, p)$ in the phase space of the system. Hamilton's equations provide the evolution law of the state, i.e., if the state at time zero is x then the state at time t is $x_t := (q_t, p_t) = \phi_H^t(x)$. An observable of the system is, by definition, a smooth real function $f = f(x)$ defined on the phase space of the system. Examples of observables are position, momentum, angular momentum, etc. Given the observable f at time zero, we define the observable at time t by

$$
f_t(x) := f(\phi_H^t(x)) .
\tag{1.156}
$$

Using the fact that $f_{t+h}(x) = f_t(\phi_H^h(x)) = f_t(x) + \nabla f_t(x) \cdot F_H(x)h + O(h^2)$, we differentiate (1.156) with respect to t and we find

$$
\frac{\partial f_t}{\partial t} = \{f_t, H\} ,
\tag{1.157}
$$

which is the evolution law for the observables of the system.

Concerning the experimental predictions, we recall that Classical Mechanics is a deterministic theory. Such a statement means that if the pure state x is known then we can predict with certainty, i.e., with probability one, that the result of a measurement on the system of any observable f at time t is $f_t(x)$. Note that the knowledge of the pure state corresponds to the maximal information available on the system.

Let us now consider the case in which only partial information on the initial state of the system are available and we can only say that there exists a probability density $\rho_0 = \rho_0(x)$, $\rho_0(x) \geq 0$, $\int dx \, \rho_0(x) = 1$, such that the mean value of any observable f at time t is

$$\langle f_t \rangle = \int dx \, f_t(x)\rho_0(x) = \int dx \, f(\phi_H^t(x))\rho_0(x) \,. \tag{1.158}$$

In this case, one says that ρ_0 represents a mixed state of the system at time zero. Observe that, with the change of variable $y = \phi_H^t(x)$, we have

$$\langle f_t \rangle = \int dy \, f(y)\rho_0(\phi_H^{-t}(y)) := \int dy \, f(y)\rho_t(y) \,. \tag{1.159}$$

Therefore, by definition,

$$\rho_t(x) := \rho_0(\phi_H^{-t}(x)) \tag{1.160}$$

is the mixed state of the system at time t. Differentiating (1.160) with respect to t we find

$$\frac{\partial \rho_t}{\partial t} = -\{\rho_t, H\} \,, \tag{1.161}$$

which is the evolution law for the mixed states of the system. We note that the mixed state ρ_t evolves according to Liouville's equation while the equation for the observable f_t differs from (1.161) only for a sign.

A further important comment is in order. We have found two ways to represent the mean value of an observable at time t, given by Eqs. (1.158) and (1.159), corresponding to two possible descriptions of the evolution problem. In the first case the observable evolves in time according to (1.157) while the mixed state do not depend on time (Hamiltonian picture). In the second case, the mixed state evolves in time according to (1.161) while the observable do not depend on time (Liouville picture). The two approaches are equivalent and the choice depends on matter of convenience. We shall see that a similar situation occurs in Quantum Mechanics.

1.8   Geometrical Optics and Mechanical Analogy

Geometrical Optics is a theory of light propagation in material media (see e.g. [2, 12, 14]). It is based on the hypothesis that light propagates as a beam of rays, with each ray characterized by a well-defined trajectory and a velocity in a generic point x given by

$$v(x) = \frac{c}{n(x)}, \tag{1.162}$$

where c is the speed of light in vacuum and $n(x)$ is a given function, named refraction index, describing the optical property of the medium.

It is important to stress that the hypothesis can be considered reasonably valid only when the refraction index is slowly varying over a distance of the order of the wavelength of the light in the medium. Therefore, such condition defines the field of applicability of the theory.

The central problem of Geometrical Optics is to determine the trajectory of the rays for a given refraction index $n(x)$. The fundamental law that allows to find the trajectory is a variational principle, known as Fermat's principle, which can be formulated as follows.

Given two points A and B in $\mathbb{R}^3$, we consider the set of curves γ in $\mathbb{R}^3$, sufficiently regular, with initial point A and final point B. On this set, we define the following functional, named optical path

$$\mathscr{C}(\gamma) = c \int_\gamma dt . \tag{1.163}$$

Therefore, the optical path is proportional to the time it takes for the light to go from A to B along the curve γ.

Let us fix a parametrization $\lambda \to x(\lambda)$, $\lambda \in [\lambda_1, \lambda_2]$, with $x(\lambda_1) = A$ and $x(\lambda_2) = B$, of the curve γ. Moreover, from (1.162) we have $dt = v(x)^{-1}ds = c^{-1}n(x)\,ds$. Then, the optical path can be more precisely written as

$$\mathscr{C}(\gamma) = \int_{\lambda_1}^{\lambda_2} d\lambda\, n(x(\lambda)) \sqrt{\sum_i \left(\frac{dx_i(\lambda)}{d\lambda}\right)^2} \tag{1.164}$$

and we have

Fermat's principle: the trajectory followed by a light ray to travel from A to B in a medium with refraction index $n(x)$ is given by the critical point of the optical path (1.164).

Remark 1.9 For a more precise formulation of the principle, it would be required to specify the regularity of the refraction index and the domain of definition of the functional. We do not insist on these technical aspects.

Note that the optical path in a homogeneous medium, where $n(x) = cost.$, reduces to the length of the curve joining A and B. Then in a homogeneous medium, the rays propagate along straight lines.

When the medium is inhomogeneous the trajectory of the ray can be found solving the Euler–Lagrange equations associated to the functional (1.164). Taking into account that the Lagrangian is $\mathscr{L}(x, \eta) = n(x)\sqrt{\sum_i \eta_i^2}$, we have

$$\frac{d}{d\lambda}\left(\frac{n(x(\lambda))}{\sqrt{\sum_i \left(\frac{dx_i(\lambda)}{d\lambda}\right)^2}}\frac{dx_k(\lambda)}{d\lambda}\right) = \frac{\partial n(x(\lambda))}{\partial x_k}\sqrt{\sum_i \left(\frac{dx_i(\lambda)}{d\lambda}\right)^2}, \qquad (1.165)$$

with $k = 1, 2, 3$. A more convenient form of the equations is obtained choosing a particular parametrization of the curve. More precisely, we define the parameter τ

$$\tau(\lambda) = \int_{\lambda_1}^{\lambda} dv \frac{1}{n(x(v))}\sqrt{\sum_i \left(\frac{dx_i(v)}{dv}\right)^2}. \qquad (1.166)$$

Note that in the vacuum, i.e., for $n = 1$, the r.h.s. of (1.166) reduces to the arc length.

The map $\lambda \to \tau(\lambda)$ is invertible and we can define

$$y(\tau) = x(\lambda(\tau)). \qquad (1.167)$$

We have

Proposition 1.12 *The trajectory of the light ray $\tau \to y(\tau)$ solves the equations*

$$\frac{d^2 y_k(\tau)}{d\tau^2} = \frac{1}{2}\frac{\partial n^2(y(\tau))}{\partial y_k}, \qquad k = 1, 2, 3 \qquad (1.168)$$

and moreover the following identity holds

$$\frac{1}{2}\sum_{i=1,2,3}\left(\frac{dy_i(\tau)}{d\tau}\right)^2 - \frac{1}{2}n^2(y(\tau)) = 0. \qquad (1.169)$$

Proof From (1.228) we have

$$\frac{dy_k}{d\tau} = \frac{dx_k}{d\lambda}\frac{d\lambda}{d\tau} = \frac{n}{\sqrt{\sum_i \left(\frac{dx_i}{d\lambda}\right)^2}}\frac{dx_k}{d\lambda} \qquad (1.170)$$

and, using (1.165) and (1.170)

$$\frac{d^2 y_k}{d\tau^2} = \frac{d}{d\tau}\left(\frac{n}{\sqrt{\sum_i \left(\frac{dx_i}{d\lambda}\right)^2}}\frac{dx_k}{d\lambda}\right) = \frac{d}{d\lambda}\left(\frac{n}{\sqrt{\sum_i \left(\frac{dx_i}{d\lambda}\right)^2}}\frac{dx_k}{d\lambda}\right)\frac{d\lambda}{d\tau} = \frac{\partial n}{\partial x_k}n$$

$$= \frac{1}{2}\frac{\partial n^2(y(\tau))}{\partial y_k}. \qquad (1.171)$$

Moreover, from (1.170) we easily obtain the identity (1.169).   □

Equation (1.168) has the form of Newton's equations for a point particle. This means that there is a formal analogy between the equations of Geometrical Optics and Classical Mechanics, which can be precisely formulated as follows:

If we identify the parameter τ with the time, the trajectory of the light ray $\tau \to y(\tau)$ in a medium with refraction index $n(y)$ formally coincides with the trajectory of a point particle with mass $m = 1$, subject to a force with potential energy $U(y) = -\frac{1}{2}n^2(y)$. Furthermore, condition (1.169) says that the energy of this point particle is zero.

It is also possible to reformulate such analogy using Hamilton–Jacobi formalism. In fact, we observe that Eq. (1.168) coincide with Hamilton's equations associated to the Hamiltonian

$$H = \frac{1}{2}p^2 - \frac{1}{2}n^2 . \tag{1.172}$$

Taking into account of (1.169), the corresponding Hamilton–Jacobi equation is

$$|\nabla W_o|^2 = n^2(x) . \tag{1.173}$$

In Optics, such equation is known as eikonal equation and it is formally identical to Eq. (1.121) written in the case of a point particle.

Proceeding as we did in that case, we shall see that from the knowledge of a complete integral of (1.173) we can find the trajectory $\tau \to y(\tau)$ of a light ray starting from a point x_0 with a given unit tangent vector $\hat{\theta}_0$ in x_0.

Indeed, given a complete integral $W_o(x, \alpha_1, \alpha_2)$ of (1.173), we consider the surface with equation $W_o(x, \alpha_1, \alpha_2) = C$, where C is an arbitrary constant. Then, we find $\alpha_1^0, \alpha_2^0, C_0$ such that x_0 belongs to the surface

$$\Sigma_0 = \left\{ x \in \mathbb{R}^3 \mid W_o(x, \alpha_1^0, \alpha_2^0) = C_0 \right\} \tag{1.174}$$

and $\hat{\theta}_0 = \frac{\nabla W_o(x_0)}{|\nabla W_o(x_0)|}$ is the unit normal to Σ_0 in x_0. Then, we define the family of surfaces

$$\Sigma_t = \left\{ x \in \mathbb{R}^3 \mid W_o(x, \alpha_1^0, \alpha_2^0) = C_0 + c\,t \right\} \tag{1.175}$$

and the curve $\tau \to y(\tau)$ obtained as solution of the problem

$$\frac{dy(\tau)}{d\tau} = \nabla W_o(y(\tau), \alpha_1^0, \alpha_2^0), \qquad y(0) = x_0 . \tag{1.176}$$

Note that such a solution is orthogonal to the family of surfaces. We have

Proposition 1.13 *The solution of problem* (1.176) *is the trajectory of the light ray starting from x_0 with unit tangent vector $\hat{\theta}_0$ in x_0.*

Proof The solution of (1.176) satisfies

$$\left.\frac{dy(\tau)}{d\tau}\right|_{\tau=0} = \nabla W_o(y(0)) = \nabla W_o(x_0) = |\nabla W_o(x_0)|\,\hat{\theta}_0 \tag{1.177}$$

where we have dropped the dependence on α_1^0, α_2^0 to simplify the notation. Moreover,

$$
\begin{aligned}
\frac{d^2 y_i(\tau)}{d\tau^2} &= \frac{d}{d\tau}\frac{\partial W_o}{\partial x_i}(y(\tau)) = \sum_j \frac{\partial^2 W_o}{\partial x_j \partial x_i}(y(\tau))\frac{dy_j(\tau)}{d\tau} \\
&= \sum_j \frac{\partial^2 W_o}{\partial x_j \partial x_i}(y(\tau))\frac{\partial W_o}{\partial x_j}(y(\tau)) = \frac{1}{2}\frac{\partial}{\partial x_i}\sum_j \left(\frac{\partial W_o}{\partial x_j}(y(\tau))\right)^2 \\
&= \frac{1}{2}\frac{\partial}{\partial x_i}n^2(y(\tau)).
\end{aligned}
\tag{1.178}
$$

Thus, $\tau \to y(\tau)$ is the trajectory of the light ray (see (1.168)) and this concludes the proof.   □

Proceeding further as in the case of the point particle (see the end of Sect. 1.6), we define the phase velocity associated to the family of surfaces (1.175), i.e., the velocity $v_f(x)$ of a point of $x \in \Sigma_t$ computed along the normal to Σ_t. In this case, the computation yields

$$
v_f(x) = \frac{c}{|\nabla W_o(x)|} = \frac{c}{n(x)}.
\tag{1.179}
$$

We conclude with the (equivalent) reformulation of the analogy between the equations of Geometrical Optics and Classical Mechanics in terms of the Hamilton–Jacobi formalism. As we shall see in Chap. 3, this form of the analogy played a crucial role in the formulation of Wave Mechanics.

Comparing Eqs. (1.121) and (1.129) for Mechanics and (1.173), (1.176) for Optics, the analogy can be reformulated in the following way:

The trajectory of a point particle with mass m, subject to a force with potential energy $V(x)$, coincides with the trajectory of a light ray propagating in a medium with refraction index $\hat{n}(x) = \sqrt{2m(E - V(x))}$.

Exercise 1.20 Verify the law of refraction of light using the Hamilton–Jacobi method.

1.9 Maxwell's Equations in Vacuum

Electrical and magnetic phenomena at macroscopic level are described by two vector fields $E(x,t)$, $H(x,t)$, called electric and magnetic field respectively, satisfying Maxwell's equations (see e.g. [2, 10, 14, 16]). In particular, light is interpreted as an electromagnetic wave, i.e., a time-dependent solution of Maxwell's equations, with given frequency, or wavelength. Such a theory of light propagation, called Wave Optics, provides a detailed description of all optical phenomena at macroscopic level, including typical undulatory effects like interference and diffraction.

In absence of material media, electric and magnetic fields are solutions of Maxwell's equations in vacuum

$$\nabla \wedge E + \frac{1}{c}\frac{\partial H}{\partial t} = 0 , \tag{1.180}$$

$$\nabla \cdot E = 4\pi\rho , \tag{1.181}$$

$$\nabla \wedge H - \frac{1}{c}\frac{\partial E}{\partial t} = \frac{4\pi j}{c} , \tag{1.182}$$

$$\nabla \cdot H = 0 , \tag{1.183}$$

where c is the speed of light in vacuum, the scalar function $\rho(x, t)$ is the density of charge and the vector function $j(x, t)$ is the density of current. The functions $\rho(x, t)$ and $j(x, t)$ are called sources. The couple of vector fields $E(x, t)$, $H(x, t)$ is called electromagnetic field.

The typical problem of electromagnetism in vacuum can be formulated as follows: assigned the sources $\rho(x, t)$, $j(x, t)$, find the electromagnetic field $E(x, t)$, $H(x, t)$.

Remark 1.10 If we compute the divergence of Eq. (1.182) and use (1.181), we find

$$\frac{\partial\rho}{\partial t} + \nabla \cdot j = 0 \tag{1.184}$$

which is the local conservation law of the electric charge. This means that the assigned sources ρ and j are not independent and they must satisfy Eq. (1.184). We recall that (1.184) implies the conservation of the total charge, i.e.,

$$\frac{d}{dt}\int dx\, \rho(x, t) = 0 . \tag{1.185}$$

Remark 1.11 Maxwell's equations are eight scalar equations in the two unknown vector functions E, H, i.e., six unknown scalar functions. Then the problem could appear overdetermined. However, one can show that Eqs. (1.181) and (1.183) are consequences of Eqs. (1.180) and (1.182) and of a constraint on the initial fields.

To prove this fact, we consider the divergence of (1.182) and use (1.184)

$$0 = \nabla \cdot \left(\nabla \wedge H - \frac{1}{c}\frac{\partial E}{\partial t} - \frac{4\pi j}{c} \right) = -\frac{1}{c}\frac{\partial}{\partial t}\nabla \cdot E + \frac{4\pi}{c}\frac{\partial\rho}{\partial t} . \tag{1.186}$$

Then

$$\frac{\partial}{\partial t}(\nabla \cdot E - 4\pi\rho) = 0 . \tag{1.187}$$

Analogously, the divergence of (1.180) gives

$$\frac{\partial}{\partial t}\nabla \cdot H = 0 . \tag{1.188}$$

Therefore, if Eqs. (1.180) and (1.182) hold and moreover the initial fields satisfy the conditions $\nabla \cdot E|_{t=0} = 4\pi\rho|_{t=0}$ and $\nabla \cdot H|_{t=0} = 0$ then Eqs. (1.181) and (1.183) are satisfied for any t.

The above remark suggests that the Cauchy problem for Maxwell's equations in vacuum can be formulated as follows (problem 1):

$$\frac{\partial E}{\partial t} = c\, \nabla \wedge H - 4\pi j\,, \tag{1.189}$$

$$\frac{\partial H}{\partial t} = -c\, \nabla \wedge E\,, \tag{1.190}$$

$$E(x,0) = E_0(x)\,, \qquad H(x,0) = H_0(x)\,, \tag{1.191}$$

where E_0, H_0 are the assigned fields at time zero, satisfying the conditions

$$(\nabla \cdot E_0)(x) = 4\pi\rho(x,0)\,, \qquad (\nabla \cdot H_0)(x) = 0 \tag{1.192}$$

and $\rho(x,t)$, $j(x,t)$ are the assigned sources satisfying (1.184).

In the next proposition, we show that the solution of problem 1 reduces to the solution of a nonhomogeneous wave equation.

Proposition 1.14 *The fields E, H are solutions of problem 1 if and only if they are solutions of the following problem 2*

$$\frac{\partial^2 E}{\partial t^2} - c^2 \Delta E = -4\pi c^2 \nabla\rho - 4\pi \frac{\partial j}{\partial t}\,, \tag{1.193}$$

$$E(x,0) = E_0(x)\,, \qquad \frac{\partial E}{\partial t}(x,0) = c\,(\nabla \wedge H_0)(x) - 4\pi j(x,0)\,, \tag{1.194}$$

$$\frac{\partial H}{\partial t} = -c\, \nabla \wedge E\,, \qquad H(x,0) = H_0(x)\,, \tag{1.195}$$

where E_0, H_0 satisfy the conditions $(\nabla \cdot E_0)(x) = 4\pi\rho(x,0)$, $(\nabla \cdot H_0)(x) = 0$ and $\rho(x,t)$, $j(x,t)$ satisfy (1.184).

Proof Let E and H be solutions of problem 1. Then, using (1.189) and (1.190), the identity $\nabla \wedge (\nabla \wedge A) = \nabla(\nabla \cdot A) - \Delta A$ and (1.181), we have

$$\frac{\partial^2 E}{\partial t^2} = c\, \nabla \wedge \left(\frac{\partial H}{\partial t}\right) - 4\pi \frac{\partial j}{\partial t} = -c^2\, \nabla \wedge (\nabla \wedge E) - 4\pi \frac{\partial j}{\partial t}$$

$$= c^2 \Delta E - c^2 \nabla(\nabla \cdot E) - 4\pi \frac{\partial j}{\partial t} = c^2 \Delta E - 4\pi c^2 \nabla\rho - 4\pi \frac{\partial j}{\partial t}\,, \tag{1.196}$$

i.e., E satisfies Eq. (1.193). Moreover (1.194) and (1.195) clearly hold and therefore E e H are solutions of problem 2.

Let E and H be solutions of problem 2. Let us compute the divergence of Eq. (1.193)

$$\frac{\partial^2}{\partial t^2}(\nabla \cdot E) - c^2 \Delta(\nabla \cdot E) = -4\pi c^2 \nabla \cdot (\nabla\rho) - 4\pi \frac{\partial}{\partial t}\nabla \cdot j$$

$$= -4\pi c^2 \Delta\rho + 4\pi \frac{\partial^2 \rho}{\partial t^2}\,. \tag{1.197}$$

The above equation implies that the function $u := \nabla \cdot E - 4\pi\rho$ is the solution of the homogeneous wave equation. Moreover $u|_{t=0} = 0$ and

$$
\left.\frac{\partial u}{\partial t}\right|_{t=0} = \left.\nabla \cdot \frac{\partial E}{\partial t}\right|_{t=0} - \left.4\pi\frac{\partial \rho}{\partial t}\right|_{t=0}
$$

$$
= c\nabla \cdot (\nabla \wedge H_0) - 4\pi\,\nabla \cdot j|_{t=0} - \left.4\pi\frac{\partial \rho}{\partial t}\right|_{t=0} = 0. \qquad (1.198)
$$

Then u is identically zero, i.e., $\nabla \cdot E = 4\pi\rho$ for any t. Analogously, from Eq. (1.195), we find $\nabla \cdot H = 0$ for any t.

It remains to show that E satisfies Eq. (1.193). We have

$$
\frac{\partial}{\partial t}\left(\frac{\partial E}{\partial t} - c\,\nabla \wedge H + 4\pi j\right) = \frac{\partial^2 E}{\partial t^2} - c\,\nabla \wedge \frac{\partial H}{\partial t} + 4\pi\frac{\partial j}{\partial t}
$$

$$
= \frac{\partial^2 E}{\partial t^2} + c^2\,\nabla \wedge (\nabla \wedge E) + 4\pi\frac{\partial j}{\partial t} = \frac{\partial^2 E}{\partial t^2} - c^2 \Delta E + c^2\nabla(\nabla \cdot E) + 4\pi\frac{\partial j}{\partial t}
$$

$$
= \frac{\partial^2 E}{\partial t^2} - c^2 \Delta E + 4\pi c^2\nabla\rho + 4\pi\frac{\partial j}{\partial t} = 0. \qquad (1.199)
$$

Taking into account of the initial condition in (1.194), we find that E satisfies (1.193) for any t and therefore E and H are solutions of problem 1. $\qquad\square$

Remark 1.12 By the above proposition, we can explicitly solve Maxwell's equations in vacuum. Indeed, the field E is found by solving the nonhomogeneous wave Eq. (1.193) with initial data (1.194) and the field H is determined from (1.195) by quadratures.

For the electromagnetic field, the important conservation laws of energy and momentum hold. Here, we briefly recall the case of the energy. Let E and H be the fields produced by a system of charges described by the sources ρ and j. We define the Lorentz force per unit volume acting on the system of charges

$$
f_L(x, t) = \rho(x, t)E(x, t) + \frac{j(x, t)}{c} \wedge H(x, t), \qquad (1.200)
$$

the corresponding power per unit volume

$$
p_L(x, t) = \frac{j(x, t)}{\rho(x, t)} \cdot f_L(x, t) = j(x, t) \cdot E(x, t), \qquad (1.201)
$$

the Poynting vector

$$
S(x, t) = \frac{c}{4\pi}\,E(x, t) \wedge H(x, t), \qquad (1.202)
$$

and the quantity

$$
u_{em}(x, t) = \frac{1}{8\pi}\left(E(x, t)^2 + H(x, t)^2\right). \qquad (1.203)
$$

We have

Proposition 1.15 *The following equation holds*

$$\frac{\partial u_{em}}{\partial t} + \nabla \cdot S + p_L = 0. \tag{1.204}$$

Moreover, let V be a bounded region in $\mathbb{R}^3$, Σ the boundary of V and $n(x)$ the unit outward normal in $x \in \Sigma$. Then

$$-\frac{d}{dt} \int_V dx \, u_{em}(x, t) = \int_V dx \, p_L(x, t) + \int_\Sigma d\sigma(x) \, S(x, t) \cdot n(x). \tag{1.205}$$

Proof Let us consider the scalar product of (1.182) by $(c/4\pi)E$ and of (1.180) by $(c/4\pi)H$ and then consider the sum

$$
\begin{aligned}
0 &= p_L + \frac{1}{4\pi}\left(E \cdot \frac{\partial E}{\partial t} + H \cdot \frac{\partial H}{\partial t}\right) + \frac{c}{4\pi}(H \cdot \nabla \wedge E - E \cdot \nabla \wedge H) \\
&= p_L + \frac{1}{8\pi}\frac{\partial}{\partial t}\left(E^2 + H^2\right) + \frac{c}{4\pi}\nabla \cdot (E \wedge H),
\end{aligned} \tag{1.206}
$$

where we used the identity $\nabla \cdot (a \wedge b) = b \cdot \nabla \wedge a - a \cdot \nabla \wedge b$. Thus, we have proved (1.204). Integrating over the region V and using divergence theorem we conclude the proof. □

Let us comment on the above proposition (also known as Poynting theorem). If the electromagnetic field is initially confined in a region strictly smaller than V and, at time t, it has not reached the boundary Σ then the last term in (1.205) is zero. In this case, (1.205) reduces to

$$-\frac{d}{dt} \int_V dx \, u_{em}(x, t) = \int_V dx \, p_L(x, t). \tag{1.207}$$

Moreover, the r.h.s. of (1.207) is equal to the work per unit time done by the field on the system of charges contained in V and therefore the l.h.s. represents the decrease per unit time of the energy associated to the field in V.

This means that the quantity u_{em} must be interpreted as energy density per unit time of the electromagnetic field in vacuum.

In the general case, when the field are not zero on Σ, it is natural to interpret the last term in (1.205) as the energy per unit time of the field flowing out through the boundary Σ of V.

In conclusion, Eq. (1.205) expresses the conservation law of energy of the electromagnetic field: the decrease per unit time of the energy associated to the field in V is equal to the work per unit time done by the field on the system of charges contained in V plus the energy per unit time of the field flowing out through the boundary Σ of V.

Remark 1.13 Let us assume that the region V is a sphere of radius R and that the system of charges is confined in a sphere of radius r, with $r \ll R$. If the fields E and H decay faster than $|x|^{-1}$ for $|x|$ large then the last term in (1.205) is negligible for R large, uniformly in time. In this case, the energy of the electromagnetic field remains confined in a bounded region.

On the other hand, if the fields E and H decay as $|x|^{-1}$ for $|x|$ large then the last term in (1.205) gives a non-vanishing contribution for any R. This means that the energy flux outgoing from the sphere is different from zero and it does not decrease with R. Fields of this kind are called radiation fields and they can be found as the solutions of Maxwell's equations. The existence of such radiation fields is the reason why it is possible to send electromagnetic signals from a region V to another one V', with V and V' separated by a large distance.

We finally briefly mention an important case of radiation fields. Let us consider a point charge moving along a given trajectory $t \to q(t)$, confined in a bounded region of $\mathbb{R}^3$. Then

$$\rho(x,t) = e\,\delta(x - q(t)), \qquad j(x,t) = e\,\dot{q}(t)\,\delta(x - q(t)). \qquad (1.208)$$

For sources of this kind, the solution of Maxwell's equations can be explicitly computed (see e.g. [10], Chap. 14). For the sake of brevity, here we only recall the asymptotic behavior at large distance of the solution. More precisely, we assume

$$\frac{\sup_t |q(t)|}{|x|} \ll 1, \qquad \frac{\sup_t |\dot{q}(t)|}{c} \ll 1. \qquad (1.209)$$

Then the asymptotic behavior of E and H for $|x| \to \infty$ is

$$E(x,t) \simeq \frac{e}{4\pi c^2 |x|}\left[\ddot{q}\left(t - \frac{|x|}{c}\right) \wedge \frac{x}{|x|}\right] \wedge \frac{x}{|x|}, \qquad (1.210)$$

$$H(x,t) \simeq \frac{e}{4\pi c^2 |x|}\,\ddot{q}\left(t - \frac{|x|}{c}\right) \wedge \frac{x}{|x|}. \qquad (1.211)$$

Remark 1.14 Equations (1.210) and (1.211) have some important consequences:
• the fields produced by an accelerated point charge decay as $|x|^{-1}$ for $|x|$ large and therefore they are radiation fields;
• for $|x|$ large the fields E and H are orthogonal and their plane is orthogonal to the propagation direction $x/|x|$, i.e., (1.210) describe a transverse wave;
• if the motion of the point charge $t \to q(t)$ is periodic with frequency ω then the emitted electromagnetic wave has the same frequency ω.

The properties of the electromagnetic field produced by a moving point charge played an important role in the first attempts to elaborate a consistent theory of the atom.

1.10 Maxwell's Equations in Inhomogeneous Media and Short Wavelength Asymptotics

In presence of material media, Maxwell's equations must be modified in order to take into account the response of the medium. In absence of sources, the equations are

$$\nabla \cdot D = 0 , \qquad \nabla \wedge H - \frac{1}{c}\frac{\partial D}{\partial t} = 0 , \qquad (1.212)$$

$$\nabla \cdot B = 0 , \qquad \nabla \wedge E + \frac{1}{c}\frac{\partial B}{\partial t} = 0 , \qquad (1.213)$$

where we have introduced two auxiliary fields $D(x,t)$ and $B(x,t)$, called electric and magnetic induction, respectively. These fields are related to electric and magnetic fields via the equations

$$D = E + 4\pi P , \qquad B = H + 4\pi M , \qquad (1.214)$$

where the vector P, M are called electric and magnetic polarization and are assigned functions of the fields E and H

$$P = P(E) , \qquad M = M(H) . \qquad (1.215)$$

Equation (1.215) are called constitutive equations. They are characteristic of the medium under consideration and are determined experimentally. In particular, in vacuum we have $D = E$ and $B = H$. In many cases of interest, the constitutive equations are linear, possibly expressed by integral operators. Here, we illustrate the line to solve Eqs. (1.212) and (1.213) in the case

$$P(x,t) = \chi_e(x)E(x,t) , \qquad M(x,t) = \chi_m H(x,t) . \qquad (1.216)$$

In (1.216), the function $\chi_e(x)$ is a scalar and positive function, called electric susceptibility, while $\chi_m > 0$, the magnetic susceptibility, is supposed constant for simplicity. The constitutive Eq. (1.216) define the response of a nondispersive and isotropic inhomogeneous medium. Denoted

$$\varepsilon(x) = 1 + 4\pi \chi_e(x) , \qquad \mu = 1 + 4\pi \chi_m , \qquad (1.217)$$

Eq. (1.214) can be written as

$$D(x,t) = \varepsilon(x)E(x,t) , \qquad B(x,t) = \mu H(x,t) \qquad (1.218)$$

and therefore Maxwell's equations read

$$\nabla \wedge E + \frac{\mu}{c} \frac{\partial H}{\partial t} = 0, \tag{1.219}$$

$$\nabla \cdot (\varepsilon(x)E) = 0, \tag{1.220}$$

$$\nabla \wedge H - \frac{\varepsilon(x)}{c} \frac{\partial E}{\partial t} = 0, \tag{1.221}$$

$$\nabla \cdot H = 0. \tag{1.222}$$

From such equations, one can derive an equation for the electric field only. Indeed, from (1.219), using the identity $\nabla \wedge (\nabla \wedge A) = \nabla(\nabla \cdot A) - \Delta A$ and (1.221), we have

$$0 = \nabla \wedge \left(\nabla \wedge E + \frac{\mu}{c} \frac{\partial H}{\partial t} \right) = \nabla \wedge (\nabla \wedge E) + \frac{\mu}{c} \frac{\partial}{\partial t} \nabla \wedge H$$

$$= \nabla(\nabla \cdot E) - \Delta E + \frac{\mu\varepsilon(x)}{c^2} \frac{\partial^2 E}{\partial t^2}. \tag{1.223}$$

Note that from Eq. (1.220), we find

$$0 = \nabla \cdot (\varepsilon(x)E) = \varepsilon(x) \nabla \cdot E + \nabla\varepsilon(x) \cdot E. \tag{1.224}$$

Using this equation in (1.223) and defining the refraction index

$$n(x) := \sqrt{\mu\varepsilon(x)}, \tag{1.225}$$

we obtain the following equation for the electric field:

$$\Delta E - \frac{n^2(x)}{c^2} \frac{\partial^2 E}{\partial t^2} + \nabla\left(E \cdot \frac{\nabla\varepsilon(x)}{\varepsilon(x)} \right) = 0. \tag{1.226}$$

Once E is found solving (1.226), the field H is obtained from Eq. (1.219) by quadratures. We remark that the vector Eq. (1.226) corresponds to three scalar equations which are coupled due to the presence of the last term in (1.226). We do not deal with the solution of Eq. (1.226) but we shall consider a simplified version of the equation, obtained by neglecting the last term in (1.226), so that the vector equation reduces to three independent scalar equations for each component of the electric field. This is called scalar Optics approximation. Such approximation is justified when $\varepsilon(x)$, or equivalently the refraction index $n(x)$, is a slowly varying function of x.

To summarize, the generic component u of the electric field E in the scalar Optics approximation is determined by solving the equation

$$\Delta u - \frac{n^2(x)}{c^2} \frac{\partial^2 u}{\partial t^2} = 0, \tag{1.227}$$

while the magnetic field H is determined by quadratures using Eq. (1.219).

In the rest of this section, we shall show how to construct a suitable approximate solution of (1.227). More precisely, given the trajectory of a light ray starting from the point x_0, with unit tangent vector $\hat{\theta}_0$ in x_0, i.e.,

$$\tau \to y(\tau), \qquad y(0) = x_0, \qquad \frac{dy}{d\tau}(0) \left| \frac{dy}{d\tau}(0) \right|^{-1} = \hat{\theta}_0, \qquad (1.228)$$

we shall find an approximate solution of (1.227) which has the form of a well concentrated wave packet propagating along the above trajectory. We shall not give a rigorous derivation but we shall only provide heuristic arguments to justify the construction (for a more detailed analysis, see e.g. [2, 12, 14, 16]). We proceed through several steps.

(i) We look for a solution of Eq. (1.227) in the form of an amplitude times an oscillating exponential

$$u_{k_1}(x, t) = A_o(x, t) \, e^{i k_1 S_o(x,t)}, \qquad S_o(x, t) = W_o(x) - c\, t, \qquad (1.229)$$

where k_1 is a positive parameter and A_o, W_o are regular functions to be determined. We assume that A_o is positive and W_o is real. By an explicit computation, one verifies that u_{k_1} is solution of Eq. (1.227) if

$$A_o \left(|\nabla W_o|^2 - n^2(x) \right) - \frac{i}{k_1} \left(2\nabla A_o \cdot \nabla W_o + A_o \Delta W_o + 2\frac{n^2(x)}{c} \frac{\partial A_o}{\partial t} \right)$$

$$+ \frac{1}{k_1^2} \left(-\Delta A_o + \frac{n^2(x)}{c^2} \frac{\partial^2 A_o}{\partial t^2} \right) = 0. \qquad (1.230)$$

Separating real and imaginary part of the above equation and neglecting the term proportional to k_1^{-2}, we find that u_{k_1} is an approximate solution of Eq. (1.227) for $k_1 \to \infty$ if W_o satisfies the eikonal equation (see Sect. 1.8)

$$|\nabla W_o|^2 = n^2(x) \qquad (1.231)$$

and A_o satisfies the equation

$$2\nabla W_o \cdot \nabla A_o + A_o \Delta W_o + 2\frac{n^2(x)}{c} \frac{\partial A_o}{\partial t} = 0. \qquad (1.232)$$

Therefore, solving Eqs. (1.231) and (1.232), we obtain the approximate solution (1.229) of the Eq. (1.227) in the limit $k_1 \to \infty$.

(ii) Let us first discuss the eikonal equation. From Sect. 1.8, we know that we can find a complete integral $W_o(x, \alpha_1^0, \alpha_2^0)$ and the family of surfaces

$$\Sigma_t = \left\{ x \in \mathbb{R}^3 \mid W_o(x, \alpha_1^0, \alpha_2^0) = C_0 + ct \right\} \tag{1.233}$$

such that $x_0 \in \Sigma_0$ and the unit normal to Σ_0 in x_0 is $\hat{\theta}_0$. We call Σ_t surface of constant phase for the approximate solution (1.229) since for $x \in \Sigma_t$ the phase of the exponential in (1.229) remains constant, i.e., we have $S_o(x, t) = C_0$.

Moreover, we know that the phase velocity is

$$v_f(x) = \frac{c}{n(x)} . \tag{1.234}$$

The frequency of the oscillations in the phase of (1.229) is

$$v_1 = \frac{c \, k_1}{2\pi} . \tag{1.235}$$

The wavelength, defined by the relation $\lambda v_1 = v_f$, is given by

$$\lambda(x) = \frac{2\pi}{n(x)k_1} \tag{1.236}$$

and the wave number, defined as $2\pi\lambda^{-1}$, is

$$k(x) = n(x)k_1 . \tag{1.237}$$

This means that the limit $k_1 \to \infty$ we are considering corresponds to the high frequency or short wavelength regime. We also observe that in vacuum, i.e., for $n(x) = 1$, the surface of constant phase Σ_t reduces to the plane

$$\pi_t = \left\{ x \in \mathbb{R}^3 \mid \hat{\theta}_0 \cdot x = \hat{\theta}_0 \cdot x_0 + ct \right\} \tag{1.238}$$

(note the analogy with the case of a free particle in exercise 1.18).

(iii) Let us consider equation for the amplitude (1.232), where W_o is the complete integral of the eikonal equation (1.231). By a direct computation one verifies that Eq. (1.232) is equivalent to the continuity equation

$$\frac{\partial I}{\partial t}(x, t) + \nabla \cdot (I(x, t)v_g(x)) = 0 , \tag{1.239}$$

where

$$I(x, t) := n^2(x)A_o^2(x, t) \tag{1.240}$$

represents the intensity of the wave and the vector

$$v_g(x) := \frac{c}{n^2(x)} \nabla W_o(x) \tag{1.241}$$

is called group velocity. To simplify the notation, in (1.241) we have dropped the dependence on α_1^0, α_2^0 of the complete integral $W_o(x, \alpha_1^0, \alpha_2^0)$. Let $I_0(x)$ be the initial datum, $\varphi^t(\bar{x})$ the solution of the problem

$$\frac{dx}{dt} = v_g(x), \qquad x(0) = \bar{x} \tag{1.242}$$

and $J^t(\bar{x}) = \det(D\varphi^t(\bar{x}))$. Then, we know that the solution of the Cauchy problem for Eq. (1.239) is

$$I(x, t) = \frac{I_0(\varphi^{-t}(x))}{J^t(\varphi^{-t}(x)} \tag{1.243}$$

and therefore the solution of (1.232) is

$$A_o(x, t) = \frac{n(\varphi^{-t}(x))}{n(x)} \frac{A_o(\varphi^{-t}(x), 0)}{\sqrt{J^t(\varphi^{-t}(x))}}. \tag{1.244}$$

(iv) Let us now consider the approximate solution $u_{k_1}(x, t)$ defined by (1.229), where $W_0(x)$ is the complete integral of (1.231) defined above and $A_o(x, t)$ is given by (1.244) with an initial datum $A_o(x, 0)$ supported only in a small neighborhood of the point x_0.

As a consequence of (1.244), at time t the approximate solution is supported only in a small neighborhood of x such that $\varphi^{-t}(x) = x_0$, i.e., in a small neighborhood of $x(t) := \varphi^t(x_0)$. In other words, the approximate solution behaves like a wave packet well localized around the moving point $x(t)$.

(v) The last step is to show that the curve $t \to x(t)$ coincides, via a change of parametrization, with the assigned trajectory $\tau \to y(\tau)$ of the light ray.

Let us define the positive and monotone increasing function

$$t(\tau) = \frac{1}{c} \int_0^\tau dv\, n^2(\varphi^{t(v)}(x_0)), \tag{1.245}$$

with $t(0) = 0$. Then we have $\varphi^{t(0)}(x_0) = x_0$ and

$$\frac{d}{d\tau}\varphi^{t(\tau)}(x_0) = \frac{d}{dt}\varphi^t(x_0)\Big|_{t=t(\tau)} \frac{dt(\tau)}{d\tau} = \frac{c}{n^2(\varphi^{t(\tau)}(x_0))} \nabla W_o(\varphi^{t(\tau)}(x_0)) \frac{dt(\tau)}{d\tau}$$
$$= \nabla W_o(\varphi^{t(\tau)}(x_0)), \tag{1.246}$$

where we have used (1.241), (1.242) and (1.245). Thus, by Proposition 1.13 in Sect. 1.8, we conclude that $\tau \to \varphi^{t(\tau)}(x_0)$ coincides with the assigned trajectory of the light ray $\tau \to y(\tau)$. Moreover, the velocity of the wave packet is given by the group velocity (1.241) and its modulus is $c/n(x)$, in agreement with the assumption made in Geometrical Optics.

Let us summarize the above construction. In the asymptotic short wavelength regime, we can find an approximate solution of Maxwell's equations which has the form of wave packet well localized around a moving point $x(t)$. Moreover, the trajectory and the velocity of this point coincide with those of a light ray. In other words, it is possible to derive the propagation law of a light ray from a suitable solution of Maxwell's equations in the short wavelength limit. In this sense, we can conclude that Geometrical Optics is a particular limiting case of Wave Optics.

We underline that this fact and the analogy between Mechanics and Geometrical Optics were the starting points for the construction of Wave Mechanics.

Remark 1.15 Note that the description of the ray propagation provided by Wave Optics is more detailed. In fact, the theory allows to derive the motion of the ray as a function of the time, with the velocity of the ray given by $c/n(x)$ in each point x of the trajectory. On the other hand, in Geometrical Optics one can only derive the trajectory of the ray, i.e., the curve described by the ray, and, moreover, the velocity of the ray in each point x is assumed to be $c/n(x)$ a priori.

References

1. Arnold, V.I.: Mathematical Methods of Classical Mechanics. Springer-Verlag, New York (1989)
2. Born, M., Wolf, E.: Principles of Optics. University Press, Cambridge (1999)
3. Buttà, P., Negrini, P.: Note del Corso di Sistemi Dinamici (2008). (in Italian). http://www1.mat.uniroma1.it/~butta/didattica
4. Dell'Antonio, G.: Elementi di Meccanica (in Italian). Liguori Editore, Napoli (1996)
5. Esposito, R.: Appunti dalle Lezioni di Meccanica Razionale (in Italian). Aracne, Roma (1999)
6. Faddeev, L.D., Yakubovskii, O.A.: Lectures on Quantum Mechanics for Mathematics Students. AMS (2009)
7. Fasano, A., Marmi, S.: Analytical Mechanics. Oxford University Press, New York (2006)
8. Gallavotti, G.: The Elements of Mechanics. Springer-Verlag, New York (1983)
9. Gantmacher, F.: Lectures in Analytical Mechanics. Mir Publishers, Moscow (1975)
10. Jackson, J.D.: Classical Electrodynamics. Wiley, New York (1962)
11. Josè, J.V., Saletan, E.J.: Classical Dynamics: A Contemporary Approach. University Press, Cambridge (1998)
12. Kravtsov, YuA, Orlov, YuI: Geometrical Optics of Inhomogeneous Media. Springer-Verlag, Berlin Heidelberg (1990)
13. Landau, L.D., Lifshitz, E.M.: Mechanics. Pergamon Press, Oxford (1976)
14. Someda, C.G.: Electromagnetic Waves. CRC Press, New York (2006)
15. Strocchi, F.: An Introduction to the Mathematical Structure of Quantum Mechanics, 2th edition. World Scientific, Singapore (2008)
16. Stroffolini, R.: Lezioni di Elettrodinamica (in Italian). Bibliopolis, Napoli (2001)

Chapter 2
From Planck's Hypothesis to Bohr's Atom

2.1 Physics at the End of the Nineteenth Century

At the end of the nineteenth century, the two most successful and fundamental physical theories, universally accepted in the scientific community, were Mechanics and Electromagnetism.

Mechanics, in the formulation given by Newton in 1687, refined during the two following centuries, describes the motion of any material body subject to assigned forces, from planets and stars to objects of our everyday life. During the two centuries, the theory was applied with great success and it was able to describe a large variety of phenomena.

Electromagnetism was formulated in its final form by Maxwell in 1873. It gives a detailed account of Electricity, Magnetism, and Optics, for the first time described in a unified theoretical framework, and it also predicts the existence of the electromagnetic waves, experimentally verified by Hertz in 1887.

Beyond Mechanics and Electromagnetism, the nineteenth century saw also the development of Thermodynamics, a powerful phenomenological physical theory, which describes the behavior of macroscopic systems without making any assumption on their microscopic structure.

The body of knowledge contained in the theories above constitutes what is usually called Classical Physics.

The success obtained by these theories led to the belief that a complete knowledge of the laws of nature had been reached and that a correct application of these laws was sufficient to account for a satisfactory description of all natural phenomena. However, starting from the beginning of the twentieth century, scientists were forced to abandon this optimistic view under the thrust of new experimental data and new theoretical difficulties which were hard to explain using classical laws. The result was a critical review of the laws of Physics which culminated with the elaboration of two new theories, Relativity and Quantum Mechanics. We shall briefly describe the historical path that led to the birth of Quantum Mechanics. In this chapter, we outline the development of ideas in the period from Planck's analysis of the black body radiation to the formulation of the so-called "Old Quantum Theory". For a detailed account of the subject, we refer to [3, 5, 6]. In Ref. [6], the reader can also find the most important original papers of the period.

© Springer International Publishing AG, part of Springer Nature 2018
A. Teta, *A Mathematical Primer on Quantum Mechanics*,
UNITEXT for Physics, https://doi.org/10.1007/978-3-319-77893-8_2

2.2 Black Body Radiation

The first difficulties of the classical theoretical framework emerged in the study of the electromagnetic radiation emitted by a black body.

A black body, according to the definition given by Kirchhoff in 1860, is a body that absorbs all the radiation incident upon it.

A concrete realization of a black body is obtained considering a hollow body with walls opaque to radiation and with a small hole connecting the internal cavity with the external environment. A radiation incident on the hole will enter the cavity and, almost surely, it will be trapped inside. At a given temperature T, the electromagnetic radiation inside the cavity reaches an equilibrium condition with the internal walls of the cavity. On the other hand, the hole emits a small amount of electromagnetic radiation which does not alter the equilibrium condition inside the cavity. Under these conditions, the hole approximately behaves as a black body. Moreover, the radiation emitted by the hole is the black body radiation and it can be directly observed.

Using purely thermodynamic considerations, one can show that the black body radiation depends only on the temperature and not on the specific properties of the cavity, like the shape or the type of material used for the walls. In other words, the behavior of the radiation is universal. This fact suggests that such a behavior is determined by some fundamental property of nature and this is the reason why the analysis of the physical problem is considered particularly relevant.

We denote by $u(v, T)$ the energy density of the radiation component with frequency v and define the energy density of the radiation at temperature T as

$$U(T) = \int_0^\infty dv \, u(v, T). \tag{2.1}$$

The experimental analysis suggested that, for any T fixed, $u(v, T)$ is an increasing function for small v and, after having reached its maximum, it rapidly decreases to zero for large v (Fig. 2.1).

Despite the attempts made by many physicists, a satisfactory theoretical explanation of this behavior was not known. In particular, in 1900, Rayleigh approached the problem using the classical laws of Mechanics and Electromagnetism and derived the formula

$$u(v, T) = A \, v^2 T, \tag{2.2}$$

where A is a constant. Formula (2.2) correctly describes the experimental behavior for small v but it is clearly unsatisfactory for v large. In particular, it makes the integral in (2.1) divergent.

A possible solution of the problem was proposed by Planck in 1900. He studied a model where the atoms of the walls were described as a charged harmonic oscillator in equilibrium with the electromagnetic radiation. However, he also introduced the following ad hoc hypothesis: the value of the energy of the harmonic oscillators cannot be arbitrary but it must be one of the possible discrete values

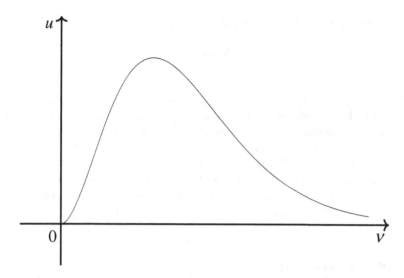

Fig. 2.1 Graph of $u(v, T)$ for T fixed

$$E_n = n\,hv, \qquad n = 1, 2, \ldots, \tag{2.3}$$

where v is the frequency of the oscillator and h is a universal constant with the physical dimension of an action, later called Planck's constant. A comparison with known experimental values of relevant quantities in the emission process gave for h the value

$$h = 6.6 \times 10^{-27} \ \mathrm{g\,cm^2\,sec^{-1}}. \tag{2.4}$$

Note that h has the physical dimension of an action. Thus, formula (2.3) means that the energy of an oscillator is an integer multiple of the elementary "quantum of energy" hv. In this case, one says that the energy is "quantized".

It is worth mentioning that Planck's hypothesis is definitely in contrast with the laws of Classical Mechanics. Indeed, from Mechanics, we know that the energy of a harmonic oscillator of mass m and frequency v is the following function of the initial conditions (x_0, v_0)

$$E(x_0, v_0) = \frac{1}{2}mv_0^2 + \frac{1}{2}m\omega^2 x_0^2, \qquad \omega = 2\pi v. \tag{2.5}$$

From (2.5) it is obvious that, varying the initial conditions, the energy can take any value of the interval $[0, \infty)$, and this fact is clearly in contradiction with the assumption (2.3).

It should also be recalled that Planck was inspired by previous works of Boltzmann, who used the same idea as a purely mathematical trick to perform some computations in problems of Statistical Mechanics.

Using the hypothesis (2.3), Planck derived the formula

$$u(\nu, T) = \frac{8\pi h \nu^3}{c^3} \frac{1}{e^{h\nu/kT} - 1}, \tag{2.6}$$

where c is the speed of light and k is the Boltzmann constant. Formula (2.6) is in good agreement with the experimental data. Note that for small values of $h\nu/kT$, it reproduces Rayleigh's formula (2.2) whereas for large value of $h\nu/kT$, it exponentially decreases to zero making the integral in (2.1) finite.

Nevertheless, we underline that Planck's derivation was not considered satisfactory by the scientific community, due to the arbitrariness of the hypothesis (2.3). Planck himself was convinced that, with a more detailed analysis of the problem, formula (2.6) could be derived without using such a hypothesis.

2.3 Photoelectric Effect

In 1905, Einstein wrote a fundamental paper (see, e.g., [6]) where Planck's hypothesis on the quantization of energy was reconsidered and further developed.

Using thermodynamic considerations, Einstein analyzed the energy of the electromagnetic radiation at low density and observed that phenomena involving exchange of energy between radiation and matter, like in the black body radiation problem, "*can be better understood on the assumption that the energy of light is distributed discontinuously in space. According to the assumption considered here, when a light ray starting from a point is propagated, the energy is not continuously distributed over an ever increasing volume, but it consists of a finite number of energy quanta, localized in space, which move without being divided and which can be absorbed or emitted only as a whole*" ([6] p. 92). In other terms, the distribution of the radiation energy at low density is quantized and therefore it is completely analogous to that of a gas of particles.

Einstein used this new idea to give an explanation of the so-called photoelectric effect, i.e., the emission of electrons (negatively charged particles) when the surface of a conductor is illuminated by a beam of light. In order to observe the phenomenon, an electromagnetic wave of frequency ν is sent on a metallic plate placed in vacuum. As a result of the interaction between the wave and the metal, some electrons are extracted from the metal, forming an electric current which can be measured. From the experiments it was known that

- the electrons are emitted only if the frequency ν of the wave is larger than a threshold value ν_0, characteristic of the metal;
- the kinetic energy of the emitted electrons has a maximum w_{max} which is proportional to the frequency ν;
- the number of emitted electrons per square centimeter per second is proportional to the intensity of the incident wave.

The explanation of these experimental facts using Maxwell theory of the electromagnetic field encounters various difficulties. Indeed, it is not clear why a wave with low frequency but high intensity does not extract electrons. Moreover, the emitted electrons should have a kinetic energy which is proportional to the intensity of the wave rather than to the frequency.

On the other hand, following Einstein's idea one arrives at a simple and intuitive explanation. Let us assume that the radiation energy is distributed in small packets, or quanta, of energy $h\nu$. Then the intensity of the wave is proportional to $nh\nu$, where n is the number of quanta per unit volume of the incident radiation. In this picture, the extraction of an electron is the result of the interaction between a single quantum of radiation and a single atom of the metal. As a consequence, an electron is extracted only if it interacts with a quantum of radiation of sufficiently high energy $h\nu$, i.e., an energy larger than the binding energy w_0 of the electron in the atom. This explains why the extraction takes place only if $\nu > w_0/h \equiv \nu_0$. Moreover, the emitted electron has a kinetic energy which is at most equal to $w_{max} = h\nu - w_0$, i.e., proportional to the frequency. Finally, if an electron is extracted by a single quantum of radiation then the number of extracted electrons is proportional to the number of quanta of the incident wave, and therefore to the intensity of the wave.

In conclusion, the analysis of Einstein provided a satisfactory explanation of the photoelectric effect.

In few years, the idea of quantization of energy was applied with success in the description of other physical situations, e.g., in the study of specific heat, and this led to the widespread belief that the assumption had a deep physical meaning. The scientific debate culminated in the first Solvay conference in 1911. In that occasion, there was a general acceptance of the idea that the exchanged energy in the interaction between radiation an matter is an integer multiple of an energy quantum proportional to the Planck's constant.

2.4 Atomic Structure of Matter

At the end of the nineteenth century, a further field of research devoted to the investigation of the microscopic constituents of matter became increasingly relevant.

We recall that atomism, as a philosophical materialistic conception, was developed by the Greeks Leucippus and Democritus between fifth and fourth century BC. It is based on the idea that matter is made of microscopic and indivisible objects, called atoms, subject to a continuous motion in vacuum. They can collide and form aggregates determining the properties of macroscopic objects that we observe. Such a metaphysical theory was recovered in the framework of the modern western philosophy and, starting from the eighteenth century, it was used as a possible basis for the scientific explanation of some natural phenomena.

In particular, the first laws of modern Chemistry, formulated by Lavoisier and Dalton between the eighteenth and the nineteenth century, were interpreted in a natural way on the basis of the atomic hypothesis.

Another important element in favor of atomism was given by the botanist Brown in 1827. He observed grains of pollen suspended in an aqueous solution using a microscope and he noted a strongly erratic motion of the grains. This motion can be explained as a consequence of the large number of collisions of the grains with the molecules of the solution which, in turn, are subject to thermal agitation. In this sense, the observed motion of the grains can be considered as an indirect proof of the existence of the molecules, i.e., small aggregates of atoms, of the solution. However, it is worth mentioning that a satisfactory theoretical explanation of this "Brownian motion" was given only in 1905 in another fundamental work of Einstein.

Despite some success obtained using the atomic hypothesis, it should be remarked that at the end of the nineteenth century, the debate on the existence of the atoms was still open. For example, two important scientists like Ostwald and Mach believed that physical theories are only a system of rules, expressed in mathematical language, useful to organize the observable phenomena in the most simple and economic way and to predict the results of the experiments. In such a view the atoms, which could not be directly observed, were to be considered as purely metaphysical entities without any physical reality. Therefore, they had to be excluded from any scientific investigation.

The intense scientific debate between atomists and their opponents was solved, in some sense, by the great technological and industrial development that took place in Europe in those years. Such development allowed to highly improve the experimental techniques and then to accumulate a considerable amount of accurate experimental data on properties and characteristics of the atoms. As a result, at the beginning of the twentieth century, the atomic structure of matter was an accepted idea in the scientific community. Without entering the details of all the experimental discoveries, we only summarize some quantitative information on the atoms acquired in that period:

- the radius of an atom is approximately 10^{-8} cm;
- the mass of the lightest atom (hydrogen) is approximately 10^{-24} g;
- in normal condition an atom is electrically neutral;
- it is possible to extract negatively charged particles (electrons) from an atom, which in turn means that the atom contains also positive charges;
- the charge e of the electron is approximately 10^{-10} $e.s.u.$ and its mass is $(1840)^{-1}$ times the mass of the hydrogen atom;
- the atoms of a gas, subject to an incident electromagnetic radiation, can absorb and emit radiation, whose wavelength can be measured and it is characteristic of the atoms.

We give some further details only on the last property. The electromagnetic radiation absorbed or emitted by the atoms of a gas was intensively studied from the experimental point of view in the framework of the so-called atomic spectroscopy. In the simple case of a hydrogen gas, it was known that the gas emits or absorbs radiation with different wavelengths, characterized by the following empirical formula (originally found by Balmer in 1885 and generalized by Rydberg in 1888):

$$\frac{1}{\lambda_n} = R_H \left(\frac{1}{4} - \frac{1}{n^2} \right), \qquad n = 3, 4, 5, \ldots, \tag{2.7}$$

where R_H (Rydberg constant) had at that time the estimated value $R_H = 109721.6$ cm^{-1}. The relevant aspect of formula (2.7) is that the possible wavelengths λ_n, and then the frequencies $\nu_n = c/\lambda_n$, form a discrete set parametrized by the integer n. It is worth noticing that this phenomenology is hard to understand inside the framework of classical Electromagnetism. Indeed, according to this theory, the frequencies of the radiation emitted or absorbed by the electric charges present in the atoms must be equal to the frequencies of the motion of the charges themselves (see Remark 1.14 in Sect. 1.9). As a consequence, one should expect that the possible frequencies could vary with continuity as a function of the initial conditions instead of being restricted to the discrete set of values given by formula (2.7).

The large amount of experimental data accumulated at the beginning of the twentieth century stimulated the formulation of the first hypothesis on the structure of the atom. One of the first proposed model was due to Thomson in 1904. He imagined the atom as a sphere made of a uniformly distributed positive charge, with a radius of order of 10^{-8} cm, with the electrons embedded in the sphere and arranged in equilibrium positions. The model was accepted for some years and it was considered as a first attempt to describe some qualitative characteristics of the atom. Nevertheless, it encountered severe difficulties in the explanation of some important experiments performed by Geiger e Marsden in 1909 on the scattering of α-particles, later identified as nuclei of helium. In these experiments, the α-particles were sent on a metal foil of thickness of order of 10^{-5} cm and their deflection with respect to the incident direction was measured. It was found that the large majority of particles passed through the foil with a negligible deflection while a small percentage, about one particle on $2 \cdot 10^4$, were deflected of an angle larger than $90°$.

The theoretical analysis of these experiments was given in 1911 by Rutherford (see, e.g., [6]). He observed that the experimental data could not be explained by the Thomson model of the atom. Indeed, if the metal foil is thought as an arrangement of Thomson atoms then all the α-particles should be deflected of a small angle and there would be no reason to see any deflection of large angles.

In order to solve the difficulty, Rutherford proposed the following different model of atom. The positive charge, with almost all the mass of the atom, is concentrated in a very small region, of order of 10^{-13} cm, which constitutes the nucleus of the atom. Moreover, the electrons are supposed to move around the nucleus at the distance of 10^{-8} cm, corresponding to the known radius of the atom. In this picture, the atom is thought as a small planetary system, with the nucleus and the electrons playing the role of the sun and the planets, respectively. Note that there is a large empty space between the orbiting electrons and the inner nucleus.

Using this model, it is possible to obtain a satisfactory explanation of the experimental data. Indeed, it is reasonable to expect that the large majority of the incident α-particles pass through the thin foil traveling in the empty space between electrons and nuclei. Therefore, they are subject to a negligible deflection. On the other hand, a small fraction of particles pass close to the nuclei, feel an intense electric field and then it is deflected at a large angle.

In his work, Rutherford made detailed computations starting from the hypothesis of the model and he obtained a quantitative prediction of the behavior of the α-particles in good agreement with the experimental data.

The success of this theoretical explanation led to the definitive acceptance of the planetary model of the atom. Moreover, it opened the way to the study of the nuclear structure and of the behavior of the electrons around the nucleus, subjects of Nuclear and Atomic Physics, respectively.

It should be stressed that Rutherford himself was aware that, according to the classical laws of Electromagnetism, there was a stability problem for the model. Indeed, the electrons moving around the nucleus are accelerated and therefore they emit electromagnetic radiation (see Remark 1.14 in Sect. 1.9). If such radiation is dispersed in the surrounding space then the electrons progressively lose their kinetic energy and, after a very short time, they fall on the nucleus. Thus, the stability of the model remained an open problem.

2.5 Bohr's Atom

The first dynamical model of an atom was proposed in 1913 by Bohr (see, e.g., [6]). His starting point was Rutherford's analysis, where the atom was described as a system of point charged particles (the electrons) moving under the attraction of the Coulomb force centered in the nucleus, considered as a fixed center of force. Bohr was aware that "*In an attempt to explain some of the properties of matter on the basis of this atom-model*" (i.e., Rutherford planetary model) *we meet, however, with difficulties of a serious nature arising from the apparent instability of the system of electrons* ([6] p. 132).

He also noted that the analysis of this instability problem could be reconsidered on the basis of the recent discussion on the new hypothesis of the quantization of the energy proposed by Planck and Einstein. He was convinced that "*The result of the discussion of these questions seems to be a general acknowledgment of the inadequacy of the classical electrodynamics in describing the behaviour of systems of atomic size*" ([6] p. 133).

Bohr's fundamental intuition was to approach the stability problem modifying the classical laws and introducing at the atomic level the hypothesis that the exchanged energy is an integer multiple of the Planck's constant h. In his view, this could also lead in a natural way to Balmer empirical formula (2.7), which is also expressed in terms of an integer. Moreover, the hypothesis introduced the new universal constant h in the theory, having the physical dimension of an action (see 2.4). This fact made possible to construct a quantity with the dimension of a length. Indeed, using the other two atomic constants, i.e., the mass of the electron m and its charge e, with $[e] = g^{1/2}\text{cm}^{3/2}\text{sec}^{-1}$, one obtains that the quantity

$$\frac{h^2}{me^2} \tag{2.8}$$

has the physical dimension of a length and, moreover, its value is of order of 10^{-8} cm. This fact made reasonable the possibility to derive the linear dimension of the atom in the new theory.

Starting from these motivations, Bohr approached the study of the hydrogen atom, i.e., the simplest atom containing just one electron, and he formulated the following three basic assumptions for his theory.

(a) For most of the time, the electron does not emit or absorb electromagnetic radiation. In this situation, one says that the atom is in a stationary state and the motion of the electron is correctly described by the laws of Classical Mechanics.

(b) In some interval of time, very short with respect to the period of the motion, the electron can make a transition from one stationary state to another one. In this process, it emits or absorbs a quantum of electromagnetic radiation with frequency

$$\nu = \frac{|E_i - E_f|}{h},\tag{2.9}$$

where E_i and E_f are the energies of the initial and final stationary states of the electron, respectively.

(c) The possible values of the energy of the electron in a stationary state are

$$E_n = -\frac{1}{2}n\,h\,\nu_e \qquad n = 1, 2, \ldots,\tag{2.10}$$

where ν_e is the frequency of the electronic motion.

Let us briefly comment the three assumptions.

Assumption (a) says that the classical laws of emission and absorption of electromagnetic radiation by accelerated charges (see Remark 1.14 in Sect. 1.9) are not valid at atomic level. This implies that the energy of the electron in a stationary state is constant, the motion is periodic, and the orbit is an ellipse, as in any standard Kepler problem.

Assumption (b) is the most important and it constitutes a first attempt toward a new theory of emission or absorption of electromagnetic radiation at atomic level. It does not describe the transition process from two stationary states but it only asserts that, when the process is triggered, the frequency of the emitted or absorbed radiation is given by (2.9). Note that such a frequency has nothing to do with the frequency of the electron motion and therefore (2.9) contradicts the laws of Classical Electromagnetism.

Assumption (c) provides the selection rule for the possible energies of the electron in a stationary state. It is chosen in complete analogy with Planck's hypothesis on the possible energies of a harmonic oscillator and, consequently, it is in contrast with the the laws of Classical Mechanics.

Let us see how Bohr's assumptions can be used to study the hydrogen atom.

When the electron is in a stationary state, by Assumption (a) we know that the motion is Keplerian. We briefly recall here the analysis of this motion (see also, e.g., [1, 2, 4]).

By the conservation law of the angular momentum, we know that the motion of the electron takes place in a plane π orthogonal to the direction of the angular momentum. We fix a reference frame with origin in the nucleus and the plane xy coinciding with π. If we denote by (r, ϕ) the polar coordinates of the electron, the conservation laws of the energy E and the z-component L of the angular momentum are

$$ E = \frac{1}{2}m(\dot{r}^2 + r^2\dot{\phi}^2) - \frac{e^2}{r}, \qquad L = mr^2\dot{\phi}, \qquad (2.11) $$

where $V(r) = -e^2/r$ is the Coulomb potential. Let us fix $L > 0$ (the case $L < 0$ is analogous and for $L = 0$ the trajectory reduces to a line). Solving for $\dot{\phi}$ the second equation in (2.11) and replacing in the first equation, we find

$$ E = \frac{1}{2}m\dot{r}^2 + V_{eff}(r), \qquad V_{eff}(r) = -\frac{e^2}{r} + \frac{L^2}{2mr^2}. \qquad (2.12) $$

Studying the graph of V_{eff} one sees that the trajectory is bounded if and only if

$$ \min_r V_{eff}(r) \leq E < 0, \qquad \min_r V_{eff}(r) = -\frac{me^4}{2L^2}. \qquad (2.13) $$

For $E = \min_r V_{eff}(r)$, one has a uniform circular motion of radius $\bar{r} = L^2/me^2$. For $\min_r V_{eff}(r) < E < 0$, the motion takes place in the region $\{r \mid r_{min} \leq r \leq r_{max}\}$, where

$$ r_{min} = \frac{e^2}{2|E|}\left(1 - \sqrt{1 - \frac{2|E|L^2}{m\,e^4}}\right), \qquad r_{max} = \frac{e^2}{2|E|}\left(1 + \sqrt{1 - \frac{2|E|L^2}{m\,e^4}}\right). $$
$$ (2.14) $$

In order to find the trajectory $r = r(\phi)$, we note that

$$ \frac{dr}{d\phi} = \frac{dr}{dt}\frac{dt}{d\phi} = \frac{\sqrt{2m}}{L}r^2\sqrt{E - V_{eff}(r)}, \qquad (2.15) $$

where we have used (2.12) and the second equation in (2.11). The differential equation (2.15) can be explicitly solved by separation of variables. Taking for simplicity $r(0) = r_{min}$, the reader can verify that the solution is the ellipse

$$ r(\phi) = \frac{p}{1 + \varepsilon\cos\phi}, \qquad p = \frac{L^2}{me^2}, \qquad \varepsilon = \sqrt{1 - \frac{2|E|L^2}{m\,e^4}}, \qquad (2.16) $$

where p and ε denote the parameter and the eccentricity of the ellipse respectively. In Cartesian coordinates, the equation reads

$$\frac{\left(x + \frac{\varepsilon p}{1-\varepsilon^2}\right)^2}{\frac{p^2}{(1-\varepsilon^2)^2}} + \frac{y^2}{\frac{p^2}{1-\varepsilon^2}} = 1 \tag{2.17}$$

and then the semimajor axis and the semiminor axis of the ellipse are

$$a = \frac{p}{1 - \varepsilon^2} = \frac{e^2}{2|E|}, \qquad b = \frac{p}{\sqrt{1 - \varepsilon^2}} = \frac{L}{\sqrt{2|E|m}}. \tag{2.18}$$

Let us compute the frequency ν_e of the periodic motion. As a consequence of the conservation of the angular momentum, the areal velocity is constant, indeed

$$\frac{dA}{dt} := \frac{1}{2}r^2\dot{\phi} = \frac{L}{2m}. \tag{2.19}$$

Taking into account that the ellipse area is πab and that $\nu_e = T_e^{-1}$, where T_e is the period of the motion, we have

$$\frac{dA}{dt} = \frac{\pi ab}{T_e} = \pi \nu_e ab = \pi \nu_e \frac{e^2 L}{(2|E|)^{3/2}\sqrt{m}}. \tag{2.20}$$

By (2.19) and (2.20) we obtain

$$\nu_e(E) = \frac{\sqrt{2}|E|^{3/2}}{\pi e^2 \sqrt{m}}. \tag{2.21}$$

Let us now consider Assumption (c). Using Eq. (2.21) we have

$$E_n = -\frac{1}{2}nh\nu_e(E_n) = -\frac{1}{2}nh\frac{\sqrt{2}|E_n|^{3/2}}{\pi e^2 \sqrt{m}}. \tag{2.22}$$

Solving with respect to E_n, we find the possible values for the energy, called energy levels, of the electron in a stationary state

$$E_n = -\frac{2\pi^2 e^4 m}{h^2 n^2}. \tag{2.23}$$

The integer n labeling the energy levels is called quantum number. By (2.18) and (2.23), we also obtain the semimajor axis of the corresponding ellipse

$$a_n = \frac{h^2 n^2}{4\pi^2 e^2 m}. \tag{2.24}$$

Finally, using Assumption (b), we find the frequency ν_{nk} of the emitted radiation in the transition from the stationary state with energy E_n to that with energy E_k, with $n > k$

$$\nu_{nk} = \frac{2\pi^2 e^4 m}{h^3} \left(\frac{1}{k^2} - \frac{1}{n^2} \right). \tag{2.25}$$

Let us comment on formulas (2.23), (2.24), (2.25), which are the most relevant results obtained in Bohr's model for the hydrogen atom.

Formula (2.23) shows that the minimum value of the energy is obtained for $n = 1$. The corresponding stationary state is called ground state. When the electron is in the ground state, it cannot emit radiation with a further decay. This means that the atom is stable. Moreover, the energy of the ground state is

$$E_1 = -\frac{2\pi^2 e^4 m}{h^2} \simeq 13.59\,\text{eV}, \tag{2.26}$$

where 1 eV (electron volt) is the kinetic energy gained or lost by an electron moving across an electric potential difference of one volt. It is remarkable that the above theoretical value is in good agreement with the known experimental value of the ionization energy for the hydrogen atom (by definition the ionization energy is the energy required to extract an electron from the atom).

Concerning formula (2.24), for $n = 1$ we have

$$a_1 = \frac{h^2}{4\pi^2 e^2 m} \simeq 0.529 \times 10^{-8}\,\text{cm}. \tag{2.27}$$

Therefore, the semimajor axis of the elliptic orbit, called Bohr's radius, is of the same order of magnitude as the atomic radius known from the experiments.

Formula (2.25) reproduces the structure of Balmer empirical formula (2.7) and it also provides a correct theoretical derivation of the Rydberg constant (verify).

To summarize, we have seen that Bohr's model of 1913, based on Assumptions (a), (b), (c), is able to describe with good accuracy the properties of the hydrogen atom known at that time.

Exercise 2.1 Show that the emitted frequency $\nu_{n+1,n}$ reduces, for n large, to the frequency ν_e of the periodic motion of the electron.

Remark 2.1 The result expressed in the previous exercise states that the new law obtained by Assumption (b) for the frequencies of the emitted radiation reduces, for n large, to the corresponding law of Classical Electromagnetism. Thus, in this case, the validity of the so-called "correspondence principle" is verified. Roughly speaking, this principle, formulated by Bohr himself, says that the results of the new atomic theory should reproduce the results of the classical theory in a suitable limit (typically, for large quantum numbers).

We underline that the correspondence principle played an important role in the long formation process of the new quantum theory as a "heuristic guiding principle" for the elaboration of new theoretical concepts.

In conclusion, we remark once again that the assumptions of Bohr's model are hard to understand on the basis of the classical laws. They are a strange mixture of classical ideas, e.g., Assumption (a) states that the motion in a stationary state is described by Classical Mechanics, with radically new ideas like Assumptions (b) and (c) in evident contrast with Classical Electromagnetism and Mechanics. In Bohr's view, the assumptions had to be only considered as the first elements of a forthcoming theory, both mechanical and electromagnetic, able to describe the behavior of the atom.

Despite these limitations, Bohr's model accurately explains some important properties af the hydrogen atom. Moreover, it shows the fundamental role that Planck's constant h and the quantization of the energy levels must play in the description of the atomic dynamics.

2.6 Bohr–Sommerfeld Quantization

The success obtained by Bohr's model encouraged experimental physicists to realize new experiments in order to evaluate the validity of the assumptions at the basis of the model. In particular, it is worth mentioning the important experiment performed by Franck and Hertz in 1914 (see [6]) which provided a convincing evidence of the existence of discrete energy levels for the atoms.

Concerning the theoretical aspect, the main problem was to extend the assumptions of the model in order to describe more complex atomic structures, e.g., an atom with many electrons possibly subject to external fields.

The most unsatisfactory point was Assumption (c). Indeed, it requires the existence of just one frequency of the electronic motion, i.e., it requires a periodic motion, and this is certainly not the case for an atom with many electrons.

The problem was approached in 1916 by Sommerfeld and Ehrenfest, who exploited the properties of the action variables of multiperiodic systems (see Sect. 1.5).

Let us consider consider a mechanical system with n degrees of freedom described by a separable Hamiltonian $H(q, p)$. Let us also assume that the hypothesis made in Proposition 1.8 of Sect. 1.5 is satisfied. Then we can introduce the action-angle variables $(\theta, I) = (\theta_1, \ldots, \theta_n, I_1, \ldots, I_n)$, the Hamiltonian in the new variables is $H(q(\theta, I), p(\theta, I)) = \hat{H}(I)$ and Hamilton's equations in the new variables are

$$\dot{I}_k = 0, \qquad \dot{\theta}_k = \frac{\partial \hat{H}}{\partial I_k} := 2\pi \nu_k(I), \tag{2.28}$$

where $\nu_1(I), \ldots, \nu_n(I)$ are the frequencies of the system (note that $\nu_k(I)$ differs for a factor 2π from the frequency defined in Proposition 1.8 for each $k = 1, \ldots, n$).

It may happen that the frequencies are not independent, i.e., there exist s relations, $0 < s \leq n - 1$, of the type

$$\sum_{l=1}^{n} a_{\tau l} \, \nu_l(I) = 0, \tag{2.29}$$

where $\tau = 1, \ldots, s$ and $a_{\tau l}$ are integer coefficients such that the matrix $a_{\tau h}$, with $\tau, h = 1, \ldots, s$, is non-singular. Therefore, the frequencies $\nu_1, \ldots, \nu_s$ can be expressed as linear combinations with rational coefficients of the remaining $\nu_{s+1}, \ldots, \nu_n$.

In this situation, we consider the further coordinate transformation $(\theta, I) \rightarrow (\phi, J)$ defined by

$$\phi_h = \sum_{l=1}^{n} a_{hl} \theta_l, \qquad I_h = \sum_{\tau=1}^{s} a_{\tau h} J_\tau, \qquad h = 1, \ldots, s,$$

$$\phi_k = \theta_k, \qquad I_k = J_k + \sum_{\tau=1}^{s} a_{\tau k} J_\tau, \qquad k = s+1, \ldots, n. \tag{2.30}$$

The transformation (2.30) is completely canonical and one can verify that the Hamiltonian in the variables (ϕ, J) is a function of $J_{s+1}, \ldots, J_n$. As a consequence, in the new variables, the first s frequencies are zero and the remaining $n - s$ are independent. The action variables $J_{s+1}, \ldots, J_n$ are called proper action variables. Note that if $s = n - 1$, there is just one frequency different from zero and we are in the case of periodic motion.

We have seen that the action variables can be introduced under sufficiently general conditions. Moreover, they are adiabatic invariants, i.e., they remain essentially constant under the action of slowly varying external perturbations. This fact was considered crucial for the applications to atomic dynamics due to the following observation. Atoms are continuously subject to slowly varying external electromagnetic fields which can slowly modify the electronic orbits without determining transitions from a stationary state to another (with emission or absorption of radiation). This suggests that the right quantities to be quantized are the ones remaining constant under such kind of perturbations, i.e., they must be adiabatic invariants.

On the basis of the above considerations, the (proper) action variables were identified as the most "natural" quantities to be quantized, in place of the energy used by Bohr. Therefore, Bohr's Assumption (c) was replaced by the following.

(c1) Consider a mechanical system with n degrees of freedom where it is possible to introduce action-angle variables and let $J_{s+1}, \ldots, J_n$ be the corresponding proper action variables. Then the possible values of $J_{s+1}, \ldots, J_n$ in a stationary state are

$$J_k = n_k h, \qquad n_k \in \mathbb{N}, \qquad k = s+1, \ldots, n. \tag{2.31}$$

Condition (2.31) is referred to as the Bohr–Sommerfeld quantization rule and the integers n_k are called quantum numbers of the system.

Exercise 2.2 Find the possible values of the energy levels for the hydrogen atom using the Bohr–Sommerfeld quantization rule.

Assumptions (a), (b), introduced by Bohr, together with the above Assumption (c1) were the basis of the so-called "Old Quantum Theory". Such a theory was applied for almost 10 years with some success to the study of various phenomena, e.g., atoms subject to electric fields (Stark effect) or magnetic fields (Zeeman effect). Nevertheless, it could not be considered satisfactory.

A first limitation was the fact that the theory could be applied only to multiperiodic systems and therefore situations in which the orbits of the system are unbounded, relevant in scattering problems, could not be described.

Another, and more fundamental, limitation was that the theory, as for Bohr's model, was a strange mixture of Classical Mechanics and quantum hypothesis, which are clearly incompatible from a conceptual point of view.

Finally, it encountered serious difficulties in the explanation of multielectron atoms, like helium.

In conclusion, physicists believed that the theory was provisional, even if with a certain degree of validity. In few years, it became clear that a coherent and correct description of atomic dynamics could ultimately be obtained only through a radical revision of the basic concepts of Classical Mechanics and Electromagnetism.

References

1. Arnold, V.I.: Mathematical Methods of Classical Mechanics. Springer, New York (1989)
2. Fasano, A., Marmi, S.: Analytical Mechanics. Oxford University Press, New York (2006)
3. Hund, F.: The History of Quantum Theory. Barnes & Noble Books (1974)
4. Landau, L.D., Lifshitz, E.M.: Mechanics. Pergamon Press, Oxford (1976)
5. Tagliaferri, G.: Storia della Fisica Quantistica (in Italian). Franco Angeli, Milano (1985)
6. Ter Haar, D.: The Old Quantum Theory. Pergamon Press, Oxford (1967)

Chapter 3
The Formulation of Wave Mechanics

3.1 Matrix Mechanics and Wave Mechanics

The Old Quantum Theory was replaced by a radically new theory of the dynamical behavior of microscopic objects, called Quantum Mechanics. Such a theory was elaborated in the years 1925–26 following two different routes.

The first approach was initiated by Heisenberg in 1925, with relevant contributions due to Born, Jordan, Dirac, and led to the formulation of the so-called Matrix Mechanics (see, e.g., [1–3]).

Heisenberg's starting point was the consideration that the rules of the Old Quantum Theory, based on classical kinematics, "*contain, as basic element, relationships between quantities that are apparently unobservable in principle, e.g., the position and period of revolution of the electron*" ([3] p. 261). This fact was a strong limitation since, in Heisenberg's view, a good physical theory should always be formulated in terms of quantities which can be concretely observed in experiments. In other words, he was convinced that the origin of the difficulties encountered by the Old Quantum Theory in the description of atomic dynamics was the use of classical kinematics, based on positions and velocities of point particles.

Therefore, his program was to propose a new kinematics based entirely on quantities that are observable at the atomic level. In his words "*it seems more reasonable to try to establish a theoretical quantum mechanics, analogous to classical mechanics, but in which only relations between observable quantities occur*" ([3] p. 262).

As observable quantities at atomic level, he chose frequencies and amplitudes of the radiation emitted or adsorbed in quantum jumps. He noticed that these quantities are characterized by two indices (specifying the initial and the final stationary states) and that their product is not commutative. As for the dynamics, the assumption was that two rules of the Old Quantum Theory, i.e., Newton's law and the selection rule for stationary states, should remain valid and that they had only to be conveniently rewritten in terms of the new observable quantities.

In 1925, Born and Jordan finalized Heisenberg's program. Using a Hamiltonian formalism, they were able to associate infinite, Hermitian matrices Q, P, obeying the canonical commutation relations $QP - PQ = i\hbar I$, to the observable quantities introduced by Heisenberg. This led to the first formulation of Matrix Mechanics that appeared to satisfactorily describe atomic phenomena. It is worth emphasizing that

© Springer International Publishing AG, part of Springer Nature 2018
A. Teta, *A Mathematical Primer on Quantum Mechanics*,
UNITEXT for Physics, https://doi.org/10.1007/978-3-319-77893-8_3

such a theory relied on the explicit rejection of the idea of a continuous motion in ordinary space to describe the dynamical evolution of a microscopic object. In other words, the classical space-time description of a physical process is replaced by a more abstract description, whose only aim is to provide quantitative predictions of the observable quantities in agreement with the experimental results.

The second approach, initiated by De Broglie in 1924 and developed by Schrödinger in 1926, led to the formulation of the so-called Wave Mechanics (see, e.g., [1, 2, 4]). The line of thought of De Broglie and Schrödinger was the following.

As we discussed in the previous chapter, Planck and Einstein had shown that electromagnetic radiation can exhibit a wave or a corpuscular behavior, depending on the physical situation. Such a twofold nature of the radiation led De Broglie to the idea that also matter can exhibit a wave behavior, besides the known corpuscular one. More precisely, he supposed that a point particle with energy E and momentum p could be associated with a wave packet with frequency $\hat{v} = E/h$ and wavelength $\hat{\lambda} = h/p$, where h is the Planck's constant, whose behavior should be governed by a new evolution equation, analogous to the ordinary wave equation. Moreover, the evolution of the wave packet should have the property to reduce to the classical evolution of a point particle for large values of the momentum p, or small values of $\hat{\lambda}$, in complete analogy with the reduction of Wave Optics to Geometrical Optics in the short wavelength limit.

Such a property was a necessary requirement since it was known that the laws of Classical Mechanics are valid when applied to the ordinary macroscopic world. In other terms, the undulatory aspect of matter should emerge only in the description of the dynamical behavior of microscopic objects.

An important reason supporting this hypothetical undulatory aspect of matter was that the new wave equation governing the evolution of the wave packet has a natural mathematical structure allowing to find the quantization of the energy levels of the atom. Indeed, it was well known that a wave equation in a bounded region, with suitable boundary conditions, produces a discrete spectrum of frequencies. By analogy, the discrete energy levels of the atom could be produced as eigenvalues of a boundary value problem for this new wave equation.

De Broglie's ideas were concretely realized by Schrödinger, who introduced the new evolution equation and gave a first formulation of Wave Mechanics. Such a theory was explicitly based on the idea that a classical space-time description of microscopic phenomena could be preserved, provided one replaces the ordinary mechanics of point particles with a new mechanics of waves.

It is worth underlining that the two approaches, Matrix and Wave Mechanics, were based on radically different physical ideas. Nevertheless, it was soon proved by Schrödinger himself in 1926 that the two approaches are equivalent from the mathematical point of view. In other words, they are two different ways to describe the same theory, from that time known as Quantum Mechanics. Finally, in 1932, von Neumann [5] provided an axiomatic and mathematically rigorous formulation of the theory essentially based on the theory of linear operators in Hilbert spaces.

In the next sections, we shall introduce the basic elements of the new theory following the line of thought of Wave Mechanics.

3.2   The Schrödinger Equation

In a series of papers published in 1926, Schrödinger developed the program of Wave Mechanics. His idea was that the motion of a microscopic object, e.g., an electron, must be *"represented by the wave process in q-space"* (i.e., in position space) *and not by the motion of representative points in this space. The study of motion of representative points, which forms the subject of classical mechanics, is only an approximate process and as such has exactly the same justification as geometrical or ray optics in comparison with the true optical processes* ([4] p. 117). Therefore, following this idea, the motion of an electron is described by means of a "wave function" $\psi(t) \equiv \psi(\cdot, t)$ defined in the whole space and satisfying a new evolution equation.

In order to construct such an equation, Schrödinger considered the analogy between Geometrical Optics and Classical Mechanics and the fact that Geometrical Optics is obtained as the short wavelength limit of Wave Optics. Letting himself be guided by these known facts, he conjectured that Classical Mechanics could be considered a suitable limit of a more general Wave Mechanics able to describe the behavior of microscopic particles. The situation could be summarized by the following diagram.

Wave Optics		"Wave Mechanics"
$\Big\downarrow \lambda \to 0$		$\Big\downarrow \hat{\lambda} \to 0$
Geometrical Optics	$\longleftrightarrow$	Classical Mechanics

Here, we present some heuristic arguments leading in a natural way to the formulation of the equation governing the evolution of the wave function.

As a preliminary step, we associate the typically wave quantities (refraction index, phase velocity, frequency, wavelength) to the particle motion exploiting the analogy between Geometrical Optics and Classical Mechanics. From Sect. 1.8, we know that the trajectory of a point particle with mass m, energy E and subject to a potential $V(x)$ coincides with the trajectory of a light ray propagating in a medium with refraction index

$$\hat{n}(x) = \sqrt{2m(E - V(x))}. \tag{3.1}$$

Moreover, we have associated the following phase velocity to the motion of the point particle

$$\hat{v}_f(x) = \frac{E}{\sqrt{2m(E - V(x))}}. \tag{3.2}$$

In order to associate a frequency to the point particle motion, we recall the discussion in Sect. 1.10 on the short wavelength limit for the solutions of Maxwell's equations. In particular, we have seen that the generic component of the electric field is the product of an amplitude times the rapidly oscillating exponential

$$e^{ik_1 S_o(x,t)}, \qquad S_o(x, t) = W_o(x) - c\,t, \tag{3.3}$$

where the wave number k_1 is proportional to the inverse of the wavelength. Moreover, the function W_o is solution of the eikonal equation with refraction index $n(x)$, characteristic of the medium.

By analogy, we assume that also the wave function $\psi(x, t)$, in a suitable limit, reduces to the product of an amplitude times a rapidly oscillating exponential

$$e^{i\frac{2\pi}{b}S(x,t)}, \qquad S(x,t) = W(x) - Et, \tag{3.4}$$

where b is a constant to be determined and W is solution of the eikonal equation with refraction index $\hat{n}(x)$. Note that the constant b is required to make the phase dimensionless and then the ratio E/b must have the dimension of a frequency. The constant b can be fixed recalling the idea of Planck and Einstein that the energy is the product of the frequency times the Planck's constant. This suggests to fix $b = h$ and therefore the frequency associated to the particle motion is

$$\hat{\nu} = \frac{E}{h}. \tag{3.5}$$

Concerning the wavelength $\hat{\lambda}$, we exploit the fact that wavelength, frequency, and phase velocity are related by the equation $\hat{\lambda}\,\hat{\nu} = \hat{\nu}_f$. Then, using (3.5) and (3.2), we find

$$\hat{\lambda} = \frac{\hat{\nu}_f}{\hat{\nu}} = \frac{h}{p}. \tag{3.6}$$

To summarize, by the above heuristic arguments, we have obtained the wave quantities (3.1), (3.2), (3.5), (3.6) naturally associated to the particle motion. In particular, the relations (3.5) and (3.6) coincide with that conjectured by De Broglie.

The next step is to find the evolution equation for the wave function $\psi(x, t)$. Following again the analogy with Optics, we recall that (see Sect. 1.10) the equation for the generic component of the electric field is

$$\Delta u - \frac{n^2(x)}{c^2} \frac{\partial^2 u}{\partial t^2} = 0. \tag{3.7}$$

Let $\tilde{u}(x, \omega)$ be the Fourier transform with respect to the time variable of $u(x, t)$. Then we also have

$$\Delta \tilde{u} + 4\pi^2 \frac{\nu^2}{\nu_f^2(x)} \tilde{u} = 0, \tag{3.8}$$

where we have used the relations $\omega = 2\pi\nu$ and $\nu_f(x) = c/n(x)$.

By analogy, we consider the Fourier transform $\tilde{\psi}(x, \hat{\omega})$ of the wave function, where $\hat{\omega} = 2\pi\hat{\nu}$, and we assume that it satisfies an equation of the form (3.8) with ν and $\nu_f(x)$ replaced by $\hat{\nu}$ and $\hat{\nu}_f(x)$, i.e.,

$$\Delta \tilde{\psi} + 4\pi^2 \frac{\hat{v}^2}{\hat{v}_f^2(x)} \tilde{\psi} = 0 .$$ (3.9)

Using (3.5) and (3.2) in (3.9), we find

$$-\frac{\hbar^2}{2m}\Delta \tilde{\psi} + V(x)\tilde{\psi} = E\,\tilde{\psi}$$ (3.10)

where we have defined

$$\hbar = \frac{h}{2\pi} .$$ (3.11)

Equation (3.10) is called stationary Schrödinger equation.

It remains to find the time-dependent equation for $\psi(x, t)$, whose stationary solutions are provided by solving (3.10). We underline that such an evolution equation is not uniquely determined and the choice made by Schrödinger corresponds to a simplicity criterion. Let us multiply Eq. (3.10) by $(2\pi)^{-1/2} e^{-i\hat{\omega}t}$ and let us integrate with respect to the variable $\hat{\omega}$. The left hand side of (3.10) reduces to

$$-\frac{\hbar^2}{2m}\Delta \psi + V(x)\psi .$$ (3.12)

To simplify the notation, in (3.12) (and often in the sequel) the wave function is simply denoted by ψ. Concerning the right-hand side of (3.10), we use the relation $E = \hbar\hat{\omega}$ and we obtain

$$\frac{1}{\sqrt{2\pi}} \int d\hat{\omega}\, E\, e^{-i\hat{\omega}t} \tilde{\psi}(x, \hat{\omega}) = i\hbar \frac{\partial}{\partial t} \frac{1}{\sqrt{2\pi}} \int d\hat{\omega}\, e^{-i\hat{\omega}t} \tilde{\psi}(x, \hat{\omega}) = i\hbar \frac{\partial}{\partial t} \psi(x, t) .$$ (3.13)

By (3.12) and (3.13), we finally arrive at the time-dependent equation for the wave function

$$i\hbar \frac{\partial \psi}{\partial t} = -\frac{\hbar^2}{2m}\Delta \psi + V(x)\psi.$$ (3.14)

Equation (3.14) is called time-dependent Schrödinger equation.

It is a time-dependent partial differential equation which, given the wave function at time $t = 0$, determines the wave function at any time $t > 0$.

We underline once again that the arguments used to derive Eq. (3.14) are heuristic and they have the only aim to make reasonable to assume, as a postulate, the validity of the equation.

Remark 3.1 An important aspect of Eq. (3.14) is that it is a linear equation. This is in contrast with the situation in Classical Mechanics where, in general, the evolution equation is nonlinear.

A second important characteristic is that it contains the imaginary unit i and therefore the solution is in general a complex valued function.

Once the validity of Eq. (3.14) is assumed, the next step is to understand in which sense the solution ψ describes the motion of a microscopic particle. The wave function itself, being complex valued, cannot have a direct physical meaning and, therefore, we have to look for a suitable real quantity associated to ψ.

To this aim, we prove the following proposition which expresses the conservation law of the the L^2-norm of the solution of Eq. (3.14) (a more general proof will be given in the next chapter).

Proposition 3.1 *Assume that there exists $\psi(x,t)$, $x \in \mathbb{R}^3$, solution of Eq. (3.14), twice differentiable in x and differentiable in t, such that for any $t > 0$*

$$|\psi(x,t)\nabla\psi(x,t)| < \frac{a}{|x|^{2+\eta}}, \tag{3.15}$$

where a, η are two positive constants. Then for any $t > 0$, we have

$$\int_{\mathbb{R}^3} dx\, |\psi(x,t)|^2 = \int_{\mathbb{R}^3} dx\, |\psi(x,0)|^2. \tag{3.16}$$

Proof Multiplying (3.14) by $\bar{\psi}$, we have

$$i\hbar\bar{\psi}\frac{\partial\psi}{\partial t} = -\frac{\hbar^2}{2m}\bar{\psi}\Delta\psi + V(x)\bar{\psi}\psi. \tag{3.17}$$

Taking the complex conjugate of Eq. (3.17), we also find

$$-i\hbar\psi\frac{\partial\bar{\psi}}{\partial t} = -\frac{\hbar^2}{2m}\psi\Delta\bar{\psi} + V(x)\psi\bar{\psi}. \tag{3.18}$$

Subtracting (3.17) and (3.18) we obtain

$$\frac{\partial}{\partial t}|\psi|^2 = \frac{i\hbar}{2m}(\bar{\psi}\Delta\psi - \psi\Delta\bar{\psi}) = \frac{i\hbar}{2m}\,\nabla\cdot\left(\bar{\psi}\nabla\psi - \psi\nabla\bar{\psi}\right). \tag{3.19}$$

Let us integrate Eq. (3.19) on the sphere B_R of radius R and center in the origin and let us apply the divergence theorem. We find

$$\frac{d}{dt}\int_{B_R} dx\, |\psi(x,t)|^2 = \frac{i\hbar}{2m}\int_{\partial B_R} dS(x)\left(\bar{\psi}\frac{\partial\psi}{\partial n} - \psi\frac{\partial\bar{\psi}}{\partial n}\right)(x,t). \tag{3.20}$$

Taking the limit $R \to \infty$ in Eq. (3.20) and taking into account of (3.15), we obtain the thesis.   □

Remark 3.2 Note that the conservation law (3.16) can also be expressed in local form. Indeed, let us introduce the density ρ and the current density j

$$\rho := |\psi|^2, \qquad j := -\frac{i\hbar}{2m}(\bar{\psi}\nabla\psi - \psi\nabla\bar{\psi}). \tag{3.21}$$

Then, Eq. (3.19) can be rewritten in the form of a local conservation law

$$\frac{\partial \rho}{\partial t} + \nabla \cdot j = 0. \tag{3.22}$$

The above proposition indicates that the Hilbert space $L^2(\mathbb{R}^3)$ of the square integrable, complex valued functions is a natural space for the solutions of Eq. (3.14).
Moreover, it suggests that $|\psi(x, t)|^2$ could be interpreted as a quantity with a physical
meaning.

A first proposal was made by Schrödinger himself. In the case of an electron, he
assumed that the electron charge is distributed in space with a charge density given
by $e|\psi(x, t)|^2$, at time t (here e is the electron charge). The conservation of the total
charge would be guaranteed by the conservation of the L^2-norm of the solutions of
the equation and by the choice of a normalized wave function at time zero.

Schrödinger expected that an electron moving in space would be represented
by solutions in the form of wave packets remaining well concentrated during time
evolution. He verified the existence of solutions of this kind in the special case of the
harmonic oscillator. However, a serious difficulty arose as soon as it was realized that,
except in the special case of the harmonic oscillator, the solutions would inevitably
spread in space as time goes by.

Schrödinger proposal was consequently soon abandoned in favor of a new interpretation formulated by Born in 1926 and based on statistical considerations.

3.3 Born's Statistical Interpretation

In Born's view, an electron, being always experimentally detected in a point of space,
must be considered a corpuscle, i.e., a point particle. On the other hand, $|\psi(x, t)|^2$
is a function defined in the whole space and then it cannot uniquely identify the
position in space of a point particle.

The problem was analogous to that approached by Einstein in 1905 for the description of the photoelectric effect. In Born's words: "*I adhere to an observation of Einstein on the relationship of wave field and light quanta; he said, for example, that the
waves are present only to show the corpuscular light quanta the way, and he spoke
in the sense of a "ghost field". This determines the probability that a light quanta,
the bearer of energy and momentum, takes a certain path; however, the field itself
has no energy and no momentum*" ([4] p. 207).

Following Einstein, Born proposed a similar idea for the motion of microscopic
particles: "*the guiding field, represented by a scalar function ψ of the coordinates of
all the particles involved and the time, propagates in accordance with Schrödinger
differential equation. Momentum and energy, however, are transferred in the same
way as if corpuscles (electrons) actually moved. The paths of these corpuscles are
determined only to the extent that the laws of energy and momentum restrict them;
otherwise, only a probability for a certain path is found, determined by the values of*

the ψ function" ([4] p. 207). In particular, Born assumed that such a probability for the particle positions is given by the quantity $|\psi|^2$.

In a more precise way, Born's statement can be summarized as follows.

Given an electron described at time t by the wave function $\psi(x, t)$, the quantity $|\psi(x, t)|^2$ is the probability density to find the electron at time t in x. Therefore, the probability to find the electron at time t in a region A is given by $\int_A dx \, |\psi(x, t)|^2$.

Due to its crucial importance, it is convenient to further clarify the above statement explaining its empirical meaning.

Let us consider a system, e.g., a particle, prepared in some initial condition at $t = 0$ and then let the system evolve in time. At time t, we measure the position of the particle. Let us assume that this experiment is repeated N times, with the system prepared at $t = 0$ always in the same initial condition. Let A be a subset of $\mathbb{R}^3$ and let $N_{A,t}$ be the number of experiments in which the particle position is found in A at time t.

Furthermore, let us assume to know the wave function $\psi_0(x)$ describing the system at time $t = 0$ and to determine $\psi(x, t)$, solution of the Schrödinger equation (3.14) at time t corresponding to the initial datum $\psi_0(x)$.

Then, according to Born's statistical interpretation, the ratio (statistical frequency)

$$\frac{N_{A,t}}{N} \tag{3.23}$$

converges to

$$\int_A |\psi(x, t)|^2 dx \tag{3.24}$$

for N large. This is precisely the sense in which one must interpret the statement: "the probability to find the particle in A at time t is given by (3.24)".

Remark 3.3 The probability to find the particle somewhere in the space must be obviously one. Therefore, the initial wave function ψ_0 must be normalized in such a way that

$$\int_{\mathbb{R}^3} dx \, |\psi_0(x)|^2 = 1. \tag{3.25}$$

By Proposition 3.1, the same normalization holds for the wave function at any later time t. In this sense, Proposition 3.1 must be interpreted as the conservation law of probability.

Born's interpretation was soon accepted by the physics community and became a crucial aspect of the new quantum theory. In other words, such a theory must be considered as a probabilistic theory in the sense that, given the knowledge of the wave function, one can only predict the probability to find the electron in a given region of space.

3.4 Observables

A further element to be clarified is the role of the observables (e.g., position, momentum, angular momentum etc.) in the new theory. More precisely, we have to understand how the observable quantities can be represented as mathematical objects within the theory.

We recall that in Classical Mechanics, the observables quantities relative to a given system are represented by real and smooth functions defined in the phase space of the system. For example, in the case of a point particle the observable "j-th component of the position" is represented by the function $(x, p) \rightarrow x_j$, the observable "j-th component of the momentum" by the function $(x, p) \rightarrow p_j$, the observable "third component of the angular momentum" by the function $(x, p) \rightarrow x_1 p_2 - x_2 p_1$ and so on. We also note that the set of the classical observables is a commutative algebra with respect to the ordinary product of functions.

Let us consider the case of the new theory. We first observe that, if $|\psi(x, t)|^2$ is the probability density to find the particle in x at time t, then the mean value, or expectation, of the j-th component of the position of the particle at time t is

$$\langle x_j \rangle (t) = \int dx \, x_j \, |\psi(x, t)|^2 \tag{3.26}$$

(if not indicated, the domain of integration is the whole space). More generally, the mean value of an arbitrary function g of the position when the particle is described by the wave function $\psi(x, t)$ will be given by

$$\langle g \rangle (t) = \int dx \, g(x) |\psi(x, t)|^2. \tag{3.27}$$

Concerning the momentum, it is natural to write

$$\langle p_j \rangle (t) = m \frac{d}{dt} \langle x_j \rangle (t). \tag{3.28}$$

Using (3.26), (3.22) and then integrating by parts, one has

$$\langle p_j \rangle (t) = m \int dx \, x_j \frac{\partial}{\partial t} |\psi(x, t)|^2 = -m \int dx \, x_j \, \nabla \cdot j(x, t)$$
$$= m \int dx \, j_j(x, t) = -\frac{i\hbar}{2} \int dx \left(\bar{\psi}(x, t) \frac{\partial}{\partial x_j} \psi(x, t) - c.c. \right)$$
$$= \frac{\hbar}{i} \int dx \, \bar{\psi}(x, t) \frac{\partial}{\partial x_j} \psi(x, t). \tag{3.29}$$

Note that the right-hand side of (3.26) and (3.29) can be rewritten as the scalar products in $L^2(\mathbb{R}^3)$

$$\langle x_j \rangle(t) = (\psi(t), \hat{x}_j \psi(t)), \tag{3.30}$$

$$\langle p_j \rangle(t) = (\psi(t), \hat{p}_j \psi(t)), \tag{3.31}$$

where $\hat{x}_j$ denotes the position operator

$$\hat{x}_j \ : \ \phi(x) \to \left(\hat{x}_j \phi\right)(x) := x_j \phi(x) \tag{3.32}$$

and $\hat{p}_j$ denotes the momentum operator

$$\hat{p}_j \ : \ \phi(x) \to \left(\hat{p}_j \phi\right)(x) := \frac{\hbar}{i} \frac{\partial \phi(x)}{\partial x_j}. \tag{3.33}$$

We note that the operators (3.32) and (3.33) are linear and symmetric in $L^2(\mathbb{R}^3)$. We recall that an operator A in a Hilbert space $\mathcal{H}$ is linear if $A(\lambda_1 \psi_1 + \lambda_2 \psi_2) = \lambda_1 A \psi_1 + \lambda_2 A \psi_2$ and symmetric if $(\psi_1, A\psi_2) = (A\psi_1, \psi_2)$ for any $\lambda_1, \lambda_2 \in \mathbb{C}$ and $\psi_1, \psi_2 \in \mathcal{H}$. Moreover, the operators (3.32) and (3.33) are well defined only on a restricted set of elements of $L^2(\mathbb{R}^3)$, called domain of the operator. For example, $\hat{x}_j$ could be defined on the domain of $\phi \in L^2(\mathbb{R}^3)$ such that $\int dx \, x_j |\phi(x)|^2 < \infty$. In this chapter, the aim is to illustrate the emergence of the concepts at the basis of the new quantum theory and therefore we do not insist on these technical points of the theory of linear operators. We shall come back on these, and others, mathematical aspects in the next chapter.

The above considerations suggest to introduce the following (tentative) rule.

The observables "position" and "momentum" in Quantum Mechanics are represented by the linear and symmetric operators (3.32), (3.33) acting in the Hilbert space $L^2(\mathbb{R}^3)$.

Remark 3.4 Using the Fourier transform, formula (3.31) can be written as

$$\langle p_j \rangle(t) = \int dk \, \hbar k_j |\tilde{\psi}(k, t)|^2 = \frac{1}{\hbar^3} \int dq \, q_j |\tilde{\psi}(\hbar^{-1}q, t)|^2. \tag{3.34}$$

In analogy with (3.26) for the position observable, such a formula suggests to interpret the quantity $\hbar^{-3}|\tilde{\psi}(\hbar^{-1}q, t)|^2$ as the probability density to find the particle at time t with momentum q. Accordingly, we obtain that

$$\frac{1}{\hbar^3} \int_A dq \, |\tilde{\psi}(\hbar^{-1}q, t)|^2 \tag{3.35}$$

is the probability to find the particle at time t with momentum in $A \subset \mathbb{R}^3$.

In general, the two operators (3.32), (3.33) do not commute. Indeed, for any smooth function ϕ, a straightforward computation shows that

$$\left([\hat{x}_k, \hat{p}_l]\phi\right)(x) := (\hat{x}_k \hat{p}_l \phi)(x) - (\hat{p}_l \hat{x}_k \phi)(x) = i\hbar \delta_{kl} \phi(x). \tag{3.36}$$

The relations (3.36) are known as canonical commutation relations. They contain the constant $\hbar$ which provides a sort of "measure" of the noncommutativity of the observables $\hat{x}_j$ and $\hat{p}_j$.

As a physically relevant consequence of (3.36), we have the so-called Heisenberg uncertainty principle

$$\Delta x_k(t) \cdot \Delta p_l(t) \geq \frac{\hbar}{2} \delta_{kl}, \tag{3.37}$$

where $\Delta x_k(t)$ and $\Delta p_l(t)$ denote the standard deviations of position and momentum respectively

$$\Delta x_k(t) := \sqrt{\left(\psi(t), (\hat{x}_k - \langle \hat{x}_k \rangle)^2 \psi(t)\right)} = \sqrt{\langle \hat{x}_k^2 \rangle(t) - \langle \hat{x}_k \rangle(t)^2}, \tag{3.38}$$

$$\Delta p_l(t) := \sqrt{\left(\psi(t), (\hat{p}_l - \langle \hat{p}_l \rangle)^2 \psi(t)\right)} = \sqrt{\langle \hat{p}_l^2 \rangle(t) - \langle \hat{p}_l \rangle(t)^2}. \tag{3.39}$$

We recall that the standard deviation, or dispersion, $\Delta \xi$ of a random variable ξ is a quantity that measures the dispersion of ξ around its mean value $\langle \xi \rangle$. Roughly speaking, a small $\Delta \xi$ means that the possible values taken by ξ in repeated experiments are well concentrated around $\langle \xi \rangle$ and, therefore, we can predict with reasonable certainty (i.e., with probability close to one) that the result of a measurement of ξ will be very close to $\langle \xi \rangle$.

Inequality (3.37) expresses the fact that the standard deviations of the position $\hat{x}_k$ and momentum $\hat{p}_k$ cannot be made both arbitrarily small. In other words, we can always choose a wave function $\psi(t)$ such that $\Delta x_k(t)$ is very small, but this is obtained at the price to have $\Delta p_k(t)$ large in order to satisfy (3.37) (and vice versa). This means that for any given $\psi(t)$, we cannot predict with probability close to one both the k-component of the position and the k-component of the momentum. It is worth to underline the radical difference with Classical Mechanics, where limitations of this kind do not occur.

Finally, we note that the relations (3.37) impose no restriction on the product $\Delta x_k(t) \cdot \Delta p_l(t)$ for $k \neq l$.

Exercise 3.1 Let us consider a particle in dimension one. For any $\varepsilon > 0$ find $\psi_\varepsilon \in L^2(\mathbb{R})$ such that the standard deviation of the position computed with the wave function ψ_ε is less than ε. Repeat the computation in the case of the momentum.

The proof of the Heisenberg uncertainty principle (3.37) is obtained as a special case of the following result.

Proposition 3.2 *Let A, B, C be linear and symmetric operators in the Hilbert space $\mathcal{H}$. Let $\psi \in \mathcal{H}$ such that*

$$[A, B] \psi = i C \psi. \tag{3.40}$$

Then

$$\Delta A \cdot \Delta B \geq \frac{1}{2} |\langle C \rangle|. \tag{3.41}$$

Proof We are tacitly assuming that ψ is fixed in such a way that the expressions involved in (3.40) and (3.41) are well defined.

The Schwarz inequality and the symmetry of A and B imply

$$\Delta A \cdot \Delta B = \|(A - \langle A \rangle)\psi\| \cdot \|(B - \langle B \rangle)\psi\| \geq \left|\left((A - \langle A \rangle)\psi, (B - \langle B \rangle)\psi\right)\right|$$
$$= \left|\left(\psi, (A - \langle A \rangle)(B - \langle B \rangle)\psi\right)\right|. \tag{3.42}$$

Let us consider the identity

$$(A - \langle A \rangle)(B - \langle B \rangle)\psi = \frac{1}{2}\left[(A - \langle A \rangle)(B - \langle B \rangle) + (B - \langle B \rangle)(A - \langle A \rangle)\right]\psi + \frac{1}{2}[A, B]\psi$$
$$:= F\psi + \frac{i}{2}C\psi \tag{3.43}$$

where F is symmetric. Using the identity in (3.42) and recalling that the mean value of a symmetric operator is real, we obtain

$$\Delta A \cdot \Delta B \geq \left|\left(\psi, F\psi + \frac{i}{2}C\psi\right)\right| = \left|\langle F \rangle + \frac{i}{2}\langle C \rangle\right| = \sqrt{\langle F \rangle^2 + \frac{1}{4}\langle C \rangle^2} \geq \frac{1}{2}|\langle C \rangle|$$
$$\tag{3.44}$$

and then the proposition is proved.   □

Note that in the case of position and momentum observables in dimension one, with $\langle x \rangle = \langle p \rangle = 0$, the inequality (3.41) reduces to

$$\left(\int dx\, |\psi'(x)|^2\right)^{1/2} \left(\int dx\, x^2|\psi(x)|^2\right)^{1/2} \geq \frac{1}{2}\|\psi\|^2. \tag{3.45}$$

Remark 3.5 By formulas (3.42), (3.44) in the above proof, we see that the equality in the uncertainty relation (3.41) holds if and only if

$$(B - \langle B \rangle)\psi = \lambda (A - \langle A \rangle)\psi, \qquad \langle F \rangle = 0 \tag{3.46}$$

where $\lambda \in \mathbb{C}$.

Exercise 3.2 Let us consider the uncertainty principle for position and momentum in dimension one. Verify that the equality holds if and only if the wave function (with norm one) has the form

$$\psi_\beta(x) = \left(\frac{\beta}{\pi\hbar}\right)^{1/4} e^{-\frac{\beta}{2\hbar}(x-x_0)^2 + \frac{i}{\hbar}p_0 x}, \qquad \beta > 0, \tag{3.47}$$

where x_0, p_0 are the mean values of position and momentum. A wave function that realizes the condition $\Delta x \, \Delta p = \hbar/2$ is called coherent state.

In the above discussion, we have identified the fundamental quantum observables of position and momentum. For the identification of a generic quantum observable, it is quite natural to introduce the following quantization rule.

If $f(x, p)$ is a classical observable, then the corresponding quantum observable is represented by the linear and symmetric operator $\hat{f}$ obtained by replacing x_j and p_j, $j = 1, 2, 3$, with the operators $\hat{x}_j$ and $\hat{p}_j$, respectively.

It is important to stress that the noncommutativity of $\hat{x}_j$ and $\hat{p}_j$ makes the above rule ambiguous in certain cases. For example, it is not clear which quantum observable can be associated with a classical observable represented by the product $x_j p_j$. We do not further discuss this point here and we only note that in the applications that we consider this ambiguity does not occur.

Finally, we note that in a classical system, the set of the possible results of a measurement on the system of an observable represented by the function f is the spectrum, or the range, of the function f. By analogy, in the quantum case, we assume that the set of the possible results of a measurement on the system of an observable represented by the operator $\hat{f}$ is the spectrum of the operator $\hat{f}$ (the notion of spectrum of an operator is a generalization of the spectrum of a matrix and it will be defined in the next section).

Exercise 3.3 Verify that for a particle in $\mathbb{R}^3$ the observable energy is represented by the Hamiltonian operator

$$\hat{H} \; : \; \phi(x) \rightarrow \left(\hat{H}\phi\right)(x) := -\frac{\hbar^2}{2m}\Delta\phi(x) + V(x)\phi(x), \qquad (3.48)$$

the observable angular momentum by

$$\hat{L}_1 \; : \; \phi(x) \rightarrow \left(\hat{L}_1\phi\right)(x) := \frac{\hbar}{i}\left[x_2\frac{\partial\phi(x)}{\partial x_3} - x_3\frac{\partial\phi(x)}{\partial x_2}\right] \qquad (3.49)$$

and analogously for the components $\hat{L}_2$, $\hat{L}_3$. Note that, due to (3.48), the Schrödinger equation can be written in the form

$$i\hbar\frac{\partial\psi(t)}{\partial t} = \hat{H}\psi(t). \qquad (3.50)$$

Exercise 3.4 Verify the commutation rule

$$[\hat{L}_1, \hat{L}_2] = i\hbar\hat{L}_3 \qquad (3.51)$$

and the others obtained from (9.80) by cyclic permutation.

3.5   A First Sketch of the Theory

Let us summarize the elements of the new theory that we have identified in the previous sections. In order to underline the novelty of the quantum description, we first recall the basic elements of the classical description. Let us consider for simplicity a system made of a particle moving in $\mathbb{R}^3$ and subject to a force of potential energy $V(x)$. In the following rules, we define states and observables, the evolution law of the states and the possible predictions of the theory.

- The (pure) state of the particle is (x, p), where x is the position and p is the momentum of the particle. The space of the states is the phase space $\mathbb{R}^6$ of the system.
- The observables of the system are represented by smooth real functions defined on the phase space.
- Given the state (x_0, p_0) at $t = 0$, the state $(x(t), p(t))$ at time t is given by the solution of the Cauchy problem for the Hamilton's equations of the system.
- Given an observable represented by the function f, the set of the possible results of a measurement on the system of the observable is the spectrum (or range) of f. Moreover, if the system is in the state (x, p), we can predict with probability one that the result of a measurement on the system of the observable represented by f is $f(x, p)$.

Let us recall that a pure state corresponds to the maximal information available on the system. Otherwise, we have the so-called mixed state where we only know the probability distribution of the possible values of position and momentum.

We also emphasize that in Classical Mechanics both the evolution law of the states and the predictions of the theory are deterministic.

In the case of the quantum description of the particle motion, we have introduced the following new tentative rules.

- The (pure) state of the particle is the wave function ψ, with $\psi \in L^2(\mathbb{R}^3)$ and $\|\psi\| = 1$. Then the space of the states is the Hilbert space $L^2(\mathbb{R}^3)$. Note that $\mathbb{R}^3$ is the configuration space of the corresponding classical system.
- The observables of the system are represented by linear and symmetric operators in $L^2(\mathbb{R}^3)$. In particular, position and momentum observables are represented by (3.32), (3.33) and the other observables are obtained using the quantization rule.
- Given the state ψ_0 at $t = 0$, the state at time t is given by the solution of the Cauchy problem for the Schrödinger equation (3.50) of the system.
- Given an observable represented by the operator $\hat{f}$, the set of the possible results of a measurement on the system of the observable is the spectrum of $\hat{f}$. Moreover, if the system is in the state ψ, in general, we can only make probabilistic predictions on the possible results of a measurement on the system of the observable. In particular, $|\psi(x, t)|^2$ is the probability density to find the particle at time t with position x and $\hbar^{-3}|\tilde{\psi}(\hbar^{-1}q, t)|^2$ is the probability density to find the particle at time t with momentum q.

Note that in the quantum description, the evolution law of the states is deterministic while the predictions of the theory are probabilistic.

The above quantum rules are surely incomplete and require substantial clarifications and insights from the mathematical point of view.

The main point concerns the mathematical representation of the observables. In particular, we note that the result of a measurement of an observable is a real number and then, according to the last rule, the spectrum of the operator representing the observable must be real. Such a property holds if the operator is self-adjoint, while it is not guaranteed if the operator is only symmetric. As we shall see in the next chapter, the notions of symmetry and self-adjointness coincide only for bounded operators and, as a matter of fact, the operators representing observables in Quantum Mechanics are usually unbounded. This means that for a correct mathematical representation of a quantum observable we must make use of the notion of self-adjoint and unbounded operator in a Hilbert space.

Moreover, for a self-adjoint and unbounded operator, one can prove the spectral theorem and define the notion of function of an operator. This will allow us to find a natural generalization of the Born's rule to a generic observable. In this way, we will establish the possible probabilistic predictions for the results of a measurement of any observable, and not only for position and momentum.

Also the correct definition of the evolution law of the state requires the notion of self-adjointness. Indeed, the solution of the Schrödinger equation (3.50) is given, at least formally, by the linear operator in $L^2(\mathbb{R}^3)$

$$U(t) \; : \; \psi_0 \to \psi(t) = e^{-i\frac{\hat{H}}{\hbar}t}\psi_0. \tag{3.52}$$

For such a map, we must require that it preserves the L^2-norm, it is invertible, it is continuous in t, and it satisfies the group property. In other words, $U(t), t \in \mathbb{R}$, must be a strongly continuous one-parameter group of unitary operators in the Hilbert space $L^2(\mathbb{R}^3)$. In the next chapter, we shall see that this property holds if and only if $\hat{H}$ is a self-adjoint operator and, therefore, the construction of the self-adjoint operator $\hat{H}$ representing the quantum Hamiltonian is the first step in the study of a dynamical problem in Quantum Mechanics.

We can conclude saying that a precise formulation of Quantum Mechanics requires the use of some notions of the theory of linear operators in Hilbert spaces which will be discussed in the next chapter. In Chap. 5, we shall use these notions to reformulate the rules of Quantum Mechanics for a single particle in a more precise way.

3.6 Classical Limit of Quantum Mechanics

The new theory has been constructed following the analogy with Optics. As a guiding heuristic principle, we have used the fact that Quantum Mechanics is a generalization of Classical Mechanics in the same way as Wave Optics generalizes Geometrical

Optics. On the other hand, Classical Mechanics works perfectly well when applied to the macroscopic world. Therefore, the problem remains to understand in which sense and under which conditions the evolution of the wave function generated by the Schrödinger equation reproduces the classical evolution. Such a problem is known as the classical limit of Quantum Mechanics.

The problem involves delicate aspects both from the mathematical and from the conceptual point of view. Here, we shall only give a brief introductory discussion.

A first relevant result connecting quantum and classical description is the following proposition, known as Ehrenfest theorem.

Proposition 3.3 *Let* $\psi(x, t)$ *be a sufficiently smooth solution of the Schrödinger equation (3.14). Then*

$$m \frac{d^2}{dt^2} \langle x_j \rangle(t) = \langle -\frac{\partial V}{\partial x_j} \rangle(t). \tag{3.53}$$

Proof Using (3.29), the Schrödinger equation and the Green's formula, we have

$$m \frac{d^2}{dt^2} \langle x_j \rangle = \frac{d}{dt} \langle p_j \rangle = -i\hbar \int dx \left(\frac{\partial \bar{\psi}}{\partial t} \frac{\partial \psi}{\partial x_j} + \bar{\psi} \frac{\partial^2 \psi}{\partial t \partial x_j} \right)$$

$$= \int dx \left(-\frac{\hbar^2}{2m} \Delta \bar{\psi} + V \bar{\psi} \right) \frac{\partial \psi}{\partial x_j} - \int dx \, \bar{\psi} \frac{\partial}{\partial x_j} \left(-\frac{\hbar^2}{2m} \Delta \psi + V \psi \right)$$

$$= \int dx \, V \bar{\psi} \frac{\partial \psi}{\partial x_j} - \int dx \, \bar{\psi} \left(\frac{\partial V}{\partial x_j} \psi + V \frac{\partial \psi}{\partial x_j} \right)$$

$$= -\int dx \, \bar{\psi} \frac{\partial V}{\partial x_j} \psi$$

$$= \langle -\frac{\partial V}{\partial x_j} \rangle. \tag{3.54}$$

□

Remark 3.6 Note that, in general,

$$\langle -\frac{\partial V}{\partial x_j} \rangle(t) \neq -\frac{\partial V}{\partial x_j} (\langle x \rangle(t)) \tag{3.55}$$

and then Ehrenfest theorem does not imply that the mean value of the position satisfies Newton's equation.

In order to discuss the conditions for the classical limit, let us consider a particle in dimension one for simplicity. Ehrenfest theorem would suggest the following line of reasoning. Let $t \to x(t)$ be the classical trajectory of a particle with initial conditions x_0, p_0. Suppose that we can construct an initial state ψ_0 with $\langle x \rangle(0) = x_0$, $\langle p \rangle(0) = p_0$ and such that the solution of the Schrödinger equation $\psi(t)$ is a wave packet with dispersion in position $\Delta x(t)$ very small for any t in a fixed time interval.

This means that $\psi(t)$ is well concentrated around the mean value $\langle x \rangle(t)$ or, roughly speaking, $|\psi(x,t)|^2 \simeq \delta(x - \langle x \rangle(t))$. In this case

$$\langle -\frac{dV}{dx} \rangle(t) \simeq -\frac{dV}{dx}(\langle x \rangle(t)) \tag{3.56}$$

and, by Ehrenfest theorem, we could affirm that $\langle x \rangle(t)$ is arbitrarily close to the classical trajectory of the particle. Therefore, the wave packet $\psi(t)$ would behave like the point particle following its classical trajectory.

Unfortunately, there is a serious difficulty in the above reasoning. To construct a wave packet $\psi(t)$ of the above kind, we have to require that the dispersion $\Delta x(0)$ at $t = 0$ is very small so that ψ_0 is well concentrated around $\langle x \rangle(0)$. But this condition is not sufficient and we should require that also $\Delta p(0)$ is very small. Indeed, if $\Delta p(0)$ is large then the probability to have an initial momentum different from $\langle p \rangle(0)$ is also large and this would imply that at a fixed later time t the dispersion of the position $\Delta x(t)$ would be significantly large.

On the other hand, by the Heisenberg uncertainty principle (3.37), we know that $\Delta x(0)$ and $\Delta p(0)$ cannot be made both arbitrarily small. Therefore, we must conclude that the uncertainty principle prevents the possibility to construct a solution $\psi(t)$ of the Schrödinger equation with an arbitrary small dispersion in position $\Delta x(t)$ for any t in a fixed time interval.

We underline that this is a characteristic aspect of the new theory and it shows the reason why the ordinary notion of trajectory seems to lose its meaning.

Nevertheless, taking into account the limitation introduced by the uncertainty principle, it is still possible to construct a wave packet whose evolution is at least "approximately classical" for a time interval "not too long".

Roughly speaking, the typical result can be formulated as follows. Let us assume that $\Delta x(0)$ and $\Delta p(0)$ are both small compatibly with the constraint $\Delta x(0) \cdot \Delta p(0) = \hbar/2$. Let us also assume that $\hbar$ is "small" with respect to the characteristic action of the system. Then, it is possible to prove that, for t not too large, the wave function behaves like a wave packet well concentrated around the classical trajectory of the particle. A proof of this statement in a simple case will be given in Appendix A.

References

1. Hund, F.: The History of Quantum Theory. Barnes & Noble Books (1974)
2. Tagliaferri, G.: Storia della Fisica Quantistica (in Italian). Franco Angeli, Milano (1985)
3. van der Waerden, B.L.: Sources of Quantum Mechanics. Dover Publications, New York (1968)
4. Ludwig, G.: Wave Mechanics. Pergamon Press, Oxford (1968)
5. von Neumann, J.: Mathematical Foundations of Quantum Mechanics. Princeton University Press, Princeton (1955)

Chapter 4
Linear Operators in Hilbert Spaces

4.1 Preliminaries

In this chapter, we discuss some elements of the theory of linear operators in Hilbert spaces. For a detailed treatment of the whole subject, the reader is referred to [1–3]. Further useful readings are [4–16].

Let us introduce some basic notation and recall few preliminary notions. We denote by $\mathscr{H}$ a Hilbert space on the field of the complex numbers $\mathbb{C}$, by $(\cdot, \cdot)$ the scalar product, antilinear in the first argument, and by $\| \cdot \|$ the norm induced by the scalar product. We always assume that $\mathscr{H}$ is infinite dimensional and separable. We denote by $\bar{z}$ the complex conjugate of $z \in \mathbb{C}$.

In general, examples and applications will concern the spaces l^2 and $L^2(\Omega)$, where $\Omega \subseteq \mathbb{R}^n$ is an open set. Unless otherwise stated, $L^2(\Omega)$ is the space of square-integrable functions in Ω with respect to Lebesgue measure.

Some useful spaces of regular functions are dense in $L^2(\Omega)$. In particular, we shall consider the space $C_0^\infty(\Omega)$ of the infinitely differentiable functions with compact support in Ω and the Schwartz space $\mathscr{S}(\mathbb{R}^n)$ of the infinitely differentiable functions $f : \mathbb{R}^n \to \mathbb{C}$ such that

$$\sup_{x \in \mathbb{R}^n} \left| x_1^{j_1} \cdots x_n^{j_n} D^\alpha f(x) \right| < \infty \tag{4.1}$$

for any choice of $j_1, \ldots, j_n \in \mathbb{N}$ and for any multi-index $\alpha = (\alpha_1, \ldots, \alpha_n) \in \mathbb{N}^n$ of order $|\alpha| = \alpha_1 + \cdots + \alpha_n$, where

$$D^\alpha f(x) := \frac{\partial^{|\alpha|} f(x)}{\partial x_1^{\alpha_1} \cdots \partial x_n^{\alpha_n}} . \tag{4.2}$$

We denote by $C^k(\Omega)$, $k \in \mathbb{N}$, the space of k-times differentiable functions in Ω.

When we write an integral we do not specify the domain of integration if such domain coincides with the domain of definition of the integration variable.

© Springer International Publishing AG, part of Springer Nature 2018
A. Teta, *A Mathematical Primer on Quantum Mechanics*,
UNITEXT for Physics, https://doi.org/10.1007/978-3-319-77893-8_4

We shall extensively use the Fourier transform of a function f from $\mathbb{R}^n$ to $\mathbb{C}$, defined by

$$\tilde{f}(k) := (\mathscr{F} f)(k) = \frac{1}{(2\pi)^{n/2}} \int dx \, e^{-ik\cdot x} f(x) \qquad (4.3)$$

with inverse

$$f(x) := (\mathscr{F}^{-1} \tilde{f})(x) = \frac{1}{(2\pi)^{n/2}} \int dk \, e^{ik\cdot x} \tilde{f}(k). \qquad (4.4)$$

The Fourier transform is a bijection from $\mathscr{S}(\mathbb{R}^n)$ to $\mathscr{S}(\mathbb{R}^n)$ and it has the fundamental property to reduce derivation to multiplication by the independent variable. More precisely, for $f \in \mathscr{S}(\mathbb{R}^n)$

$$\left(\mathscr{F} \frac{\partial f}{\partial x_j}\right)(k) = ik_j \, (\mathscr{F} f)(k). \qquad (4.5)$$

Fourier transform can also be defined for an element of $L^2(\mathbb{R}^n)$. In fact, for any $f \in L^2(\mathbb{R}^n)$ we consider

$$\tilde{f}_N(k) = \frac{1}{(2\pi)^{n/2}} \int_{|x|<N} dx \, e^{-ik\cdot x} f(x). \qquad (4.6)$$

By Plancherel theorem, there exists a unique $\tilde{f} \in L^2(\mathbb{R}^n)$ such that $\|\tilde{f}_N - \tilde{f}\| \to 0$ for $N \to \infty$ and moreover

$$\int dk \, |\tilde{f}(k)|^2 = \int dx \, |f(x)|^2. \qquad (4.7)$$

By definition, $\tilde{f}$ is the Fourier transform of f.

We also recall the notions of weak derivative and Sobolev space (for details see, e.g., [17, 18]). Given $u \in L^1_{loc}(\Omega)$ and a multi-index α, we say that $v \in L^1_{loc}(\Omega)$ is the α-th weak partial derivative of u if

$$\int_\Omega dx \, u \, D^\alpha \phi = (-1)^{|\alpha|} \int_\Omega dx \, v \, \phi, \qquad \text{for any} \quad \phi \in C_0^\infty(\Omega) \qquad (4.8)$$

and we write $D^\alpha u = v$. The weak derivative is uniquely defined (up to a set of measure zero) and it reduces to the standard partial derivative if $u \in C^{|\alpha|}(\Omega)$. The Sobolev space $H^m(\Omega)$, $m \in \mathbb{N}$, is the set of $u \in L^2(\Omega)$ such that for any multi-index α, $|\alpha| \leq m$, there exists the weak derivative $D^\alpha u$ and it belongs to $L^2(\Omega)$. It turns out that $H^m(\Omega)$, equipped with the scalar product

$$(u, v)_{H^m(\Omega)} = \int_\Omega dx \, \overline{u} \, v + \sum_{1 \leq |\alpha| \leq m} \int_\Omega dx \, \overline{D^\alpha u} \, D^\alpha v, \qquad (4.9)$$

is a separable Hilbert space.

We remark that for $u \in H^1(\mathscr{I})$, where $\mathscr{I}$ is an open interval of the real line, there exists $u_1 \in C^0(\bar{\mathscr{I}})$ such that $u = u_1$ a.e. in $\mathscr{I}$ and $u_1(x) - u_1(y) = \int_y^x ds\, u'(s)$ for $x, y \in \bar{\mathscr{I}}$ (then we shall identify $u \in H^1(\mathscr{I})$ with its continuous representative u_1).

Finally, in the case $\Omega = \mathbb{R}^n$ the Sobolev space can be equivalently defined via Fourier transform, i.e.,

$$H^m(\mathbb{R}^n) = \left\{ f \in L^2(\mathbb{R}^n) \mid \int dk\, |k|^{2m} |\tilde{f}(k)|^2 < \infty \right\}. \tag{4.10}$$

4.2 Bounded and Unbounded Operators

Let us introduce the notion of linear operator in a Hilber space.

Definition 4.1 A linear operator in a Hilbert space $\mathscr{H}$ is a map $A : D(A) \subseteq \mathscr{H} \to \mathscr{H}$ such that $A(\alpha f + \beta g) = \alpha Af + \beta Ag$ for any $\alpha, \beta \in \mathbb{C}$ and for any $f, g \in D(A)$.

The subset $D(A)$ of $\mathscr{H}$ is called domain of A and we always assume that it is a linear manifold (i.e., any finite linear combination of elements of $D(A)$ belongs to $D(A)$). Unless otherwise stated, we shall also assume that $D(A)$ is dense in $\mathscr{H}$. The kernel of A is $\{f \in D(A) \mid Af = 0\}$ and it is denoted by $Ker(A)$. Obviously, the null vector belongs to $Ker(A)$. The range of A is $\{f \in \mathscr{H} \mid f = Ag,\ g \in D(A)\}$ and it is denoted by $Ran(A)$. The identity operator is denoted by I. Often, the operator αI, $\alpha \in \mathbb{C}$, will be simply denoted by α. In the following, by operator we always mean linear operator in $\mathscr{H}$.

Definition 4.2 The operator $A, D(A)$ is bounded if

$$\|A\| := \sup_{f \in D(A), f \neq 0} \frac{\|Af\|}{\|f\|} = \sup_{f \in D(A), \|f\|=1} \|Af\| < \infty \tag{4.11}$$

and $\|A\|$ denotes the norm of A (with an abuse of notation the norm of an operator is denoted with the same symbol used for the norm of a vector in the Hilbert space). In the following, we shall often use the fact that $\|Af\| = \sup_{g \in \mathscr{H}, \|g\|=1} |(g, Af)|$.

In certain cases an operator, initially defined on a given domain, can be extended to a larger domain.

Definition 4.3 The operator B defined on $D(B)$ is an extension of A defined on $D(A)$ if $D(B) \supseteq D(A)$ and $Bf = Af$ for any $f \in D(A)$.

We shall use the notation $B \supseteq A$; in the case $D(B) \neq D(A)$, we shall write $B \supset A$. A bounded operator A defined on $D(A)$ can be uniquely extended by continuity to the whole space $\mathscr{H}$. Then, for a bounded operator A we always assume

$D(A) = \mathcal{H}$. On the other hand, for an unbounded operator it is crucial to specify the domain.

For a sequence of bounded operators, we distinguish norm, strong, and weak convergence.

Definition 4.4 Let A_k, $k \in \mathbb{N}$, and A bounded operators in $\mathcal{H}$. Then
$n - \lim_{k \to \infty} A_k = A$ if $\lim_{k \to \infty} \|A_k - A\| = 0$ (norm sense);
$s - \lim_{k \to \infty} A_k = A$ if $\lim_{k \to \infty} \|A_k f - Af\| = 0$ for any $f \in \mathcal{H}$ (strong sense);
$w - \lim_{k \to \infty} A_k = A$ if $\lim_{k \to \infty} (A_k f, g) = (Af, g)$ for any $f, g \in \mathcal{H}$ (weak sense).

It is a simple exercise to verify that norm convergence implies strong convergence and that the latter implies weak convergence.

Note that the set of all bounded operators in $\mathcal{H}$, equipped with the norm (4.11), is a Banach space and it will be denoted by $\mathcal{B}(\mathcal{H})$ (see, e.g., [1], Theorem III.2).

Example 4.1 Let $\mathcal{H} = L^2(\mathbb{R}^n)$ and $\varphi \in L^\infty(\mathbb{R}^n)$. Let us define the multiplication operator by the bounded function φ

$$(M_\varphi f)(x) = \varphi(x) f(x) . \tag{4.12}$$

The operator is bounded, since $\|M_\varphi f\| \le \|\varphi\|_{L^\infty} \|f\|$.

Example 4.2 Let us consider an integral operator in $L^2(\mathbb{R})$

$$(Kf)(x) = \int dy\, k(x, y) f(y) , \tag{4.13}$$

where $k(x, y)$ denotes the corresponding integral kernel. Moreover, let us assume that $k \in L^2(\mathbb{R}^2)$. Using the Schwarz inequality, one has

$$\|Kf\|^2 = \int dx \left| \int dy\, k(x, y) f(y) \right|^2 \le \|f\|^2 \int dx \int dy\, |k(x, y)|^2 \tag{4.14}$$

and then the operator is bounded.

Example 4.3 In $\mathcal{H} = L^2(\mathbb{R})$, we consider the multiplication operator by the independent variable
$$D(Q_0) = \mathscr{S}(\mathbb{R}), \quad (Q_0 f)(x) = x f(x) . \tag{4.15}$$

The operator in unbounded. In fact, let us define the sequence

$$f_n(x) = \pi^{-1/4} e^{-\frac{(x-n)^2}{2}} . \tag{4.16}$$

The function f_n is a Gaussian with $\|f_n\| = 1$ and it reaches its maximum in $x_n = n$. Moreover

$$\| Q_0 f_n \|^2 = \frac{1}{\sqrt{\pi}} \int dx \, x^2 e^{-(x-n)^2} = \frac{1}{\sqrt{\pi}} \int dz \, (z^2 + n^2) e^{-z^2} \to \infty \qquad (4.17)$$

for $n \to \infty$.

Example 4.4 In $\mathscr{H} = L^2(\mathbb{R})$, we consider the operator

$$D(P_0) = \mathscr{S}(\mathbb{R}), \quad (P_0 f)(x) = -i f'(x). \qquad (4.18)$$

If we define $f_n(x) = (\mathscr{F}^{-1} \tilde{f}_n)(x)$, with

$$\tilde{f}_n(k) = \pi^{-1/4} e^{-\frac{(k-n)^2}{2}} \qquad (4.19)$$

we have

$$\| P_0 f_n \|^2 = \| \mathscr{F}(P_0 f_n) \|^2 = \frac{1}{\sqrt{\pi}} \int dk \, k^2 e^{-(k-n)^2} \qquad (4.20)$$

and then the operator is unbounded.

The next proposition provides a useful sufficient condition to prove boundedness of an integral operator.

Proposition 4.1 *Let A be an integral operator in $L^2(\Omega)$, with Ω open set of $\mathbb{R}^n$, and let $a(x, y)$ be the corresponding measurable integral kernel. If*

$$\sup_{x \in \Omega} \int dy \, |a(x, y)| := a_1 < \infty \qquad \sup_{y \in \Omega} \int dx \, |a(x, y)| := a_2 < \infty \qquad (4.21)$$

then A is bounded and $\|A\| \le \sqrt{a_1 a_2}$.

Proof By the Schwarz inequality

$$|(Af)(x)| \le \int dy \, |f(y)||a(x, y)| = \int dy \, |f(y)||a(x, y)|^{1/2} |a(x, y)|^{1/2}$$

$$\le \left(\int dy \, |f(y)|^2 |a(x, y)| \right)^{1/2} \left(\int dy \, |a(x, y)| \right)^{1/2} \le \sqrt{a_1} \left(\int dy \, |f(y)|^2 |a(x, y)| \right)^{1/2} \qquad (4.22)$$

and then

$$\|Af\|^2 = \int dx \, |(Af)(x)|^2 \le a_1 \int dy \, |f(y)|^2 \int dx \, |a(x, y)| \le a_1 a_2 \|f\|^2. \qquad (4.23)$$

$\square$

Exercise 4.1 Verify that the following operator in $L^2(0, 1)$ is bounded:

$$(Vf)(x) = \int_0^x dy \, \frac{f(y)}{\sqrt{x - y}}. \qquad (4.24)$$

Exercise 4.2 Verify that the following operator in $L^2(\mathbb{R})$ is bounded:

$$(G_1 f)(x) = \int dy \, f(y) \, e^{-|x-y|} \,. \tag{4.25}$$

Exercise 4.3 Verify that the following operator in $L^2(\mathbb{R}^3)$ is bounded:

$$(G_3 f)(x) = \int dy \, f(y) \, \frac{e^{-|x-y|}}{|x-y|} \,. \tag{4.26}$$

Exercise 4.4 Verify that the following operator in $L^2(-1, 1)$ is not bounded:

$$D(\delta) = C^0(-1, 1)\,, \qquad (\delta f)(x) = f(0)\,. \tag{4.27}$$

We recall that the inverse of an operator A, $D(A)$ is defined if and only if $Ker\,(A) = \{0\}$, moreover $D(A^{-1}) = Ran\,(A)$, $Ran\,(A^{-1}) = D(A)$.

We conclude this section observing that if C is a bounded operator in $\mathscr{H}$, with $\|C\| < 1$, then the Neumann series

$$\sum_{n=0}^{\infty} (-1)^n C^n \tag{4.28}$$

is convergent and it defines a bounded operator in $\mathscr{H}$. Indeed, denoted $S_N = \sum_{n=0}^{N} (-1)^n C^n$, we have for $N < M$

$$\left\| S_N - S_M \right\| = \left\| \sum_{n=N+1}^{M} (-1)^n C^n \right\| \le \sum_{n=N+1}^{M} \|C^n\| \le \sum_{n=N+1}^{M} \|C\|^n \,, \tag{4.29}$$

where we have used the property $\|AB\| \le \|A\|\|B\|$, for $A, B \in \mathscr{B}(\mathscr{H})$. Hence, S_N is a Cauchy sequence and therefore converges to an element of $\mathscr{B}(\mathscr{H})$. Furthermore

$$(I + C) \sum_{n=0}^{\infty} (-1)^n C^n = \sum_{n=0}^{\infty} (-1)^n C^n - \sum_{n=0}^{\infty} (-1)^{n+1} C^{n+1} = I \,. \tag{4.30}$$

Analogously, we have $\sum_{n=0}^{\infty} (-1)^n C^n \, (I+C) = I$ and we conclude that the Neumann series (4.28) defines the operator $(I + C)^{-1}$.

4.3 Adjoint Operators

Let A be a linear bounded operator in $\mathscr{H}$. For any fixed $f \in \mathscr{H}$, we define the linear functional in $\mathscr{H}$

$$\mathfrak{F}_f \; : \; g \in \mathscr{H} \rightarrow (f, Ag). \tag{4.31}$$

The boundedness of A implies that the functional $\mathfrak{F}_f$ is also bounded for any $f \in \mathscr{H}$. Then, by Riesz theorem, there exists a unique $h \in \mathscr{H}$ such that $\mathfrak{F}_f(g) = (f, Ag) = (h, g)$ for any $g \in \mathscr{H}$.

Definition 4.5 Let A be a bounded operator. The adjoint of A is the map $A^* : f \rightarrow h$ such that $(f, Ag) = (h, g)$ for any $g \in \mathscr{H}$.

One has that A^* is a linear operator (verify), it is bounded and $\|A^*\| = \|A\|$. Indeed,

$$\|A^* f\| = \sup_{g \in \mathscr{H}, \|g\|=1} |(g, A^* f)| = \sup_{g \in \mathscr{H}, \|g\|=1} |(Ag, f)|$$
$$\leq \|f\| \sup_{g \in \mathscr{H}, \|g\|=1} \|Ag\| = \|f\| \|A\| \tag{4.32}$$

so that $\|A^*\| \leq \|A\|$. In the same way, one sees that $\|A\| \leq \|A^*\|$.

Example 4.5 The adjoint of the operator M_φ defined in (4.12) is given by $(M_{\bar\varphi} f)(x) = \overline{\varphi(x)} f(x)$.

Example 4.6 Let K be the integral operator defined in (4.13). The adjoint K^* is the integral operator with kernel $k^*(x, y) = \overline{k(y, x)}$. Indeed,

$$(g, K^* f) = (Kg, f) = \int dy \int dx\, g(x) k(y, x)\, f(y) = \int dx\, \overline{g(x)} \int dy\, f(y) \overline{k(y, x)} \tag{4.33}$$

and then $(K^* f)(x) = \int dy\, f(y) \overline{k(y, x)}$.

Exercise 4.5 Find the adjoint of the operator V defined in (4.24).

For an unbounded operator, the definition of adjoint is more delicate, since the functional $\mathfrak{F}_f$ is not bounded for any $f \in \mathscr{H}$.

Definition 4.6 Let $A, D(A)$ be an unbounded linear operator. We define the linear manifold $D(A^*)$ of $\mathscr{H}$ as

$$D(A^*) = \left\{ f \in \mathscr{H} \mid \mathfrak{F}_f \; : \; g \rightarrow (f, Ag) \text{ is bounded on } D(A) \right\}$$
$$= \left\{ f \in \mathscr{H} \mid \sup_{g \in D(A), \|g\|=1} |(f, Ag)| < \infty \right\}. \tag{4.34}$$

By definition, for any $f \in D(A^*)$ the functional $\mathfrak{F}_f$ defined on $D(A)$ is bounded. Since $D(A)$ is dense in $\mathscr{H}$, the functional has a unique extension to a bounded and linear functional in $\mathscr{H}$. By Riesz theorem, there exists a unique $h \in \mathscr{H}$ such that $\mathfrak{F}_f(g) = (f, Ag) = (h, g)$. Therefore

$$D(A^*) = \{ f \in \mathcal{H} \mid \exists h \in \mathcal{H} \text{ such that } (f, Ag) = (h, g), \ \forall g \in D(A) \}. \tag{4.35}$$

Then, we have have the following.

Definition 4.7 Let $A, D(A)$ be an unbounded operator. The adjoint of $A, D(A)$ is the map $A^*, D(A^*)$ defined as

$$A^* : f \in D(A^*) \to h \text{ such that } (f, Ag) = (h, g), \ \forall g \in D(A). \tag{4.36}$$

As in the previous case, one has that $A^*, D(A^*)$ is a linear operator.

Remark 4.1 If $D(A)$ is not dense, then the functional $\mathfrak{F}_f$ cannot be uniquely extended to $\mathcal{H}$ and therefore the adjoint operator is not uniquely defined.

Example 4.7 Let us consider the unbounded operator $Q_0, D(Q_0)$ defined in (4.15) and let us show that

$$D(Q_0^*) = \left\{ f \in L^2(\mathbb{R}) \mid \int dx\, x^2 |f(x)|^2 < \infty \right\}, \qquad (Q_0^* f)(x) = x f(x). \tag{4.37}$$

Given $f \in D(Q_0^*)$, we define the sequence of integrable functions $|\xi_N(x)\, xf(x)|^2$, where $\xi_N \in C_0^\infty(\mathbb{R})$, $N \in \mathbb{N}$, is nonnegative and $\xi_N(x) = 1$ for $x \in [-N, N]$ (see Exercise 4.6). The sequence is monotone increasing and

$$\|Q_0^* f\| = \lim_N \|\xi_N Q_0^* f\| = \lim_N \sup_{g \in D(Q_0), \|g\|=1} |(g, \xi_N Q_0^* f)|$$

$$= \lim_N \sup_{g \in D(Q_0), \|g\|=1} |(Q_0 \xi_N g, f)| = \lim_N \sup_{g \in D(Q_0), \|g\|=1} \left| \int dx\, \overline{g(x)}\, \xi_N(x)\, xf(x) \right|$$

$$= \lim_N \left(\int dx\, |\xi_N(x)\, xf(x)|^2 \right)^{1/2}. \tag{4.38}$$

Applying the monotone convergence theorem, we find

$$\int dx\, x^2 |f(x)|^2 < \infty. \tag{4.39}$$

On the other hand, if f satisfies condition (4.39), we set $h(x) = xf(x)$ and then we have $(f, Q_0 g) = (h, g)$ for any $g \in D(Q_0)$, so that $f \in D(Q_0^*)$ and $(Q_0^* f)(x) = xf(x)$. Note that the adjoint $Q_0^*, D(Q_0^*)$ is an extension of $Q_0, D(Q_0)$.

Example 4.8 Let us consider the unbounded operator $P_0, D(P_0)$ defined in (4.18). Given $f \in D(P_0^*)$, we have

$$\|P_0^* f\| = \|\mathscr{F}(P_0^* f)\| = \lim_N \|\xi_N \mathscr{F}(P_0^* f)\|. \tag{4.40}$$

Proceeding as in the previous example, we find (verify)

$$D(P_0^*) = \left\{ f \in L^2(\mathbb{R}) \mid \int dk \, k^2 |\tilde{f}(k)|^2 < \infty \right\}, \quad (P_0^* f)(x) = \frac{1}{\sqrt{2\pi}} \int dk \, e^{ikx} \, k \tilde{f}(k)$$

$$\tag{4.41}$$

and thus $D(P_0^*) = H^1(\mathbb{R})$ and $(P_0^* f)(x) = -if'(x)$. Also in this case the adjoint $P_0^*, D(P_0^*)$ is an extension of $P_0, D(P_0)$.

Exercise 4.6 Consider the function

$$\zeta(x) = \begin{cases} \dfrac{2 \, e^{\frac{1}{4x^2-1}}}{\int_{-1}^{1} dy \, e^{\frac{1}{y^2-1}}} & \text{if} \quad |x| < \dfrac{1}{2} \\[4mm] 0 & \text{if} \quad |x| > \dfrac{1}{2} \end{cases} \tag{4.42}$$

and verify that

$$\xi_N(x) = \int_{-\infty}^{x/N} dy \left[\zeta\left(y + \frac{3}{2}\right) - \zeta\left(y - \frac{3}{2}\right) \right] \tag{4.43}$$

belongs to $C_0^\infty(\mathbb{R})$, is nonnegative and $\xi_N(x) = 1$ for $x \in [-N, N]$.

Exercise 4.7 Find the adjoint of the operator δ defined in (4.27).

Exercise 4.8 Consider the operator in $L^2(0, 1)$ defined by

$$D(p) = \{ u \in H^1(0, 1) \mid u(0) = u(1) = 0 \}, \quad (pu)(x) = -iu'(x) \tag{4.44}$$

and verify that the adjoint is

$$D(p^*) = H^1(0, 1), \quad (p^* u)(x) = -iu'(x). \tag{4.45}$$

4.4 Symmetric and Self-adjoint Operators

The notion of symmetric operator in a Hilbert space is analogous to that of a symmetric matrix.

Definition 4.8 A linear operator $A, D(A)$ is symmetric if

$$(Af, g) = (f, Ag) \qquad \forall f, g \in D(A). \tag{4.46}$$

The next proposition shows that the adjoint of a symmetric operator $A, D(A)$ is an extension of $A, D(A)$.

Proposition 4.2 *If $A, D(A)$ is symmetric, then $A^* \supseteq A$.*

Proof If $f \in D(A)$, we have

$$\sup_{g \in D(A), \|g\|=1} |(f, Ag)| = \sup_{g \in D(A), \|g\|=1} |(Af, g)| = \|Af\|. \qquad (4.47)$$

From the definition of $D(A^*)$, it follows that $f \in D(A^*)$ and then $D(A) \subseteq D(A^*)$. Moreover, for $f \in D(A)$

$$(Af, g) = (f, Ag) = (A^* f, g) \qquad \forall g \in D(A). \qquad (4.48)$$

Thus $Af = A^* f$ and the proposition is proved. $\square$

When the domain of a symmetric operator coincides with the domain of its adjoint, we have a self-adjoint operator.

Definition 4.9 An operator $A, D(A)$ is self-adjoint if $A = A^*$, i.e., if A is symmetric and $D(A) = D(A^*)$.

Example 4.9 The operator M_φ defined in (4.12) is self-adjoint if and only if φ is real.

Example 4.10 The operator K defined in (4.13) is self-adjoint if and only if $k(x, y) = \overline{k(y, x)}$.

Example 4.11 The operators $Q_0, D(Q_0)$ and $Q_0^*, D(Q_0^*)$ defined in (4.15), (4.37) are symmetric. Note that $Q_0, D(Q_0)$ is not self-adjoint, since $Q_0^* \supset Q_0$. The same is true for $P_0, D(P_0)$ defined in (4.18).

Example 4.12 The operator $p, D(p)$ defined in (4.44) is symmetric but not self-adjoint. Its adjoint (4.45) is not symmetric.

For a bounded operator, the notions of symmetry and self-adjointness coincide. In the case of an unbounded operator, the two notions are distinct and the problem arises to establish if a self-adjoint operator can be constructed starting from a symmetric one. To this aim, it is useful the following proposition.

Proposition 4.3 *Let $A, D(A)$ be symmetric and let $B, D(B)$ be an extension of $A, D(A)$. Then, $A^*, D(A^*)$ is an extension of $B^*, D(B^*)$.*

Proof Consider $f \in D(B^*)$. By definition of $D(B^*)$, we have

$$\sup_{g \in D(B), \|g\|=1} |(f, Bg)| < \infty \qquad (4.49)$$

and in particular

$$\sup_{g \in D(A), \|g\|=1} |(f, Bg)| = \sup_{g \in D(A), \|g\|=1} |(f, Ag)| < \infty. \qquad (4.50)$$

From (4.50), it follows that $f \in D(A^*)$ and therefore $D(B^*) \subseteq D(A^*)$. Let us consider $f \in D(B^*)$ and $g \in D(A)$. We have

$$(B^* f, g) = (f, Bg) = (f, Ag) = (A^* f, g). \tag{4.51}$$

Since $D(A)$ is dense, we conclude that $B^* f = A^* f$ for any $f \in D(B^*)$ and the proof is complete. $\qquad\square$

The previous proposition shows that if we enlarge the domain of a symmetric operator then we restrict the domain of the adjoint. Thus, the construction of a self-adjoint operator from a symmetric one reduces to find a suitable extension of the symmetric operator such that its domain and the domain of the adjoint coincide. We underline that many different situations can occur, i.e., such self-adjoint extensions may not exist or, if they exist, they can be finite or infinite.

In most applications to Quantum Mechanics, the strategy is to start with an operator defined on a domain of smooth functions, which is symmetric but not self-adjoint. Then, one looks for and classifies the possible self-adjoint extensions. If a symmetric operator has more than one self-adjoint extension, the choice depends on the specific physical problem under consideration.

Definition 4.10 A symmetric operator is essentially self-adjoint if it admits a unique self-adjoint extension.

Proposition 4.4 *Let $A, D(A)$ be a symmetric operator such that the adjoint $A^*, D(A^*)$ is self-adjoint. Then, $A, D(A)$ is essentially self-adjoint and its unique self-adjoint extension coincides with $A^*, D(A^*)$.*

Proof By Proposition 4.2, the adjoint $A^*, D(A^*)$ is a self-adjoint extension of $A, D(A)$. Let $B, D(B)$ be another self-adjoint extension of $A, D(A)$. Then, $B := B^* \subseteq A^*$. On the other hand, if $B \subseteq A^*$ then $(A^*)^* \subseteq B^* := B$. Since, by hypotheses, $A^* := (A^*)^*$, we have

$$B := B^* \subseteq A^* := (A^*)^* \subseteq B^* := B. \tag{4.52}$$

This implies that B coincides with A^*. $\qquad\square$

Exercise 4.9 Let us consider the operator in $L^2(0, 1)$

$$D(p_\alpha) = \left\{ u \in H^1(0, 1) \mid u(0) = \alpha\, u(1),\ \alpha \in \mathbb{C},\ |\alpha| = 1 \right\},$$
$$(p_\alpha u)(x) = -i u'(x). \tag{4.53}$$

Verify that the operator is self-adjoint. Note that the operator (4.53) is a self-adjoint extension of the symmetric operator $p, D(p)$ defined in (4.44). Thus, $p, D(p)$ has infinite self-adjoint extensions.

4.5 Self-adjointness Criterion and Kato–Rellich Theorem

In order to verify that a given symmetric operator $A, D(A)$ is self-adjoint, one has to prove that $D(A^*) \subseteq D(A)$. However, in many cases, the explicit characterization of $D(A^*)$ is not easy. It is therefore useful the following criterion.

Proposition 4.5 (Self-adjointness criterion)
Let $A, D(A)$ be a symmetric operator. If there exists $z \in \mathbb{C}$, with $\Im z \neq 0$ such that

$$Ran\,(A - z) = Ran\,(A - \bar{z}) = \mathscr{H}, \qquad (4.54)$$

then $A, D(A)$ is self-adjoint.

Proof Since A is symmetric, we know that $D(A^*) \supseteq D(A)$ and it is sufficient to prove $D(A^*) \subseteq D(A)$. Let $f \in D(A^*)$ and consider the vector $(A^* - \bar{z})f$. By hypothesis, there exists $g \in D(A)$ such that the vector $(A^* - \bar{z})f$ can be written as

$$(A^* - \bar{z})f = (A - \bar{z})g . \qquad (4.55)$$

Then, for any $\phi \in D(A)$ we have

$$(f, (A - z)\phi) = ((A^* - \bar{z})f, \phi) = ((A - \bar{z})g, \phi) = (g, (A - z)\phi) \qquad (4.56)$$

or, equivalently

$$(f - g, (A - z)\phi) = 0 . \qquad (4.57)$$

Since (4.57) holds for any $\phi \in D(A)$ and moreover, by hypothesis, $Ran\,(A - z) = \mathscr{H}$, it turns out that $f - g$ is orthogonal to any vector of $\mathscr{H}$ and then it is the null vector. This means that f coincides with $g \in D(A)$. Therefore, $f \in D(A)$ and the proposition is proved. □

Example 4.13 The operator $Q_0^*, D(Q_0^*)$ defined in (4.37) is symmetric. Let us show that it is self-adjoint. Given $f \in L^2(\mathbb{R})$, we define

$$u_f^{\pm}(x) = \frac{f(x)}{x \pm i} . \qquad (4.58)$$

It is easy to see that $u_f^{\pm} \in D(Q_0^*)$ and $(Q_0^* \pm i)u_f^{\pm} = f$, so that $Ran\,(Q_0^* \pm i) = L^2(\mathbb{R})$. By the self-adjointness criterion, we conclude that $Q_0^*, D(Q_0^*)$ is self-adjoint. Moreover, Proposition 4.4 implies that $Q_0, D(Q_0)$ is essentially self-adjoint and its unique self-adjoint extension, denoted as $Q, D(Q)$, coincides with $Q_0^*, D(Q_0^*)$.

Example 4.14 The operator $P_0^*, D(P_0^*)$ defined in (4.41) is symmetric. Moreover, given $f \in L^2(\mathbb{R})$, we define

$$v_f^{\pm}(x) = \frac{1}{\sqrt{2\pi}} \int dk\, e^{ikx}\, \frac{\tilde{f}(k)}{k \pm i} \qquad (4.59)$$

and we have $v_f^{\pm} \in D(P_0^*)$ e $(P_0^* \pm i)v_f^{\pm} = f$. Also in this case we conclude that $P_0^*, D(P_0^*)$ is self-adjoint and it coincides with the unique self-adjoint extension $P, D(P)$ of $P_0, D(P_0)$.

Exercise 4.10 Let us consider the operator in l^2

$$D(A_k) = \left\{ x = (x_1, \ldots, x_n, \ldots) \in l^2 \mid \sum_{n=1}^{\infty} n^{2k} |x_n|^2 < \infty \right\}, \qquad (A_k x)_n = n^k x_n$$

(4.60)

for any $k \in \mathbb{N}$. Verify that it is self-adjoint.

The sufficient condition for self-adjointness proved in Proposition 4.5 is also a necessary condition. To prove this fact, we first introduce the notion of closed operator.

Definition 4.11 The operator $A, D(A)$ is closed if for any sequence $\{f_n\}$ such that

$$f_n \in D(A), \qquad f_n \to f, \qquad A f_n \to g \tag{4.61}$$

one has

$$f \in D(A), \qquad A f = g. \tag{4.62}$$

Note that a bounded operator is closed.

Proposition 4.6 *The adjoint of an operator is a closed operator. In particular, a self-adjoint operator is closed.*

Proof Let $A, D(A)$ be an operator and let $A^*, D(A^*)$ be the adjoint. Consider a sequence $\{f_n\}$ such that $f_n \in D(A^*)$, $f_n \to f$ and $A^* f_n \to g$. For any $h \in D(A)$ we have

$$(f_n, A h) = (A^* f_n, h). \tag{4.63}$$

Taking the limit for $n \to \infty$, we find

$$(f, A h) = (g, h). \tag{4.64}$$

The above formula implies that $f \in D(A^*)$ and $A^* f = g$, concluding the proof. $\square$

We can now prove the following proposition.

Proposition 4.7 *Let $A, D(A)$ be a self-adjoint operator. Then, for any $z \in \mathbb{C}$, with $\Im z \neq 0$, one has*

$$Ker\,(A - z) = \{0\}, \qquad Ran\,(A - z) = \mathscr{H}. \tag{4.65}$$

Moreover, the operator $(A - z)^{-1} : \mathscr{H} \to D(A)$ is bounded and satisfies

$$\|(A - z)^{-1}\| \le \frac{1}{|\Im z|} \,. \tag{4.66}$$

Proof Let us fix $z \in \mathbb{C}$, with $\Im z \ne 0$. For any $f \in Ker\,(A - z)$, we can write

$$z(f, f) = (f, zf) = (f, Af) = (Af, f) = (zf, f) = \bar{z}(f, f) \,. \tag{4.67}$$

From (4.67) it follows that f is the null vector and therefore $Ker\,(A - z) = \{0\}$.

Let us show that $Ran\,(A - z)$ is a dense set. Let us assume that there exists $f \in (Ran\,(A - z))^{\perp}$, with $f \ne 0$, i.e., we have $((A-z)g, f) = 0$ for any $g \in D(A)$. In particular, this implies that $f \in D(A^*) = D(A)$ and therefore $(g, (A - \bar{z})f) = 0$ for any $g \in D(A)$. Then, it follows that $f \in Ker\,(A - \bar{z})$ and this is absurd since $Ker\,(A - z) = \{0\}$. Thus, we have proved that $Ran\,(A - z)$ is a dense set.

Let us prove that $Ran\,(A - z)$ is a closed set. Let $\{g_n\}$ be a sequence, with $g_n \in Ran\,(A - z)$ and $g_n \to g$ for $n \to \infty$. Then, there exists $f_n \in D(A)$ such that $(A - z)f_n = g_n$ and $(A - z)f_n \to g$. For any $h \in D(A)$ we have

$$\begin{aligned} \|(A - z)h\|^2 &= \left((A - \Re z)h - i\,\Im z\,h\,,\,(A - \Re z)h - i\,\Im z\,h \right) \\ &= \|(A - \Re z)h\|^2 + (\Im z)^2 \|h\|^2 \ge (\Im z)^2 \|h\|^2 \,. \end{aligned} \tag{4.68}$$

The estimate (4.68) implies that $\{f_n\}$ is a Cauchy sequence and then it converges to a vector f. Since $A - z$ is a closed operator, we conclude that $f \in D(A)$ and $(A - z)f = g$. This means that $g \in Ran\,(A - z)$ and this shows that $Ran\,(A - z)$ is a closed set.

Being $Ran\,(A - z)$ a dense and closed set, we conclude that $Ran\,(A - z) = \mathscr{H}$.

Finally, we note that by (4.65) the inverse operator $(A - z)^{-1}$ from $\mathscr{H}$ to $D(A)$ is well defined. Moreover, by (4.68), such inverse is bounded and it satisfies (4.66). This concludes the proof. $\square$

We conclude this section with an important result on the stability of self-adjointness with respect to a small perturbation. We first introduce the notion of small perturbation in the sense of operators.

Definition 4.12 Let $A, D(A)$ be a self-adjoint operator and let $B, D(B)$ a symmetric operator with $D(A) \subseteq D(B)$. We say that B is a small perturbation with respect to A if there exist $a \in (0, 1)$ and $b > 0$ such that

$$\|B\phi\| \le a\|A\phi\| + b\|\phi\| \tag{4.69}$$

for any $\phi \in D(A)$.

Note that if B is bounded then it is a small perturbation with respect to any self-adjoint operator. The following stability result is proved using Propositions 4.5 and 4.7.

Proposition 4.8 (Kato–Rellich theorem)
Let $A, D(A)$ be a self-adjoint operator and let $B, D(B)$ be a symmetric operator, with $D(A) \subseteq D(B)$. If B is a small perturbation with respect to A, then the operator $A + B, D(A)$ is self-adjoint.

Proof The operator $A + B, D(A)$ is symmetric and then it is sufficient to prove that there exists a constant $\mu > 0$ such that $Ran\,(A + B \pm i\mu) = \mathcal{H}$. By Proposition 4.7, the operator $(A + i\mu)^{-1}$ exists and it is bounded and therefore for any $\phi \in D(A)$ we can write

$$(A + B + i\mu)\phi = \left(I + B(A + i\mu)^{-1} \right)(A + i\mu)\phi. \tag{4.70}$$

Let us show that the norm of the operator $B(A + i\mu)^{-1}$ is less than one for μ sufficiently large. For any $\psi \in \mathcal{H}$ we have

$$\|B(A + i\mu)^{-1}\psi\| \leq a\|A(A + i\mu)^{-1}\psi\| + b\|(A + i\mu)^{-1}\psi\|$$
$$\leq a\|A(A + i\mu)^{-1}\psi\| + \frac{b}{\mu}\|\psi\|, \tag{4.71}$$

where we have used the hypothesis and the estimate (4.66). Moreover, proceeding as in (4.68), we find

$$\|\psi\|^2 = \|(A + i\mu)(A + i\mu)^{-1}\psi\|^2 = \|A(A + i\mu)^{-1}\psi\|^2 + \mu^2\|(A + i\mu)^{-1}\psi\|^2$$
$$\geq \|A(A + i\mu)^{-1}\psi\|^2. \tag{4.72}$$

Using (4.72) in (4.71) and taking $\mu > \frac{b}{1-a}$, we obtain

$$\|B(A + i\mu)^{-1}\psi\| \leq \left(a + \frac{b}{\mu} \right)\|\psi\| < \|\psi\|. \tag{4.73}$$

As a consequence, the operator $I + B(A + i\mu)^{-1}$ is invertible, with bounded inverse defined by the Neumann series. Given $f \in \mathcal{H}$, we define

$$\phi_f = (A + i\mu)^{-1}\left(I + B(A + i\mu)^{-1} \right)^{-1} f. \tag{4.74}$$

It is clear that $\phi_f \in D(A)$ and, using (4.70), $(A + B + i\mu)\phi_f = f$. This shows that $Ran\,(A + B + i\mu) = \mathcal{H}$. In the same way, one proves that $Ran\,(A + B - i\mu) = \mathcal{H}$ and therefore the proof is complete. $\square$

Exercise 4.11 Let us consider the operator $P, D(P)$ in $L^2(\mathbb{R})$ introduced in Example 4.14 and let $V \in L^2(\mathbb{R})$. Prove that $P + V, D(P)$ is self-adjoint.

Exercise 4.12 Let $A, D(A)$ be self-adjoint and let $B, D(B)$ be symmetric and small perturbation with respect to A. Verify that $B(A - z)^{-1}$ and $B(A + B - z)^{-1}$ are bounded operators for $z \in \rho(A)$ and $z \in \rho(A + B)$, respectively.

4.6 Resolvent and Spectrum

For a matrix A, i.e., a linear operator in $\mathbb{C}^n$, the spectrum is the set of $z \in \mathbb{C}$ such that there exists $x \in \mathbb{C}^n$, $x \neq 0$, which satisfies $Ax = z\,x$. Such values z are called eigenvalues and the corresponding vectors x are called eigenvectors. In other words, z is an eigenvalue of A if and only if the operator $A - z$ is not invertible.

For an operator $A, D(A)$ in a generic Hilbert space $\mathscr{H}$ the definition of eigenvalue and eigenvector is analogous.

Definition 4.13 The number $z \in \mathbb{C}$ is an eigenvalue of the operator A, $D(A)$ if there exists $\psi \in D(A)$, $\psi \neq 0$, such that $A\psi = z\,\psi$. The vector ψ is called eigenvector. Equivalently, $z \in \mathbb{C}$ is an eigenvalue if $Ker\,(A-z) \neq \{0\}$ and any $\psi \in Ker\,(A-z)$, $\psi \neq 0$, is an eigenvector.

As in the finite-dimensional case, if z is not an eigenvalue then the operator $A - z$ is invertible but it can happen that the inverse is not defined on the whole Hilbert space and is not bounded. This consideration suggests the following definition.

Definition 4.14 Let $A, D(A)$ be a closed operator. The resolvent set $\rho(A)$ is the set of $z \in \mathbb{C}$ such that $A - z$ is invertible and $(A - z)^{-1}$ is a bounded operator defined on $\mathscr{H}$. The operator $(A - z)^{-1}$, $z \in \rho(A)$, is the resolvent operator, or simply the resolvent, of $A, D(A)$.

Definition 4.15 The spectrum of the closed operator $A, D(A)$ is $\sigma(A) := \mathbb{C} \setminus \rho(A)$.

From the definition, we see that an eigenvalue belongs to the spectrum but, in general, there are points in the spectrum which are not eigenvalues.

For a matrix, the spectrum is made of isolated points and therefore it is a closed set. The same property is valid in the infinite-dimensional case.

Proposition 4.9 *Let A, $D(A)$ be a closed operator and let $z, w \in \rho(A)$. Then, the following first resolvent identity holds*

$$(A - z)^{-1} - (A - w)^{-1} = (z - w)\,(A - z)^{-1}(A - w)^{-1}. \qquad (4.75)$$

Furthermore, $\rho(A)$ is an open set and $\sigma(A)$ is a closed set.

Proof For the proof of (4.75), it is sufficient to observe

$$(A - z)^{-1} - (z - w)\,(A - z)^{-1}(A - w)^{-1} = (A - z)^{-1}\big[I - (z - w)(A - w)^{-1}\big]$$
$$= (A - z)^{-1}\big\{I + [(A - z - (A - w)](A - w)^{-1}\big\} = (A - w)^{-1}. \quad (4.76)$$

Let us fix $z_0 \in \rho(A)$ and the neighborhood $D_0 = \{z \in \mathbb{C}\,|\,|z - z_0|\,\|(A - z_0)^{-1}\| < 1\}$. If $z \in D_0$, then the sequence of bounded operators

$$R_n = \sum_{k=0}^{n}(z - z_0)^k\big[(A - z_0)^{-1}\big]^{k+1} \qquad (4.77)$$

converges in norm to a bounded operator R. Let us show that R coincides with the resolvent $(A - z)^{-1}$. For any $f \in \mathcal{H}$, we consider the sequence of vectors $R_n f$. We have $R_n f \in D(A)$ and $R_n f \to R f$ for $n \to \infty$. Moreover

$$AR_n f = (A - z_0) R_n f + z_0 R_n f = \sum_{k=0}^{n} (z - z_0)^k \big[(A - z_0)^{-1}\big]^k f + z_0 R_n f$$

$$= f + (z - z_0) R_{n-1} f + z_0 R_n f. \tag{4.78}$$

Taking the limit $n \to \infty$, we obtain $AR_n f \to f + zRf$. Since A is closed we have $Rf \in D(A)$ and $ARf = f + zRf$, i.e., $(A - z)Rf = f$. Analogously, given $f \in D(A)$ we have

$$R_n A f = R_n (A - z_0) f + z_0 R_n f = \sum_{k=0}^{n} (z - z_0)^k \big[(A - z_0)^{-1}\big]^k f + z_0 R_n f$$

$$= f + (z - z_0) R_{n-1} f + z_0 R_n f. \tag{4.79}$$

Taking the limit $n \to \infty$, we have $RAf = f + zRf$, i.e., $R(A - z)f = f$. We conclude that for $z \in D_0$ we have $R = (A - z)^{-1}$ and $z \in \rho(A)$. This means that $\rho(A)$ is an open set and $\sigma(A)$ is a closed set. $\qquad\square$

The following proposition is often useful.

Proposition 4.10 *Let $A, D(A)$ be self-adjoint and let $B, D(B)$ be symmetric and small perturbation with respect to A. For $z \in \rho(A) \cap \rho(A+B)$, the following second resolvent identity holds:*

$$(A + B - z)^{-1} = (A - z)^{-1} - (A - z)^{-1} B (A + B - z)^{-1} \tag{4.80}$$

$$= (A - z)^{-1} - (A + B - z)^{-1} B (A - z)^{-1}. \tag{4.81}$$

Proof Note that the operators at the right-hand side of (4.80), (4.81) are well defined (in fact, from Exercise 4.12 we know that $B(A - z)^{-1}$ and $B(A + B - z)^{-1}$ are bounded operators). Moreover

$$B(A + B - z)^{-1} = (A + B - z)(A + B - z)^{-1} - (A - z)(A + B - z)^{-1}$$

$$= I - (A - z)(A + B - z)^{-1}. \tag{4.82}$$

Applying to both sides the operator $(A - z)^{-1}$, we find (4.80). For the proof of (4.81), one proceeds analogously. $\qquad\square$

Remark 4.2 The second resolvent identity provides information on the resolvent of $A + B$, given the resolvent of A. In particular, one can obtain a perturbative expansion for the resolvent of $A + B$ in terms of the resolvent of A when the operator $B(A - z)^{-1}$ has norm less than one. Indeed, from (4.81) one has

$$(A + B - z)^{-1} = (A - z)^{-1}(I + B(A - z)^{-1})^{-1}$$
$$= (A - z)^{-1}\left(\sum_{n=0}^{\infty}(-1)^n(B(A - z)^{-1})^n\right). \qquad (4.83)$$

In the next proposition, we characterize eigenvalues and eigenvectors of a symmetric operator.

Proposition 4.11 *Let $A, D(A)$ be a symmetric operator. Then, the eigenvalues are real and the eigenvectors corresponding to different eigenvalues are orthogonal.*

Proof In the proof of Proposition 4.7, we have seen that if $f \in Ker (A - z)$, with $\Im z \neq 0$, then $f = 0$. This implies that the eigenvalues are real. Moreover, let us assume that $Af = \lambda f$ and $Ag = \mu g$, with $\mu \neq \lambda$. Then

$$\lambda(g, f) = (g, Af) = (Ag, f) = \mu(g, f) \qquad (4.84)$$

and we conclude that $(g, f) = 0$. ☐

We note that the spectrum of a closed and symmetric operator is not a subset of the real axis, in general. The property holds for a self-adjoint operator and it is a crucial property for the applications in Quantum Mechanics.

Proposition 4.12 *Let $A, D(A)$ be self-adjoint. Then $\sigma(A) \subseteq \mathbb{R}$.*

Proof If $A, D(A)$ is self-adjoint, then we know that the operator $(A - z)^{-1}$, with $\Im z \neq 0$, exists and it is bounded on $\mathcal{H}$ (see (4.66) in Proposition 4.7). This means that any $z \in \mathbb{C}$ with $\Im z \neq 0$ belongs to $\rho(A)$ and then $\sigma(A)$ is a subset of the real axis. ☐

The points of the spectrum of a self-adjoint operator can be characterized through sequences of quasi eigenvectors, known as Weyl sequences.

Definition 4.16 Let $A, D(A)$ be self-adjoint and let $\lambda \in \sigma(A)$. A Weyl sequence relative to λ is a sequence of vectors $\{f_n\}$ such that $f_n \in D(A)$, $\|f_n\| = 1$ and $\|(A - \lambda)f_n\| \to 0$ for $n \to \infty$.

Proposition 4.13 *Let $A, D(A)$ be self-adjoint. Then $\lambda \in \sigma(A)$ if and only if there exists a Weyl sequence relative to λ.*

Proof Let us assume that there exists a Weyl sequence $\{f_n\}$ relative to λ and let us suppose, by absurd, that $\lambda \in \rho(A)$. Then $(A - \lambda)^{-1}$ exists, it is bounded on $\mathcal{H}$ and

$$1 = \|f_n\| = \|(A - \lambda)^{-1}(A - \lambda)f_n\| \leq \|(A - \lambda)^{-1}\|\|(A - \lambda)f_n\|. \qquad (4.85)$$

Taking the limit for $n \to \infty$, we obtain a contradiction, and therefore we conclude that $\lambda \in \sigma(A)$.

Let $\lambda \in \sigma(A) \subseteq \mathbb{R}$ and fix $z_n = \lambda + i\,n^{-1}$. We have $z_n \in \rho(A)$ and $|z_n - \lambda| \to 0$ for $n \to \infty$. Moreover, the following inequality holds for any n:

$$|z_n - \lambda| \, \|(A - z_n)^{-1}\| \geq 1. \tag{4.86}$$

Indeed, if there exists $\bar{n}$ such that $|z_{\bar{n}} - \lambda| \, \|(A - z_{\bar{n}})^{-1}\| < 1$ then, proceeding as in Proposition 4.9, we would find $\lambda \in \rho(A)$ and this is absurd.

Inequality (4.86) implies that $\|(A - z_n)^{-1}\| \to \infty$ for $n \to \infty$. Then, there is a sequence $\{g_n\}$ such that $\|g_n\| = 1$ and $\|(A - z_n)^{-1} g_n\| \to \infty$ for $n \to \infty$. Let us consider the sequence $\{f_n\}$ defined by

$$f_n = \frac{(A - z_n)^{-1} g_n}{\|(A - z_n)^{-1} g_n\|}. \tag{4.87}$$

We have $\|f_n\| = 1$, $f_n \in D(A)$ and

$$(A - \lambda) f_n = (A - z_n) f_n + (z_n - \lambda) f_n = \frac{g_n}{\|(A - z_n)^{-1} g_n\|} + (z_n - \lambda) f_n \to 0 \tag{4.88}$$

for $n \to \infty$. Therefore, $\{f_n\}$ is a Weyl sequence relative to λ and the proposition is proved. $\square$

Example 4.15 Let us show that the spectrum of the self-adjoint operator Q, $D(Q) := Q_0^*$, $D(Q_0^*)$ defined in (4.37) coincides with the whole real axis. Given $\lambda \in \mathbb{R}$, we define the sequence

$$f_n = \sqrt{\frac{n}{2}} \, \chi_{(\lambda - n^{-1}, \lambda + n^{-1})} \tag{4.89}$$

where χ_Ω is the characteristic function of the set Ω. We have $\|f_n\| = 1$, $f_n \in D(Q)$ and

$$\|(Q - \lambda) f_n\|^2 = \frac{n}{2} \int_{\lambda - n^{-1}}^{\lambda + n^{-1}} dx \, (x - \lambda)^2 = \frac{1}{3n^2} \to 0 \tag{4.90}$$

for $n \to \infty$. Then, $\{f_n\}$ is a Weyl sequence relative to λ and this proves that $\lambda \in \sigma(Q)$.

On the other hand, the operator has no eigenvalues. Indeed, by absurd, let us assume that for $\lambda \in \mathbb{R}$ there exists $\phi_\lambda \in D(Q)$ such that $Q\phi_\lambda = \lambda\phi_\lambda$. Then

$$\|(Q - \lambda)\phi_\lambda\|^2 = \int dx \, (x - \lambda)^2 |\phi_\lambda(x)|^2 = 0. \tag{4.91}$$

Formula (4.91) implies that $\phi_\lambda = 0$ a.e. and therefore λ is not an eigenvalue. As we shall see, this is an example of self-adjoint operator with purely continuous spectrum.

It is also easy to see that for any $z \in \mathbb{C}$, with $\Im z \neq 0$, the resolvent operator is the multiplication operator

$$((Q - z)^{-1} f)(x) = \frac{f(x)}{x - z}. \tag{4.92}$$

Example 4.16 Proceeding in the same way, one shows that the self-adjoint operator $P, D(P) := P_0^*, D(P_0^*)$ defined in (4.41) has the spectrum coinciding with the real axis and it has no eigenvalues. In fact

$$f_n = \mathscr{F}^{-1}\left(\sqrt{\frac{n}{2}}\, \chi_{(\lambda-n^{-1},\,\lambda+n^{-1})}\right) \tag{4.93}$$

defines a Weyl sequence relative to $\lambda \in \mathbb{R}$. Moreover, if λ is an eigenvalue with eigenvector ϕ_λ then

$$\|(P - \lambda)\phi_\lambda\|^2 = \int dk\,(k - \lambda)^2 |\tilde{\phi}_\lambda(k)|^2 = 0 \tag{4.94}$$

and this implies $\phi_\lambda = 0$ a.e..

Exercise 4.13 Compute the resolvent of the operator $P, D(P)$.

Exercise 4.14 Let us consider the self-adjoint operator $p_\alpha, D(p_\alpha)$ in $L^2(0, 1)$ defined in (4.53). Verify that the spectrum is the set of the eigenvalues $\lambda_k = \theta + 2\pi k$, $k \in \mathbb{Z}$, where $\alpha = e^{i\theta}$.

4.7 Isometric Operators and Unitary Operators

We consider two (in general different) Hilbert spaces $\mathscr{H}$ and $\mathscr{H}'$ and we denote by $(\cdot, \cdot)'$, $\|\cdot\|'$, I' the scalar product, the norm, and the identity operator in $\mathscr{H}'$. If A is a bounded operator from $\mathscr{H}$ to $\mathscr{H}'$, then its adjoint A^* is a bounded operator from $\mathscr{H}'$ to $\mathscr{H}$ and it is defined by $(f', Ag)' = (A^* f', g)$ for any $f' \in \mathscr{H}'$ and $g \in \mathscr{H}$. An isometric operator (or isometry) is an operator from $\mathscr{H}$ to $\mathscr{H}'$ which preserves the scalar product. A surjective isometric operator is called unitary.

Definition 4.17 An operator $T : \mathscr{H} \to \mathscr{H}'$ is isometric if

$$(Tf, Tg)' = (f, g), \qquad \forall f, g \in \mathscr{H} \tag{4.95}$$

or, equivalently, $T^*T = I$.

Definition 4.18 An operator $U : \mathscr{H} \to \mathscr{H}'$ is unitary if it is isometric and $Ran\,(U) = \mathscr{H}'$.

Note that if T is isometric then $\|Tf\| = \|f\|$, i.e., T is bounded with norm equal to one, and $Ker\,(T) = \{0\}$. This implies that the inverse operator T^{-1} exists but, in general, it is not defined on the whole space $\mathscr{H}'$. If U is unitary then the inverse operator U^{-1} is bounded on $\mathscr{H}'$. More precisely, we have the following proposition.

Proposition 4.14 *A bounded operator* $U : \mathscr{H} \to \mathscr{H}'$ *is unitary if and only if* $U^*U = I$ *and* $UU^* = I'$ *or, equivalently,* $U^* = U^{-1}$.

Proof If U is unitary then, in particular, it is isometric and $U^*U = I$. Moreover, let $f', g' \in \mathscr{H}'$ and let $g \in \mathscr{H}$ such that $g' = Ug$. We have

$$(f', g')' = (f', Ug)' = (U^*f', g) = (UU^*f', Ug)' = (UU^*f', g')' \quad (4.96)$$

and therefore $UU^* = I'$. Conversely, if $U^*U = I$ then U is isometric. Moreover, let $f' \in \mathscr{H}'$. Then $f' = UU^*f' = Uf$, where $f := U^*f' \in \mathscr{H}$. This implies $Ran\,(U) = \mathscr{H}'$ and the proof is complete. □

In the next proposition, we characterize eigenvalues and eigenvectors of an isometric operator in $\mathscr{H}$.

Proposition 4.15 *If $U : \mathscr{H} \to \mathscr{H}$ is isometric, then the eigenvalues have modulus one and the eigenvectors corresponding to different eigenvalues are orthogonal.*

Proof Let $Uf = \lambda f$. Then $\|f\| = \|Uf\| = \|\lambda f\| = |\lambda| \|f\|$ and therefore $|\lambda| = 1$. Let $Ug = \mu g$, with $\mu \neq \lambda$. Then

$$(f, g) = (Uf, Ug) = (\lambda f, \mu g) = \bar{\lambda}\mu(f, g) \quad (4.97)$$

and we conclude that $(f, g) = 0$. □

When T is isometric but it is not unitary, it is useful to characterize the operator TT^*.

Proposition 4.16 *If $T : \mathscr{H} \to \mathscr{H}'$ is isometric with $Ran\,(T) \neq \mathscr{H}'$, then $Ran\,(T)$ is closed and*

$$TT^*f' = f' \quad if \quad f' \in Ran\,(T) \quad and \quad TT^*f' = 0 \quad if \quad f' \in Ran\,(T)^{\perp} \quad (4.98)$$

or, equivalently, $T^*|_{Ran\,(T)} = T^{-1}$ and $T^*|_{Ran\,(T)^{\perp}} = 0$.

Proof Let $f'_n \in Ran\,(T)$ and $f'_n \to f'$ for $n \to \infty$. This implies that for each n there exists $g_n \in \mathscr{H}$ such that $Tg_n = f'_n$ and $\|g_n - g_m\| = \|Tg_n - Tg_m\|' = \|f'_n - f'_m\|' \to 0$ for $n, m \to \infty$. Then, there exists $g \in \mathscr{H}$ such that $g_n \to g$ and $\|Tg_n - Tg\|' = \|g_n - g\| \to 0$. Thus, we have

$$f' = \lim_n f'_n = \lim_n Tg_n = Tg \quad (4.99)$$

and this means that $Ran\,(T)$ is closed.

Let $f' \in Ran\,(T)$ and let $g \in \mathscr{H}$ such that $Tg = f'$. Using the identity $T^*Tg = g$, we have

$$TT^*f' = TT^*Tg = Tg = f'. \quad (4.100)$$

If $f' \in Ran\,(T)^{\perp}$, then for any $g \in \mathscr{H}$ we find

$$(T^*f', g) = (f', Tg)' = 0. \quad (4.101)$$

It follows that $T^* f' = 0$ and therefore also $T T^* f' = 0$.   □

Example 4.17 Let us consider the operator in l^2

$$T : (x_1, \ldots, x_n, \ldots) \to (0, x_1, \ldots, x_n \ldots). \qquad (4.102)$$

T is isometric but it is not unitary, since $Ran(T) \subset l^2$. The adjoint operator is

$$T^* : (x_1, \ldots, x_n, \ldots) \to (x_2, \ldots, x_n, \ldots). \qquad (4.103)$$

T^* is not isometric and we have $T T^* = I - (x^{(1)}, \cdot) x^{(1)}$, where $x^{(1)} = (1, 0, \ldots, 0, \ldots)$.

Example 4.18 The following two operators in $L^2(\mathbb{R})$ are unitary:

$$(U_a f)(x) = e^{iax} f(x), \quad a \in \mathbb{R}, \qquad (4.104)$$
$$(V_b f)(x) = f(x + b), \quad b \in \mathbb{R} \qquad (4.105)$$

and one easily sees that $U_a^* = U_{-a}$ and $V_b^* = V_{-b}$.

Example 4.19 By Plancherel theorem, the Fourier transform $\mathscr{F}$ is a unitary operator from $L^2(\mathbb{R}^n, dx)$ to $L^2(\mathbb{R}^n, dk)$.

Example 4.20 Let $\{\phi_n\}$ be an orthonormal and complete system in the Hilbert space $\mathscr{H}$. Then, the operator $U : \mathscr{H} \to l^2$ defined by

$$(U f)_n := f_n = (\phi_n, f) \qquad (4.106)$$

is unitary, with $U^{-1}\{f_n\} = \sum_n f_n \phi_n$.

In the next proposition, it is shown that unitary operators preserve self-adjointness and leave the spectrum invariant.

Proposition 4.17 *Let A, $D(A)$ be self-adjoint in $\mathscr{H}$ and let $U : \mathscr{H} \to \mathscr{H}'$ be unitary. Then the operator in $\mathscr{H}'$*

$$D(A') = \{f' \in \mathscr{H}' \mid f' = Uf, \ f \in D(A)\}, \qquad A'f' = UAU^{-1}f' \quad (4.107)$$

is self-adjoint, $\sigma(A') = \sigma(A)$ and the eigenvalues of A and A' coincide.

Moreover, if A, $D(A)$ is essentially self-adjoint then also A', $D(A')$ is essentially self-adjoint.

Proof Let us prove that A' is symmetric. If $f', g' \in D(A')$, using the symmetry of A we have

$$(g', A'f')' = (g', UAU^{-1}f')' = (U^{-1}g', AU^{-1}f') = (AU^{-1}g', U^{-1}f')$$
$$= (UAU^{-1}g', f')' = (A'g', f')'. \qquad (4.108)$$

Let us fix $z \in \rho(A)$ and $f' \in \mathcal{H}'$ and let us define $u' = U(A - z)^{-1}U^{-1}f'$. We have $u' \in D(A')$ and $(A' - z)u' = f'$, i.e., $Ran\,(A' - z) = \mathcal{H}'$, and the self-adjointness criterion implies that (4.107) is self-adjoint.

Moreover, let us observe that for $z \in \rho(A)$ the operator $U(A - z)^{-1}U^{-1}$ from $\mathcal{H}'$ to $D(A')$ is bounded and it coincides with $(A' - z)^{-1}$. So $\rho(A) \subseteq \rho(A')$. Changing the role of A and A', we conclude that $\rho(A) = \rho(A')$ and therefore $\sigma(A) = \sigma(A')$.

The remaining part of the proof is left as an exercise.   □

In some simple cases, the previous proposition can be used to construct self-adjoint extensions of a symmetric operator A_0. In fact, let us assume that we have a unitary operator U which diagonalizes A_0, i.e., such that $A'_0 = U A_0 U^{-1}$ is a multiplication operator in some Hilbert space. Let us also assume that A'_0 is essentially self-adjoint (as in the case of the operator (4.15)). Under these conditions it can be easy to construct the self-adjoint extension A'. Then, by Proposition 4.17, one has that A_0 is essentially self-adjoint, its self-adjoint extension is $A = U^{-1}A'U, \sigma(A) = \sigma(A')$ and the eigenvalues of A and A' coincide.

This procedure will be used to construct the Hamiltonian for the free particle and for the harmonic oscillator. Here, we discuss two examples.

Example 4.21 Let us reconsider the self-adjoint extension of $P_0, D(P_0)$ in $L^2(\mathbb{R})$ defined in (4.18). We first observe that in the Fourier space the operator $\mathscr{F} P_0 \mathscr{F}^{-1}$ is the multiplication operator by k. Such an operator is essentially self-adjoint and we know how to construct its self-adjoint extension $\tilde{P}$. Moreover, $\sigma(\tilde{P}) = \mathbb{R}$ and there are no eigenvalues. By Proposition 4.17, we conclude that $\mathscr{F}^{-1}\tilde{P}\mathscr{F}$ is the only self-adjoint extension of $P_0, \sigma(P) = \mathbb{R}$ and there are no eigenvalues.

Example 4.22 Let us consider the operator in $L^2(0, \pi)$

$$(\dot{H}_D f)(x) = -f''(x), \qquad D(\dot{H}_D) = \{f \in C^2[0, \pi] \mid f(0) = f(\pi) = 0\}.$$
$$(4.109)$$

It is a symmetric operator but it is not self-adjoint (verify). Moreover, it has an orthonormal and complete system of eigenvectors

$$\phi_n(x) = \sqrt{\frac{2}{\pi}} \sin nx, \qquad n = 1, 2, \ldots \qquad (4.110)$$

with eigenvalues $\lambda_n = n^2$. The unitary operator $\mathscr{F}_D : L^2(0, \pi) \to l^2$ defined by

$$(\mathscr{F}_D f)_n = \sqrt{\frac{2}{\pi}} \int_0^\pi dx \, \sin nx \, f(x) \qquad (4.111)$$

reduces $\dot{H}_D$ to a multiplication operator by the eigenvalue in the space l^2

$$\left(\mathscr{F}_D \dot{H}_D \mathscr{F}_D^{-1}\{f_m\}\right)_n = \frac{2}{\pi} \int_0^\pi dx \, \sin nx \left(-\frac{d^2}{dx^2}\right) \sum_{m=1}^\infty f_m \sin mx = n^2 f_n. (4.112)$$

The operator (4.112) is essentially self-adjoint and its self-adjoint extension is (verify)

$$\left(H_D'\{f_m\}\right)_n = n^2 f_n, \qquad D(H_D') = \left\{\{f_m\} \in l^2 \mid \sum_{n=1}^{\infty} n^4 |f_n|^2 < \infty\right\}. \quad (4.113)$$

By Proposition 4.17, we conclude that the only self-adjoint extension of (4.109) is

$$H_D = \mathscr{F}_D^{-1} H_D' \mathscr{F}_D, \qquad D(H_D) = \left\{f \in L^2(0, \pi) \mid \sum_{n=1}^{\infty} n^4 |f_n|^2 < \infty\right\}. \quad (4.114)$$

In Quantum Mechanics, the operator H_D, $D(H_D)$ represents the Hamiltonian of a particle moving in the interval $[0, \pi]$ with Dirichlet boundary conditions at the end points (when we set $\hbar = 1$, $m = 1/2$). The spectrum consists of the eigenvalues $\lambda_n = n^2$.

Exercise 4.15 Let us consider the symmetric operator in $L^2(0, \pi)$

$$(\dot{H}_N f)(x) = -f''(x), \qquad D(\dot{H}_N) = \left\{f \in C^2[0, \pi] \mid f'(0) = f'(\pi) = 0\right\}. \quad (4.115)$$

Construct the (unique) self-adjoint extension and compute its spectrum.

4.8 Spectral Theorem

From linear algebra, it is known that a symmetric matrix can be reduced to diagonal form by a unitary transformation. In this section, we discuss the generalization of this spectral theorem to the case of a self-adjoint operator in a Hilbert space. We underline that the spectral theorem plays a crucial role both in the formulation and in the applications of Quantum Mechanics. In order to formulate the theorem, we first discuss some preliminary notions.

4.8.1 Stieltjes Measures

Here, we recall some elements of measure theory on the real line (for more details, we refer to [1, 5, 11]).

Let $\mathscr{A}$ be a σ-algebra of subsets of $\mathbb{R}$, i.e., a collection of subsets of $\mathbb{R}$ such that

(i) $\mathbb{R}, \emptyset \in \mathscr{A}$,
(ii) if $A \in \mathscr{A}$ then $\mathbb{R} \setminus A \in \mathscr{A}$,
(iii) if $\{A_k\}$ is a sequence of elements of $\mathscr{A}$ then $\cup_{k=1}^{\infty} A_k \in \mathscr{A}$.

The pair $(\mathbb{R}, \mathscr{A})$ is called a measurable space.

Let us denote by $\mathscr{B}$ the σ-algebra of Borel sets of $\mathbb{R}$, i.e., the smallest σ-algebra of subsets of $\mathbb{R}$ containing the open sets. A Borel measure m in $\mathbb{R}$ is a measure on the measurable space $(\mathbb{R}, \mathscr{B})$, i.e., $m : \mathscr{B} \to \mathbb{R}^+ \cup \{+\infty\}$ such that

(i) $m(\emptyset) = 0$,
(ii) $m(\cup_{k=1}^{\infty} B_k) = \sum_{k=1}^{\infty} m(B_k)$ if $B_k \cap B_n = \emptyset$ for $k \neq n$.

The Borel measure is finite if $m(\mathbb{R}) < \infty$ and it is a probability measure if $m(\mathbb{R}) = 1$. We list without proof some properties of a finite Borel measure.

Proposition 4.18 *Denoted by* $\{B_k\}$ *a sequence of Borel sets, we have*

(i) $m(\cup_k B_k) \leq \sum_k m(B_k)$;
(ii) $m(\cup_k B_k) = \lim_{n \to \infty} m(\cup_{k=1}^{n} B_k)$;
(iii) if $B_k \subset B_{k+1}$ *then* $m(\cup_k B_k) = \lim_{k \to \infty} m(B_k)$;
(iv) if $B_{k+1} \subset B_k$ *then* $m(\cap_k B_k) = \lim_{k \to \infty} m(B_k)$.

Definition 4.19 A Borel measure m on $\mathbb{R}$ is

(i) pure point if there exists a countable set of points $P_i, i \in \mathbb{N}$, such that $m(\{P_i\}) \neq 0$ for any i and $m(B) = \sum_{i, P_i \in B} m(\{P_i\})$ for $B \in \mathscr{B}$;
(ii) continuous if $m(\{P\}) = 0$ for any $P \in \mathbb{R}$;
(iii) absolutely continuous w.r.t. Lebesgue measure if $m(A) = 0$ for any set $A \in \mathscr{B}$ of Lebesgue measure zero;
(iv) singular continuous w.r.t. Lebesgue measure if it is continuous and there exists a set $M \in \mathscr{B}$ of Lebesgue measure zero such that $m(\mathbb{R} \setminus M) = 0$.

Let us also recall the Lebesgue decomposition theorem.

Proposition 4.19 *For any Borel measure m on $\mathbb{R}$, we have $m = m_{pp} + m_{ac} + m_{sc}$ in a unique way, where m_{pp} is pure point, m_{ac} is absolutely continuous w.r.t. Lebesgue measure, and m_{sc} is singular continuous w.r.t. Lebesgue measure.*

Let $F : \mathbb{R} \to \mathbb{R}$ be a distribution function, i.e., F is nondecreasing, it is continuous from the right and

$$\lim_{\lambda \to -\infty} F(\lambda) = 0, \qquad \lim_{\lambda \to +\infty} F(\lambda) = \alpha, \qquad \text{con} \quad 0 < \alpha < +\infty. \qquad (4.116)$$

From a given distribution function F, we can construct a finite Borel measure on $\mathbb{R}$, called Stieltjes measure. More precisely, there exists a unique finite Borel measure m^F such that

$$m^F((a, b]) = F(b) - F(a) \qquad (4.117)$$

for any $a, b \in \mathbb{R}$, with $a < b$.

Note that if m is a finite Borel measure on $\mathbb{R}$, then the function

$$F^m(\lambda) = m((-\infty, \lambda]) \qquad (4.118)$$

is a distribution function and $m^{F^m} = m$.

In the following, we use the notation $F(a - 0) = \lim_{x \to a^-} F(x)$.

Exercise 4.16 Using the above properties of a finite Borel measure, verify that

(i) $m^F(\mathbb{R}) = \alpha$;
(ii) $m^F(\{a\}) = F(a) - F(a - 0)$;
(iii) $m^F([a, b]) = F(b) - F(a - 0)$;
(iv) $m^F([a, b)) = F(b - 0) - F(a - 0)$;
(v) $m^F((a, b)) = F(b - 0) - F(a)$.

We give some examples of measures on $\mathbb{R}$ constructed from a distribution function.

Example 4.23 Let $F(\lambda) = \chi_{[0,\infty)}(\lambda)$ (characteristic function of the interval $[0, \infty)$ or Heaviside function). The associated Stieltjes measure is the pure point (or Dirac) measure concentrated at the origin

$$\delta_0(B) = 0 \quad \text{if } 0 \notin B, \quad \delta_0(B) = 1 \quad \text{if } 0 \in B. \tag{4.119}$$

More generally, let us consider $a_i \in \mathbb{R}$ and $f_i > 0$, with $i = 1, 2, \ldots, n, \ldots$, and assume that $\sum_{i=1}^{\infty} f_i < \infty$. The function

$$F(x) = \sum_{i: a_i \le x} f_i \tag{4.120}$$

is a distribution function. Moreover, it is piecewise constant, with discontinuities in $a_1, \ldots, a_n, \ldots$ and $f_i = F(a_i) - F(a_i - 0)$. The corresponding Stieltjes measure is

$$m^F(B) = \sum_{i: a_i \in B} f_i \tag{4.121}$$

and it is a pure point measure.

Example 4.24 Let us consider an absolutely continuous distribution function F. Then

$$m^F((a, b]) = F(b) - F(a) = \int_a^b d\lambda \, F'(\lambda) \tag{4.122}$$

and for a generic Borel set

$$m^F(B) = \int_B d\lambda \, F'(\lambda). \tag{4.123}$$

In this case, the Stieltjes measure is absolutely continuous w.r.t. Lebesgue measure. If F is continuous but it is not absolutely continuous, then one can verify that the corresponding Stieltjes measure is singular continuous w.r.t. Lebesgue measure (see [1], section I.4).

Exercise 4.17 Let

$$F(\lambda) = \frac{1}{2} + \frac{1}{\pi}\arctan\lambda + \chi_{[0,\infty)}(\lambda).$$ (4.124)

Construct the associated Stieltjes measure, decomposed in its pure point and absolutely continuous part, and compute the measure of $[0, 1)$ and $(0, 1]$.

Given a distribution function F and the associated Stieltjes measure m^F, one can define the integral w.r.t. such a measure following the standard Lebesgue or Riemann methods. Correspondingly, one obtains the Lebesgue–Stieltjes integral and the Riemann–Stieltjes integral. Given a complex valued-integrable function Φ, we shall denote the integral as follows:

$$\int_B \Phi(\lambda)\,dF(\lambda).$$ (4.125)

In particular, for F given by (4.120) and Φ continuous we have

$$\int_B \Phi(\lambda)\,dF(\lambda) = \sum_{i:a_i\in B} \Phi(a_i)f_i$$ (4.126)

and for F absolutely continuous we have

$$\int_B \Phi(\lambda)\,dF(\lambda) = \int_B d\lambda\,\Phi(\lambda)\,F'(\lambda).$$ (4.127)

We also recall that the validity of standard results on Lebesgue integral (monotone convergence theorem, dominated convergence theorem, Fatou's lemma, and Fubini's theorem) can be extended to the Lebesgue–Stieltjes integral.

Exercise 4.18 Compute the integral

$$\int_{[0,1]} \lambda^2\,dF(\lambda),$$ (4.128)

where F is given by (4.124).

4.8.2 Orthogonal Projectors

Let $\mathcal{H}$ be a Hilbert space and let $M \subseteq \mathcal{H}$ be a closed subspace. We recall that the orthogonal complement of M is

$$M^\perp = \{f \in \mathcal{H} \mid (f, g) = 0, \quad \forall g \in M\}$$ (4.129)

and any $f \in \mathcal{H}$ can be uniquely written in the form $f = f_M + f_M^\perp$, with $f_M \in M$ and $f_M^\perp \in M^\perp$.

Definition 4.20 The map $P : \mathcal{H} \to \mathcal{H}$ is an orthogonal projector if there exists a closed subspace $M \subseteq \mathcal{H}$, $M \neq \{0\}$, such that $Pf = f_M$, for any $f \in \mathcal{H}$.

Example 4.25 Let $\phi \in \mathcal{H}$, with $\|\phi\| = 1$. Then the map $P_\phi : f \to (\phi, f)\phi$ is an orthogonal projector.

A useful characterization of an orthogonal projector is the following.

Proposition 4.20 *P is an orthogonal projector if and only if P is a linear, bounded, symmetric, and idempotent (i.e., $P^2 = P$) operator.*

Proof Let P be an orthogonal projector. The proof that P is a linear and bounded operator is left as an exercise. For the symmetry, we have

$$(Pf, g) = (f_M, g_M + g_M^\perp) = (f_M, g_M) = (f_M + f_M^\perp, g_M) = (f, Pg). \quad (4.130)$$

Moreover, $(Pf, g) = (f_M, g_M) = (Pf, Pg) = (P^2 f, g)$ and therefore $P^2 = P$.

Let us show the other implication. Let us consider the linear manifold $N = \{f \in \mathcal{H} \mid Pf = f\}$. If $f_n \in N$ and $f_n \to f$, then

$$\|f_n - Pf\| = \|Pf_n - Pf\| \leq \|P\|\|f_n - f\| \quad (4.131)$$

which implies $f_n \to Pf$. It follows that $Pf = f$, and therefore N is a closed subspace. For any $f \in \mathcal{H}$, we write

$$f = Pf + (f - Pf). \quad (4.132)$$

Since $P^2 f = Pf$, we have $Pf \in N$. Moreover, if $g \in N$, we have $(f - Pf, g) = (f - Pf, Pg) = (Pf - P^2 f, g) = 0$ and then $f - Pf \in N^\perp$. Equation (4.132) can be written as

$$f = f_N + f_N^\perp \quad (4.133)$$

and $Pf = Pf_N + Pf_N^\perp = f_N + P(f - Pf) = f_N$. This concludes the proof. $\quad\square$

Exercise 4.19 Verify that if P is an orthogonal projector then $\|P\| = 1$, $(Pf, f) \geq 0$ for any $f \in \mathcal{H}$ (i.e., it is positive) and $\sigma(P) = \{0, 1\}$, where 0 and 1 are eigenvalues.

4.8.3 Spectral Family

A spectral family is a collection of projectors with properties similar to those of a distribution function in $\mathbb{R}$.

Definition 4.21 A spectral family in $\mathcal{H}$ is a collection of orthogonal projectors $\{E(\lambda)\}$, $\lambda \in \mathbb{R}$, satisfying the following properties for any $f \in \mathcal{H}$

$$(i) \quad \lim_{\lambda \to -\infty} E(\lambda)f = 0 \quad \text{and} \quad \lim_{\lambda \to \infty} E(\lambda)f = f, \qquad (4.134)$$

$$(ii) \quad (E(\lambda)f, f) \leq (E(\mu)f, f) \quad \text{if} \quad \lambda \leq \mu, \qquad (4.135)$$

$$(iii) \quad \lim_{\varepsilon \to 0^+} E(\lambda + \varepsilon)f = E(\lambda)f. \qquad (4.136)$$

Example 4.26 Let us consider an orthonormal basis $\{\phi_n\}$, $n \in \mathbb{N}$, in $\mathcal{H}$ and let us define the family of orthogonal projectors

$$E(\lambda)f = \sum_{n:n\leq\lambda} (\phi_n, f)\, \phi_n. \qquad (4.137)$$

Property (i) is evident. For (ii), we observe that $(E(\lambda)f, f) = \sum_{n:n\leq\lambda} |(\phi_n, f)|^2$. Finally

$$\|E(\lambda + \varepsilon)f - E(\lambda)f\|^2 = \sum_{n:\lambda<n\leq\lambda+\varepsilon} |(\phi_n, f)|^2 \to 0 \qquad (4.138)$$

for $\varepsilon \to 0^+$. Then (4.137) is a spectral family.

Example 4.27 Let us consider the multiplication operator in $L^2(\mathbb{R})$

$$(E(\lambda)f)(x) = \chi_{(-\infty,\lambda]}(x)f(x), \qquad (4.139)$$

where $\chi_{(-\infty,\lambda]}$ is the characteristic function of the interval $(-\infty, \lambda]$. For any $\lambda \in \mathbb{R}$, the operator (4.139) is bounded, symmetric, and idempotent and then it is an orthogonal projector. Moreover, properties (i) and (ii) are evident and

$$\|E(\lambda + \varepsilon)f - E(\lambda)f\|^2 = \int_{\lambda}^{\lambda+\varepsilon} dx\, |f(x)|^2 \to 0 \qquad (4.140)$$

for $\varepsilon \to 0^+$. Thus, (4.139) is a spectral family.

Given $f \in \mathcal{H}$ and a spectral family $\{E(\lambda)\}$, we define

$$F^f(\lambda) = (E(\lambda)f, f). \qquad (4.141)$$

It is easy to verify that F^f is a distribution function. Then, we construct the associated Stieltjes measure on $\mathbb{R}$, denoted by m^f.

Definition 4.22 For any $f \in \mathcal{H}$, the measure m^f in $\mathbb{R}$ generated by the distribution function (4.141) is called spectral measure associated to the vector f and the spectral family $\{E(\lambda)\}$.

Using the polarization identity, we can also define a complex spectral measure. Indeed, for any $f, g \in \mathcal{H}$ one has (verify)

$$(E(\lambda)f, g) = \frac{1}{4}(E(\lambda)(f+g), f+g) - \frac{1}{4}(E(\lambda)(f-g), f-g)$$
$$+ \frac{i}{4}(E(\lambda)(f-ig), f-ig) - \frac{i}{4}(E(\lambda)(f+ig), f+ig). \quad (4.142)$$

The scalar products in the r.h.s. of (4.142) are distribution functions and therefore they define four spectral measures $m^{f+g}, m^{f-g}, m^{f-ig}, m^{f+ig}$. We have the following definition.

Definition 4.23 For any $f, g \in \mathscr{H}$

$$m^{f,g} = \frac{1}{4}m^{f+g} - \frac{1}{4}m^{f-g} + \frac{i}{4}m^{f-ig} - \frac{i}{4}m^{f+ig} \quad (4.143)$$

is the complex spectral measure associated to f, g and the spectral family $\{E(\lambda)\}$.

Given the spectral measure and the complex spectral measure, we can define the corresponding Riemann–Stieltjes integrals. Such integrals are denoted by

$$\int \Phi(\lambda) \, d(E(\lambda)f, f), \qquad \int \Phi(\lambda) \, d(E(\lambda)f, g). \quad (4.144)$$

Example 4.28 The distribution function defined by the spectral family (4.137) is

$$F^f(\lambda) = \sum_{n: n \leq \lambda} |(\phi_n, f)|^2. \quad (4.145)$$

Such a function is piecewise constant, with discontinuity points in $\lambda = n$ and $F^f(n) - F^f(n-0) = |(\phi_n, f)|^2$. The corresponding spectral measure is a pure point measure and

$$m^f(B) = \sum_{n: n \in B} |(\phi_n, f)|^2, \qquad m^{f,g}(B) = \sum_{n: n \in B} \overline{(\phi_n, f)}(\phi_n, g). \quad (4.146)$$

The integrals w.r.t. such measures are

$$\sum_{n \in \mathbb{N}} \Phi(n)|(\phi_n, f)|^2, \qquad \sum_{n \in \mathbb{N}} \Phi(n)\overline{(\phi_n, f)}(\phi_n, g). \quad (4.147)$$

Example 4.29 The distribution function defined by the spectral family (4.139) is

$$F^f(\lambda) = \int_{-\infty}^{\lambda} dx \, |f(x)|^2. \quad (4.148)$$

It is an absolutely continuous function, with derivative $|f(\lambda)|^2$. Then, the spectral measure is absolutely continuous w.r.t. Lebesgue and

$$m^f(B) = \int_B d\lambda \, |f(\lambda)|^2, \qquad m^{f,g}(B) = \int_B d\lambda \, \overline{f(\lambda)} g(\lambda). \qquad (4.149)$$

The integrals w.r.t. such measures are

$$\int d\lambda \, \Phi(\lambda)|f(\lambda)|^2, \qquad \int d\lambda \, \Phi(\lambda)\overline{f(\lambda)}g(\lambda). \qquad (4.150)$$

Here, we list some properties of a spectral family, whose proof is left as an exercise.

Exercise 4.20

(i) $f \in Ran \, E(\lambda)$ if and only if $f = E(\lambda)f$;
(ii) if $\lambda \leq \mu$ then $Ran \, E(\lambda) \subseteq Ran \, E(\mu)$;
(iii) if $\lambda \leq \mu$ then $E(\lambda)E(\mu) = E(\mu)E(\lambda) = E(\lambda)$.

Definition 4.24 For $a \leq b$, we define the operators

$$E((a,b]) = E(b) - E(a), \qquad E((a,b)) = E(b-0) - E(a), \qquad (4.151)$$
$$E([a,b]) = E(b) - E(a-0), \qquad E([a,b)) = E(b-0) - E(a-0), \qquad (4.152)$$
$$E(\{a\}) = E(a) - E(a-0). \qquad (4.153)$$

We have the following proposition.

Proposition 4.21

(i) *The above operators are orthogonal projectors;*
(ii) *for $f \in Ran \, E((a,b])$ we have*

$$(E(\lambda)f, f) = \begin{cases} 0 & \text{if} \quad \lambda \leq a \\ (E(a,\lambda]f, f) & \text{if} \quad a < \lambda \leq b \\ \|f\|^2 & \text{if} \quad \lambda > b \end{cases} \qquad (4.154)$$

and for $f \in Ran \, E(\{a\})$ we have $(E(\lambda)f, f) = \|f\|^2 \chi_{[a,\infty)}(\lambda)$;
(iii) *if $(a,b] \cap (c,d] = \emptyset$ then $E((a,b])E((c,d]) = E((c,d])E((a,b]) = 0$.*

Proof We only prove (ii). If $f \in Ran \, E((a,b])$ then $f = E((a,b])f$ and

$$(E(\lambda)f, f) = (E(\lambda)E((a,b])f, f) = (E(\lambda)(E(b) - E(a))f, f)$$
$$= (E(\lambda)E(b)f, f) - (E(\lambda)E(a)f, f). \qquad (4.155)$$

Using (iii) of Exercise 4.20, we find (4.154). For $f \in Ran \, E(\{a\})$, one proceeds analogously. $\qquad \square$

4.8.4 Formulation of the Spectral Theorem

Here, we formulate the spectral theorem. For the proof, we refer to [1] (for further reference see also [4–7, 10, 14, 15]).

Theorem 4.1 (Spectral theorem)
Let $A, D(A)$ be a self-adjoint operator in $\mathcal{H}$. Then, there exists a unique spectral family $\{E_A(\lambda)\}$ such that

$$D(A) = \left\{ u \in \mathcal{H} \mid \int \lambda^2 \, d(E_A(\lambda)u, u) < \infty \right\}, \tag{4.156}$$

$$(Au, v) = \int \lambda \, d(E_A(\lambda)u, v), \qquad \forall u \in D(A), \quad \forall v \in \mathcal{H}. \tag{4.157}$$

Moreover

$$\|Au\|^2 = \int \lambda^2 \, d(E_A(\lambda)u, u), \qquad \forall u \in D(A) \tag{4.158}$$

and the support of the integration measure in the above integrals is $\sigma(A)$.

The spectral theorem states that any self-adjoint operator can be reduced to a multiplication operator. Nevertheless, it should be underlined that the proof is not constructive, i.e., it does not provide a criterion for the construction of the spectral family. Only in some simple cases, the so-called solvable models, the spectral family, and the spectral measure can be explicitly exhibited. Below we give some examples. The explicit construction can also be given in some other interesting cases that we will study in the next chapters, i.e., for the Hamiltonian of a free particle, of a particle subject to a point interaction, and of the harmonic oscillator.

Example 4.30 Let us consider the self-adjoint operator Q, $D(Q) := Q_0^*$, $D(Q_0^*)$ in $L^2(\mathbb{R})$ defined in (4.37) and let us reconsider the spectral family already introduced in (4.139)

$$(E_Q(\lambda)f)(x) = \chi_{(-\infty, \lambda]}(x) f(x). \tag{4.159}$$

We find

$$\int \lambda^2 \, d(E_Q(\lambda)u, u) = \int dx \, x^2 |u(x)|^2, \tag{4.160}$$

$$\int \lambda \, d(E_Q(\lambda)u, v) = \int dx \, x \, \overline{u(x)} \, v(x) \tag{4.161}$$

and then (4.156) and (4.157) are verified. One can also immediately see that (4.158) holds and that the support of the measure is $\mathbb{R} = \sigma(Q)$. The corresponding spectral measures coincide with (4.149), i.e.,

$$m_Q^f(B) = \int_B dx \, |f(x)|^2, \qquad m_Q^{f,g}(B) = \int_B dx \, \overline{f(x)} \, g(x). \tag{4.162}$$

Example 4.31 For the self-adjoint operator $P, D(P) := P_0^*, D(P_0^*)$ in $L^2(\mathbb{R})$ defined in (4.41), the spectral family is (verify)

$$(E_P(\lambda)f)(x) = \left(\mathscr{F}^{-1}(\chi_{(-\infty,\lambda]}\tilde{f})\right)(x) \tag{4.163}$$

and the corresponding spectral measures are

$$m_P^f(B) = \int_B dk\, |\tilde{f}(k)|^2\,, \qquad m_P^{f,g}(B) = \int_B dk\, \overline{\tilde{f}(k)}\, \tilde{g}(k)\,. \tag{4.164}$$

Example 4.32 Let us consider the self-adjoint operator $H_D, D(H_D)$ in $L^2(0,\pi)$ defined in (4.114). The associated spectral family is

$$(E_{H_D}(\lambda)f)(x) = \sum_{n:\, n^2 \le \lambda} f_n\, \phi_n(x), \tag{4.165}$$

where $\phi_n(x) = \sqrt{\frac{2}{\pi}} \sin nx$ and $f_n = (\phi_n, f)$. The spectral measures are

$$m^f(B) = \sum_{n:\, n^2 \in B} |(\phi_n, f)|^2\,, \qquad m^{f,g}(B) = \sum_{n:\, n^2 \in B} \overline{(\phi_n, f)}(\phi_n, g)\,. \tag{4.166}$$

Exercise 4.21 Let $A, D(A)$ be self-adjoint. Using the spectral theorem, verify that

(i) if $f \in Ran\, E_A((a, b])$ then $f \in D(A)$;
(ii) $\left[A, E_A((a, b])\right]f = 0$ for any $f \in D(A)$, where $[\cdot, \cdot]$ denotes the commutator.

Semibounded self-adjoint operators play an important role in Quantum Mechanics.

Definition 4.25 A self-adjoint operator $A, D(A)$ is bounded from below if there exists a constant γ such that

$$(u, Au) \ge \gamma \|u\|^2\,, \qquad \forall u \in D(A)\,. \tag{4.167}$$

The number $\inf_{u \in D(A), \|u\|=1}(u, Au)$ is called the infimum of A. If, in particular, $(u, Au) \ge 0$ for any $u \in D(A)$ then the operator is positive.

As an application of the spectral theorem, we prove a useful characterization of the infimum of a self-adjoint operator in terms of the infimum of the spectrum.

Proposition 4.22 *Let $A, D(A)$ be self-adjoint. Then, A is bounded from below if and only if $\sigma(A)$ is bounded from below. Moreover*

$$\inf \sigma(A) = \inf_{u \in D(A), \|u\|=1}(u, Au)\,. \tag{4.168}$$

Proof Let us denote $\gamma_\sigma = \inf \sigma(A)$, $\gamma_A = \inf_{u \in D(A), \|u\|=1} (u, Au)$ and let us suppose that A is bounded from below. For any $\lambda < \gamma_A$ and $f \in D(A)$, using the Schwarz inequality, we have

$$\|(A - \lambda)f\| \, \|f\| \geq \big((A - \lambda)f, f\big) = \big((A - \gamma_A)f, f\big) + \big((\gamma_A - \lambda)f, f\big)$$
$$\geq (\gamma_A - \lambda)\|f\|^2. \tag{4.169}$$

Let us suppose that $\lambda \in \sigma(A)$. Then, there exists a Weyl sequence relative to λ and this contradicts inequality (4.169). Therefore, we conclude that $\lambda \in \rho(A)$ for any $\lambda < \gamma_A$. This means that $\sigma(A)$ is bounded from below and $\gamma_\sigma \geq \gamma_A$.

Let $\sigma(A)$ be bounded from below. Then

$$(f, Af) = \int \lambda \, d(E(\lambda)f, f) = \int_{\lambda \geq \gamma_\sigma} \lambda \, d(E(\lambda)f, f) \geq \gamma_\sigma \|f\|^2. \tag{4.170}$$

Thus, A is bounded from below and $\gamma_A \geq \gamma_\sigma$. $\qquad\qquad\qquad\qquad\qquad\qquad\square$

4.8.5 Functional Calculus

The spectral theorem provides a natural way to define a function of a self-adjoint operator.

Proposition 4.23 *Let $A, D(A)$ be a self-adjoint operator and let $\phi : \mathbb{R} \to \mathbb{C}$ be a continuous and bounded function. Then, there exists a bounded operator $\phi(A)$ such that*

$$(\phi(A)u, v) = \int \overline{\phi(\lambda)} \, d(E_A(\lambda)u, v), \quad \forall u, v \in \mathscr{H}. \tag{4.171}$$

Moreover
$$\|\phi(A)u\|^2 = \int |\phi(\lambda)|^2 \, d(E_A(\lambda)u, u). \tag{4.172}$$

Proof For any fixed $u \in \mathscr{H}$, we define the linear and bounded functional in $\mathscr{H}$

$$\mathfrak{G}_u : v \to \int \overline{\phi(\lambda)} d(E_A(\lambda)u, v). \tag{4.173}$$

Then, there exists a unique $h \in \mathscr{H}$ such that $\mathfrak{G}_u(v) = (h, v)$ and the map $\phi(A) : u \to h$ is a linear and bounded operator in $\mathscr{H}$ satisfying (4.171).

The proof of (4.172) is left as an exercise. $\qquad\qquad\qquad\qquad\qquad\qquad\square$

Exercise 4.22 Let ϕ, ϕ_1, ϕ_2 be continuous and bounded functions. Verify that

(i) $[E_A(\mu), \phi(A)] = 0$ for any $\mu \in \mathbb{R}$;

(ii) $\|\phi(A)\| = \sup_{\lambda \in \sigma(A)} |\phi(\lambda)|$;

(iii) $\phi(A)^* = \bar{\phi}(A)$;

(iv) $\phi(A)\phi(A)^* = \phi(A)^*\phi(A)$;

(v) if $\phi(\lambda) = \alpha\phi_1(\lambda) + \beta\phi_2(\lambda)$ then $\phi(A) = \alpha\phi_1(A) + \beta\phi_2(A)$, where $\alpha, \beta \in \mathbb{C}$;

(vi) if $\phi(\lambda) = \phi_1(\lambda)\phi_2(\lambda)$ then $\phi(A) = \phi_1(A)\phi_2(A)$.

In the applications, it is often useful the following proposition, whose proof is left as an exercise.

Proposition 4.24 *Let A, $\tilde{A}$ be two self-adjoint operators in the Hilbert spaces $\mathcal{H}$, $\mathcal{H}'$, respectively, such that $A = U^{-1}\tilde{A}U$, where $U : \mathcal{H} \to \mathcal{H}'$ is unitary. Then,*

$$E_A(\lambda) = U^{-1}E_{\tilde{A}}(\lambda)U \tag{4.174}$$

and

$$\phi(A) = U^{-1}\phi(\tilde{A})U \tag{4.175}$$

for any $\phi : \mathbb{R} \to \mathbb{C}$ continuous and bounded.

As a function of the self-adjoint operator A, $D(A)$, we have already defined the resolvent $(A-z)^{-1}$, which is bounded for $\Im z \neq 0$. Given the bounded and continuous function $\phi_z(\lambda) = (\lambda-z)^{-1}$, we can construct the operator $\phi_z(A)$ using the functional calculus and this operator coincides with the resolvent of A (verify).

Making use of the functional calculus, one can prove a useful formula expressing the spectral family of a self-adjoint operator in terms of its resolvent.

Exercise 4.23 *(Stone's formula)*
Let A, $D(A)$ be a self-adjoint operator and let $E_A(\lambda)$ be its spectral family. Prove that

$$\frac{1}{2\pi i} \lim_{\varepsilon \to 0} \int_a^b d\lambda \left((A - \lambda - i\varepsilon)^{-1}f - (A - \lambda + i\varepsilon)^{-1}f, g \right)$$

$$= \left(E((a, b))f, g \right) + \frac{1}{2}\left(E(\{a\})f, g \right) + \frac{1}{2}\left(E(\{b\})f, g \right) \tag{4.176}$$

for any $f, g \in \mathcal{H}$.

(Hint: verify that the l.h.s. of (4.176) equals

$$\frac{1}{\pi} \int_a^b d\lambda \int \frac{\varepsilon}{(\mu - \lambda)^2 + \varepsilon^2} d(E_A(\mu)f, f), \tag{4.177}$$

interchange the order of integration and use dominated convergence theorem).

4.9 Unitary Groups

Functional calculus can be used to define the exponential function of a self-adjoint operator. This fact allows us to solve the Schrödinger equation and to define the time evolution of a state of a generic quantum system.

Definition 4.26 The family of linear operators $U(t)$, $t \in \mathbb{R}$, is a strongly continuous one-parameter group of unitary operators in $\mathcal{H}$ if

(i) $U(t)$ is a unitary operator in $\mathcal{H}$ for any $t \in \mathbb{R}$,
(ii) $U(t + s) = U(t)U(s)$ for any $s, t \in \mathbb{R}$,
(iii) $\lim_{t \to 0} \|U(t)f - f\| = 0$, for any $f \in \mathcal{H}$.

Note that from *(ii)* it follows $U(0) = I$ and $U(-t) = U^{-1}(t)$. Moreover, $U(t)$ is strongly continuous also for $t_0 \neq 0$. Indeed,

$$\|U(t)f - U(t_0)f\| = \|U^{-1}(t_0)U(t)f - f\| = \|U(-t_0)U(t)f - f\|$$
$$= \|U(t - t_0)f - f\| . \tag{4.178}$$

Example 4.33 The families of operators U_a and V_b defined in (4.104) and (4.105) are two strongly continuous one-parameter groups of unitary operators (verify).

The next proposition shows that a self-adjoint operator defines a strongly continuous one-parameter group of unitary operators. Such group is also differentiable when it is restricted to the domain of the operator.

Proposition 4.25 *Let A, $D(A)$ be self-adjoint in $\mathcal{H}$. Then $U(t) := e^{itA}$, $t \in \mathbb{R}$, is a strongly continuous one-parameter group of unitary operators. Moreover, if $f \in D(A)$ then $e^{itA} f \in D(A)$ and we have*

$$\frac{d}{dt} e^{itA} f \Big|_{t=0} = \lim_{t \to 0} \frac{1}{t} \left(e^{itA} f - f \right) = i A f . \tag{4.179}$$

Proof For any $t \in \mathbb{R}$, let us consider the continuous and bounded function $\phi_t(\lambda) = e^{it\lambda}$. By the functional calculus, we know that $U(t) = e^{itA}$ is a bounded operator. Using (iii) of Exercise 4.22, one has $U(t)^* = U(-t)$. Moreover, by the relation $1 = e^{it\lambda} e^{-it\lambda}$ and iv) of Exercise 4.22, we obtain $I = U(t)U(-t) = U(-t)U(t)$. Hence, we conclude that $U(t)^* = U(t)^{-1}$, i.e., $U(t)$ is unitary. By the property of the exponential function $\phi_{t+s}(\lambda) = \phi_t(\lambda)\phi_s(\lambda)$, we also obtain the group property $U(t + s) = e^{i(t+s)A} = e^{itA} e^{isA} = U(t)U(s)$. For the continuity, we observe that

$$\|e^{itA} f - f\|^2 = \int |e^{it\lambda} - 1|^2 d(E_A(\lambda)f, f) \tag{4.180}$$

and we apply the dominated convergence theorem.

It is easy to see that $f \in D(A)$ implies $e^{itA} f \in D(A)$ (verify). Furthermore

$$\left\| \frac{1}{t}\left(e^{itA}f - f\right) - iAf \right\|^2 = \int \left| \frac{1}{t}\left(e^{it\lambda} - 1\right) - i\lambda \right|^2 d(E_A(\lambda)f, f). \qquad (4.181)$$

Note that

$$\left| \frac{1}{t}\left(e^{it\lambda} - 1\right) - i\lambda \right|^2 = \left| i\int_0^\lambda dv \left(e^{itv} - 1\right) \right|^2 \le 4\lambda^2. \qquad (4.182)$$

The function λ^2 is integrable, and therefore we can apply the dominated convergence theorem and we obtain (4.179). □

Remark 4.3 The converse of the above proposition (Stone's theorem) is also true, i.e., given a strongly continuous one-parameter group of unitary operators $U(t)$, $t \in \mathbb{R}$, there exists a unique self-adjoint operator A, $D(A)$ such that $U(t) = e^{itA}$ (see, e.g., [1]).

Exercise 4.24 Let Q, P be the position and momentum operators and let U_a and V_b be the operators defined in (4.104), (4.105).

 (i) Verify that $U_a = e^{iaQ}$ and $V_b = e^{ibP}$.
 (ii) Show that Weyl's relation holds

$$V_b U_a = e^{iab} U_a V_b. \qquad (4.183)$$

(iii) Verify that for any $f \in \mathscr{S}(\mathbb{R})$ we have

$$\frac{d}{da}\frac{d}{db}(V_b U_a f)(x)\Big|_{a=b=0} = -(PQf)(x), \qquad (4.184)$$

$$\frac{d}{da}\frac{d}{db}(e^{iab} U_a V_b f)(x)\Big|_{a=b=0} = -(QPf)(x) + if(x). \qquad (4.185)$$

Therefore, by differentiation w.r.t. the parameters, Weyl's relation implies the canonical commutation relation for Q and P

$$[Q, P]f = if, \qquad f \in \mathscr{S}(\mathbb{R}). \qquad (4.186)$$

Using Proposition 4.25, we define the evolution of a state in Quantum Mechanics. Let H, $D(H)$ be the self-adjoint Hamiltonian of the system defined in some Hilbert space $\mathscr{H}$ and let $\psi_0 \in \mathscr{H}$ be the initial state. Then, the state at time t is defined by

$$\psi(t) = e^{-i\frac{t}{\hbar}H}\psi_0, \qquad (4.187)$$

where the unitary operator $e^{-i\frac{t}{\hbar}H}$ is called propagator.

We obviously have $\psi(0) = \psi_0$ and $\|\psi(t)\| = \|\psi_0\|$. Moreover, if we fix $T > 0$ and $\psi_0 \in D(H)$ then (4.187) belongs to $C^0([0, T], D(H)) \cap C^1([0, T], \mathscr{H})$ and it is solution of the Cauchy problem for the Schrödinger equation

$$i\hbar\frac{\partial\psi(t)}{\partial t} = H\psi(t), \qquad \psi(0) = \psi_0 \in D(H). \qquad (4.188)$$

For the uniqueness of the solution, we observe that if $\psi_1(t)$ and $\psi_2(t)$ are solutions with initial state ψ_0, then $\psi_1(t) - \psi_2(t)$ is solution with initial datum zero. Since the norm is preserved, we have $\|\psi_1(t) - \psi_2(t)\| = 0$, i.e., $\psi_1(t) = \psi_2(t)$.

Exercise 4.25 Verify that e^{-itH_D}, with H_D, $D(H_D)$ defined in (4.114), is given by

$$\left(e^{-itH_D} f\right)(x) = \sum_{n=1}^{\infty} e^{-itn^2} f_n \, \phi_n(x), \qquad f_n = (\phi_n, f). \tag{4.189}$$

4.10 Point, Absolutely Continuous, and Singular Continuous Spectrum

Here, we characterize different parts of the spectrum of a self-adjoint operator A, $D(A)$ in the Hilbert space $\mathscr{H}$ in terms of the corresponding spectral family $E_A(\lambda)$ and of the spectral measure m_A^f generated by the distribution function $(E_A(\lambda)f, f)$. We first characterize $\rho(A)$ and $\sigma(A)$.

Proposition 4.26
$\lambda \in \rho(A)$ *if and only if* $\exists\, \varepsilon_0 > 0$ *such that* $E_A(\lambda + \varepsilon_0) - E_A(\lambda - \varepsilon_0) = 0$.
Equivalently,
$\lambda \in \sigma(A)$ *if and only if* $\forall\, \varepsilon > 0$ *one has* $E_A(\lambda + \varepsilon) - E_A(\lambda - \varepsilon) \neq 0$.

Proof Let us assume that $\exists\, \varepsilon_0 > 0$ such that $E_A(\lambda + \varepsilon_0) - E_A(\lambda - \varepsilon_0) = 0$ and let $f \in D(A)$. Then

$$\|(A - \lambda)f\|^2 = \int (\lambda - \mu)^2 d(E_A(\mu)f, f) = \int_{|\lambda - \mu| \geq \varepsilon_0} (\lambda - \mu)^2 d(E_A(\mu)f, f)$$

$$\geq \varepsilon_0^2 \int_{|\lambda - \mu| \geq \varepsilon_0} d(E_A(\mu)f, f) = \varepsilon_0^2 \int d(E_A(\mu)f, f) = \varepsilon_0^2 \|f\|^2. \tag{4.190}$$

The above inequality implies that $\lambda \in \rho(A)$ (arguing as in the proof of Proposition 4.22).

Let us assume that $\lambda \in \rho(A)$. By absurd, we suppose that there is a sequence $\{\varepsilon_n\}$, with $\varepsilon_n > 0$ and $\varepsilon_n \to 0$ for $n \to \infty$, such that $E_A(\lambda + \varepsilon_n) - E_A(\lambda - \varepsilon_n) \neq 0$ for any n. Let us fix $f \in \mathscr{H}$ and $f_n = \left(E_A(\lambda + \varepsilon_n) - E_A(\lambda - \varepsilon_n)\right)f$. Then $f_n \in D(A)$, the spectral measure generated by $(E_A(\mu)f_n, f_n)$ is supported in $(\lambda - \varepsilon_n, \lambda + \varepsilon_n]$ and we have

$$\|(A - \lambda)f_n\|^2 = \int (\lambda - \mu)^2 d(E_A(\mu)f_n, f_n) = \int_{(\lambda - \varepsilon_n, \lambda + \varepsilon_n]} (\lambda - \mu)^2 d(E_A(\mu)f_n, f_n)$$

$$\leq \varepsilon_n^2 \int_{(\lambda - \varepsilon_n, \lambda + \varepsilon_n]} d(E_A(\mu)f_n, f_n) = \varepsilon_n^2 \|f_n\|^2. \tag{4.191}$$

By inequality (4.191), it follows that $A - \lambda$ does not have a bounded inverse and this contradicts the hypothesis. □

An eigenvalue of A is a point of discontinuity of the spectral family.

Proposition 4.27

(i) λ *is an eigenvalue of* A, $D(A)$ *if and only if* $E_A(\{\lambda\}) \neq 0$;
(ii) f *is an eigenvector of* A, $D(A)$ *corresponding to the eigenvalue* λ *if and only if* $f \in Ran\, E_A(\{\lambda\})$;

Proof Let $Af = \lambda f$, $f \in D(A)$. Then

$$0 = \|(A - \lambda)f\|^2 = \int (\lambda - \mu)^2 d(E_A(\mu)f, f). \tag{4.192}$$

In particular, for any $\varepsilon > 0$ we have

$$0 = \int_{(\lambda+\varepsilon,\infty)} (\lambda - \mu)^2 d(E_A(\mu)f, f) \geq \varepsilon^2 \int_{(\lambda+\varepsilon,\infty)} d(E_A(\mu)f, f)$$
$$= \varepsilon^2 \big(E_A(\lambda + \varepsilon, \infty)f, f\big) = \varepsilon^2 \|E_A(\lambda + \varepsilon, \infty)f\|^2, \tag{4.193}$$

i.e., $E_A(\lambda+\varepsilon,\infty)f = f - E_A(\lambda+\varepsilon)f = 0$. Similarly, we find $E_A(\lambda - \varepsilon)f = 0$. Hence,

$$\big(E_A(\lambda + \varepsilon) - E_A(\lambda - \varepsilon)\big)f = f. \tag{4.194}$$

Taking the limit $\varepsilon \to 0$, we find

$$\big(E_A(\lambda) - E_A(\lambda - 0)\big)f = E_A(\{\lambda\})f = f. \tag{4.195}$$

Thus, we have proved that $E_A(\{\lambda\}) \neq 0$ and $f \in Ran\, E_A(\{\lambda\})$.
Let us assume that $E_A(\{\lambda\}) \neq 0$ and let $f \in Ran\, E_A(\{\lambda\})$. Then, we have $f \in D(A)$ and $\|f\|^2 = (f, f) = (E_A(\{\lambda\})f, f)$. Therefore, the spectral measure generated by $(E_A(\{\lambda\})f, f)$ is supported in $\{\lambda\}$ and

$$\|(A - \lambda)f\|^2 = \int (\lambda - \mu)^2 d(E_A(\mu)f, f) = \int_{\{\lambda\}} (\lambda - \mu)^2 d(E_A(\mu)f, f) = 0. \tag{4.196}$$

which means $Af = \lambda f$. □

Exercise 4.26 Prove that $Af = \lambda f$, $f \in D(A)$, if and only if $m_A^f = \|f\|^2 \delta_\lambda$.

Definition 4.27 The point spectrum $\sigma_p(A)$ of A is the set of all eigenvalues of A. $\mathscr{H}_p(A)$ is the (closed) subspace spanned by all the eigenvectors of A.

We note that $\sigma_p(A)$ is a countable set. Indeed, eigenvectors associated to different eigenvalues are orthogonal and there is at most a countable set of mutually orthogonal vectors in a separable Hilbert space.

Moreover, we have $\mathscr{H}_p(A) = \left\{ f \in \mathscr{H} \mid m_A^f \text{ is pure point} \right\}$.

Let us introduce other closed subspaces of $\mathscr{H}$ characterized by different behaviors of the spectral measure.

Definition 4.28

$\mathscr{H}_c(A) = \left\{ f \in \mathscr{H} \mid m_A^f \text{ is continuous} \right\}$,

$\mathscr{H}_{ac}(A) = \left\{ f \in \mathscr{H} \mid m_A^f \text{ is absolutely continuous w.r.t. Lebesgue measure} \right\}$,

$\mathscr{H}_{sc}(A) = \left\{ f \in \mathscr{H} \mid m_A^f \text{ is singular continuous w.r.t. Lebesgue measure} \right\}$.

Correspondingly, we obtain the following decomposition of the Hilbert space.

Proposition 4.28 *Let $A, D(A)$ be self-adjoint in $\mathscr{H}$. Then*

$$\mathscr{H} = \mathscr{H}_p(A) \oplus \mathscr{H}_{ac}(A) \oplus \mathscr{H}_{sc}(A) = \mathscr{H}_p(A) \oplus \mathscr{H}_c(A). \qquad (4.197)$$

Moreover, the restriction of A to each subspace defines a self-adjoint operator in the subspace and

$$\sigma(A) = \sigma(A|_{\mathscr{H}_p(A)}) \cup \sigma(A|_{\mathscr{H}_{ac}(A)}) \cup \sigma(A|_{\mathscr{H}_{sc}(A)}). \qquad (4.198)$$

For the proof we refer, e.g., to [5].

The above decomposition suggests to define the following parts of the spectrum of a self-adjoint operator.

Definition 4.29

Continuous spectrum $\sigma_c(A) = \sigma(A|_{\mathscr{H}_c(A)})$,

absolutely continuous spectrum $\sigma_{ac}(A) = \sigma(A|_{\mathscr{H}_{ac}(A)})$,

singular continuous spectrum $\sigma_{sc}(A) = \sigma(A|_{\mathscr{H}_{sc}(A)})$.

Note that $\sigma_c(A), \sigma_{ac}(A)$ and $\sigma_{sc}(A)$ are spectra of self-adjoint operators and therefore are closed subsets of the real axis.

On the other hand, an accumulation point of $\sigma_p(A)$ is not in general an eigenvalue, i.e., $\sigma_p(A)$ could not be closed and we have $\sigma(A|_{\mathscr{H}_p(A)}) = \overline{\sigma_p(A)}$. Then, we can write

$$\sigma(A) = \overline{\sigma_p(A)} \cup \sigma_c(A) = \overline{\sigma_p(A)} \cup \sigma_{ac}(A) \cup \sigma_{sc}(A). \qquad (4.199)$$

Finally, we observe that the above decomposition of the spectrum plays an important role for the applications in Quantum Mechanics. In particular, we shall see that the properties of the spectrum of the Hamiltonian of a quantum system determine the characteristics of the dynamics of the system.

4.11 Discrete and Essential Spectrum

The different parts of the spectrum of a self-adjoint operator defined in the previous section could not be disjoint, e.g., an eigenvalue could be embedded in the continuous spectrum. This fact can make difficult the characterization of the spectrum. It is therefore convenient to introduce the following decomposition of the spectrum into two disjoint subsets.

Definition 4.30 Let A, $D(A)$ be a self-adjoint operator in $\mathcal{H}$.

The discrete spectrum $\sigma_d(A)$ is the set of the eigenvalues λ of A such that λ has finite multiplicity and it is an isolated point of $\sigma(A)$.

The essential spectrum is $\sigma_{ess}(A) = \sigma(A) \setminus \sigma_d(A)$.

From the definition, it follows that $\lambda \in \sigma(A)$ belongs to the essential spectrum if $\lambda \in \sigma_c(A)$, or if λ is an eigenvalue with infinite multiplicity, or if λ is an eigenvalue with finite multiplicity embedded in the continuous spectrum, or if λ is an accumulation point of the eigenvalues.

In terms of the spectral family, discrete and essential spectrum can be described as follows (verify):

$\lambda \in \sigma_d(A)$ if and only if $\exists \varepsilon_0 > 0$ such that the dimension of the subspace $Ran\, E_A((\lambda - \varepsilon, \lambda + \varepsilon))$ is finite and different from zero for any $\varepsilon \in (0, \varepsilon_0)$;

$\lambda \in \sigma_{ess}(A)$ if and only if $\forall \varepsilon > 0$ the dimension of the subspace $Ran\, E_A((\lambda - \varepsilon, \lambda + \varepsilon))$ is infinite.

A point of the essential spectrum can be characterized by the existence of a singular Weyl sequence.

Proposition 4.29 $\lambda \in \sigma_{ess}(A)$ *if and only if there exists a singular Weyl sequence relative to* λ, *i.e., a sequence of vectors* $\{f_n\}$ *such that* $f_n \in D(A)$, $\|f_n\| = 1$, $f_n \rightharpoonup 0$ *for* $n \to \infty$ *and* $\|(A - \lambda)f_n\| \to 0$ *for* $n \to \infty$.

Proof Let $\lambda \in \sigma_{ess}(A)$. This implies that the dimension of the subspace $Ran\, E_A(I_n)$ is infinite for any n, where $I_n = (\lambda - \frac{1}{n}, \lambda + \frac{1}{n})$. Let us fix $f_1 \in Ran\, E_A(I_n)$, $\|f_1\| = 1$. Then, we fix $f_2 \in Ran\, E_A(I_n)$, $\|f_2\| = 1$ with $(f_2, E_A(I_n)f_1) = (f_2, f_1) = 0$. Iterating, for any n we fix $f_n \in Ran\, E_A(I_n)$, $\|f_n\| = 1$, with $(f_n, E_A(I_n)f_1) = (f_n, E_A(I_n)f_2) = \cdots = (f_n, E_A(I_n)f_{n-1}) = 0$, i.e., $(f_n, f_m) = 0$, for any $m < n$. Thus, we have constructed an infinite orthonormal sequence $\{f_n\}$ which is weakly convergent to zero. Moreover, $f_n \in D(A)$ and

$$\|(A - \lambda)f_n\|^2 = \int_{I_n} (\lambda - \mu)^2 d(E_A(\mu)f_n, f_n) \le \frac{1}{n^2} \int d(E_A(\mu)f_n, f_n)$$
$$= \frac{1}{n^2} \|f_n\|^2 = \frac{1}{n^2}. \tag{4.200}$$

Therefore, $\{f_n\}$ is a singular Weyl sequence.

Let us assume that $\{f_n\}$ is a singular Weyl sequence relative to λ. We know that $\lambda \in \sigma(A)$ and it is sufficient to show that $\lambda \notin \sigma_d(A)$. By absurd, let us suppose that there exists $\varepsilon > 0$ such that the subspace $Ran\ E_A((\lambda-\varepsilon, \lambda+\varepsilon))$ is finite-dimensional and let $e_1, \ldots, e_N \in D(A)$ be an orthonormal basis in the subspace. Then

$$E_A((\lambda - \varepsilon, \lambda + \varepsilon))f_n = \sum_{i=1}^{N}(e_i, E_A((\lambda - \varepsilon, \lambda + \varepsilon))f_n)\,e_i = \sum_{i=1}^{N}(e_i, f_n)\,e_i \to 0$$

$$(4.201)$$

for $n \to \infty$. Moreover

$$\begin{aligned}
\|(A - \lambda)f_n\|^2 &\geq \int_{-\infty}^{\lambda-\varepsilon}(\mu - \lambda)^2\,d(E_A(\mu)f_n, f_n) + \int_{\lambda+\varepsilon}^{\infty}(\mu - \lambda)^2\,d(E_A(\mu)f_n, f_n) \\
&\geq \varepsilon^2\left(\int_{-\infty}^{\lambda-\varepsilon}d(E_A(\mu)f_n, f_n) + \int_{\lambda+\varepsilon}^{\infty}d(E_A(\mu)f_n, f_n)\right) \\
&= \varepsilon^2\left(1 - \int_{\lambda-\varepsilon}^{\lambda+\varepsilon}d(E_A(\mu)f_n, f_n)\right) \\
&= \varepsilon^2\big(1 - \|E_A((\lambda - \varepsilon, \lambda + \varepsilon))f_n\|^2\big) \to 0 \qquad (4.202)
\end{aligned}$$

since $\{f_n\}$ is a Weyl sequence. Therefore, we have $\|E_A((\lambda - \varepsilon, \lambda + \varepsilon))f_n\| \to 1$ for $n \to \infty$. This contradicts (4.201) and so we have proved that $\lambda \in \sigma_{ess}(A)$. $\square$

Using the above proposition, we prove an important stability property of the essential spectrum w.r.t. a compact perturbation.

For the convenience of the reader, we recall that a bounded operator A in a Hilbert space $\mathscr{H}$ is a compact operator if for every bounded sequence $\{f_n\}$ in $\mathscr{H}$, $\{Af_n\}$ has a subsequence convergent in $\mathscr{H}$. Moreover, we shall use the fact that a compact operator maps weakly convergent sequences into norm convergent sequences (for further details, we refer, e.g., to [1], Chapter VI).

Proposition 4.30 (Weyl's theorem)

Let A, $D(A)$ and B, $D(B)$ be self-adjoint operators in the Hilbert space $\mathscr{H}$ and let $R_A(z) = (A - z)^{-1}$, $R_B(z) = (B - z)^{-1}$ be the corresponding resolvent operators.

If there exists $z \in \rho(A) \cap \rho(B)$ such that the operator $R_A(z) - R_B(z)$ is compact, then $\sigma_{ess}(A) = \sigma_{ess}(B)$.

Proof Let $\lambda \in \sigma_{ess}(A)$ and let $\{\psi_n\}$ be a singular Weyl sequence for A relative to λ, i.e.,

$$\psi_n \in D(A), \quad \|\psi_n\| = 1, \quad \psi_n \to 0, \quad \|(A - \lambda)\psi_n\| \to 0. \qquad (4.203)$$

We note that

$$\begin{aligned}
\left(R_A(z) - \frac{1}{\lambda - z}\right)\psi_n &= \left(R_A(z) - \frac{1}{\lambda - z}R_A(z)(A - z)\right)\psi_n \\
&= R_A(z)\left(I - \frac{A - z}{\lambda - z}\right)\psi_n = \frac{R_A(z)}{z - \lambda}(A - \lambda)\psi_n \,(4.204)
\end{aligned}$$

and then

$$\left\|\left(R_A(z) - \frac{1}{\lambda - z}\right)\psi_n\right\| = \left\|\frac{R_A(z)}{z - \lambda}(A - \lambda)\psi_n\right\| \le \frac{\|R_A(z)\|}{|z - \lambda|}\|(A - \lambda)\psi_n\| \to 0. \tag{4.205}$$

Let us define $\phi_n = R_B(z)\psi_n$. We have $\phi_n \in D(B)$ and $\phi_n \rightharpoonup 0$. Moreover

$$\phi_n = R_B(z)\psi_n = (R_B(z) - R_A(z))\psi_n + \left(R_A(z) - \frac{1}{\lambda - z}\right)\psi_n + \frac{1}{\lambda - z}\psi_n. \tag{4.206}$$

By the compactness of $(R_B(z) - R_A(z))$, we have $(R_B(z) - R_A(z))\psi_n \to 0$. Taking also into account of (4.205), we find

$$\lim_n \|\phi_n\| = \frac{1}{|\lambda - z|} > 0. \tag{4.207}$$

Finally,

$$\|(B - \lambda)\phi_n\| = \|(z - \lambda)\phi_n + (B - z)\phi_n\| = |z - \lambda|\left\|\left(R_B(z) - \frac{1}{\lambda - z}\right)\psi_n\right\|$$
$$\le |z - \lambda|\left(\|R_B(z) - R_A(z))\psi_n\| + \left\|\left(R_A(z) - \frac{1}{\lambda - z}\right)\psi_n\right\|\right) \to 0. \tag{4.208}$$

We conclude that the sequence $\{\hat{\phi}_n\}$, with

$$\hat{\phi}_n = \frac{\phi_n}{\|\phi_n\|}, \tag{4.209}$$

is a singular Weyl sequence for B relative to λ and this implies $\lambda \in \sigma_{ess}(B)$. We have proved that $\sigma_{ess}(A) \subseteq \sigma_{ess}(B)$. Exchanging the role of A and B, we also find $\sigma_{ess}(B) \subseteq \sigma_{ess}(A)$ and then the proposition is proved. $\qquad\square$

References

1. Reed, M., Simon, B.: Methods of Modern Mathematical Physics, I: Functional Analysis. Academic Press, New York (1980)
2. Reed, M., Simon, B.: Methods of Modern Mathematical Physics, II: Fourier Analysis, Self-Adjointness. Academic Press, New York (1975)
3. Reed, M., Simon, B.: Methods of Modern Mathematical Physics, IV: Analysis of Operators. Academic Press, New York (1978)
4. Akhiezer, N.I., Glazman, I.M.: Theory of Linear Operators in Hilbert Space. Dover Publ, New York (1993)
5. Amrein, W.O.: Hilbert Space Methods in Quantum Mechanics. EPFL Press, Lausanne (2009)
6. Blanchard, P., Brüning, E.: Mathematical Methods in Physics. Birkhäuser, Boston (2015)

7. Birman, M.S., Solomjak, M.Z.: Spectral Theory of Self-Adjoint Operators in Hilbert Space. D. Reidel Publ. Co., Dordrecht (1987)
8. Dell'Antonio G.: Lectures on the Mathematics of Quantum Mechanics. Atlantis Press (2015)
9. Gustafson, S.J., Sigal, I.M.: Mathematical Concepts of Quantum Mechanics. Springer, Berlin (2011)
10. Kato, T.: Perturbation Theory for Linear Operators. Springer, Berlin (1980)
11. Kolmogorov, A.N., Fomin, S.V.: Elements of the Theory of Functions and Functional Analysis. Dover Publ, New York (1999)
12. de Oliveira, C.R.: Intermediate Spectral Theory and Quantum Dynamics. Birkhäuser, Basel Boston Berlin (2009)
13. Prugovecki, E.: Quantum Mechanics in Hilbert Space. Academic Press, New York (1981)
14. Schmüdgen, K.: Unbounded Self-adjoint Operators on Hilbert Space. Springer, Dordrecht (2012)
15. Teschl, G.: Mathematical Methods in Quantum Mechanics. American Mathematical Society, Providence (2009)
16. Thirring, W.: Quantum Mechanics of Atoms and Molecules. Springer, New York (1981)
17. Brezis, H.: Functional Analysis, Sobolev Spaces and Partial Differential Equations. Springer (2011)
18. Evans, L.C.: Partial Differential Equations. American Mathematical Society, Providence, Rhode Island (1998)

Chapter 5
Rules of Quantum Mechanics

5.1 Formulation of the Rules

Quantum Mechanics is a theory of extraordinary success which provides an extremely accurate description of microscopic systems, i.e., systems with a typical action of the order of the Planck's constant $\hbar$. It is also worth observing that Quantum Mechanics is formulated as a universal theory (in the nonrelativistic regime) and this means that it can be equally applied to describe macroscopic systems.

In this chapter, we use the mathematical language described in Chap. 4 to give an axiomatic and mathematically rigorous formulation of (nonrelativistic) Quantum Mechanics, in agreement with the original approach developed by von Neumann in [1]. We avoid an excess of generality and neglect some technical difficulties. We only aim to provide the basic theoretical instruments to approach elementary quantum mechanical problems from the point of view of Mathematical Physics. Some of these problems will be discussed in the next chapters. For other, more general and complete, axiomatic formulations of the theory, we refer to [2–7].

In the rest of the section, we list the rules, while in the next section we will give some comments and explanations.

The first four rules identify the notion of state and observable, define the evolution of the state and characterize the possible probabilistic predictions of the theory.

I. (State) At any instant of time, the state of the system is described by a vector ψ with $\|\psi\| = 1$, also called wave function, of a given complex Hilbert space $\mathcal{H}$. The space $\mathcal{H}$ is called state space of the system.

Given two systems S_1, S_2 with state spaces $\mathcal{H}_1$ and $\mathcal{H}_2$, the state space of the composite system $S_1 \cup S_2$ is the tensor product $\mathcal{H}_1 \otimes \mathcal{H}_2$ (for the definition and the main properties of the tensor product of Hilbert spaces see, e.g., [8], Sect. II.4[1]).

---

[1]Roughly speaking, the space $\mathcal{H}_1 \otimes \mathcal{H}_2$ is constructed as follows. Let $\{\phi_j\}$, $\{\xi_k\}$ be two orthonormal bases in $\mathcal{H}_1$ and $\mathcal{H}_2$ and consider the set of pairs $\{\Phi_{jk}\} = \{\phi_j \otimes \xi_k\}$. Then, a generic element of $\mathcal{H}_1 \otimes \mathcal{H}_2$ is $\Phi = \sum_{jk} c_{jk} \phi_j \otimes \xi_k$, where $c_{jk} \in \mathbb{C}$ and $\sum_{jk} |c_{jk}|^2 < \infty$.

© Springer International Publishing AG, part of Springer Nature 2018
A. Teta, *A Mathematical Primer on Quantum Mechanics*,
UNITEXT for Physics, https://doi.org/10.1007/978-3-319-77893-8_5

II. (Observable) The physical quantities $\mathscr{A}$, $\mathscr{B}$, ... associated to the system, called observables, are described by self-adjoint operators $A, B, \ldots$ acting in the state space $\mathscr{H}$.

III. (Evolution of the state) Assigned the state of the system ψ_0 at time zero, the state at any time t is given by

$$\psi_t = e^{-i\frac{t}{\hbar}H}\psi_0, \tag{5.1}$$

where H is the Hamiltonian operator, i.e., the self-adjoint operator describing the observable energy of the system.

IV. (Predictions) Let ψ_t be the state of the system at time t, let $\mathscr{A}$ be an observable relative to the system with A the corresponding self-adjoint operator, and let I $\subseteq \mathbb{R}$ be an interval. Let us denote by $\mathscr{P}(\mathscr{A} \in \text{I}; \psi_t)$ the probability that the result of a measurement of $\mathscr{A}$ performed at time t on the system in the state ψ_t is a value belonging to the interval I. Then

$$\mathscr{P}(\mathscr{A} \in \text{I}; \psi_t) = (\psi_t, E_A(\text{I})\psi_t), \tag{5.2}$$

where $\{E_A(\lambda)\}$ denotes the spectral family associated to A.

The next rule provides a criterion for the construction of the self-adjoint operator describing an observable associated with a given quantum system starting from the corresponding classical system (quantization procedure). We shall limit ourselves to an elementary formulation, sufficient for the applications discussed in the next chapters, i.e., for a single particle in $\mathbb{R}^d$, $d = 1, 2, 3$.

V. (Quantization procedure) For a particle in Classical Mechanics, let $(x, p) \rightarrow x_k$, $k = 1, \ldots, d$, be the function defined in the phase space R^{2d} representing the observable kth component of the position and let $(x, p) \rightarrow p_k$ the function representing the observable kth component of the momentum.

For the corresponding quantum particle, the state space is $L^2(\mathbb{R}^d)$. Moreover, position and momentum observables are represented by the (self-adjoint) position and momentum operators in $L^2(\mathbb{R}^d)$

$$(Q_k\phi)(x) = x_k\phi(x), \qquad D(Q_k) = \left\{\phi \in L^2(\mathbb{R}^d) \mid \int dx \, |x_k\phi(x)|^2 < \infty\right\} \tag{5.3}$$

$$(P_k\phi)(x) = \frac{\hbar}{i}\frac{\partial\phi}{\partial x_k}(x), \qquad D(P_k) = \left\{\phi \in L^2(\mathbb{R}^d) \mid \int dp \, |p_k\tilde{\phi}(p)|^2 < \infty\right\}. \tag{5.4}$$

More generally, let $f : \mathbb{R}^{2d} \rightarrow \mathbb{R}$ be the function representing a classical observable. The corresponding quantum observable is represented by the self-adjoint operator $f(Q, P)$ in $L^2(\mathbb{R}^d)$ defined on a suitable domain.

It is important to stress that this last statement is ambiguous due to the non-commutativity of the operators (5.3), (5.4). As a matter of fact, such ambiguity does not occur in the applications discussed in this book (for further details on this point see, e.g., [7], Sect. 3.1).

Finally, there is a further rule concerning the modification of the state as a consequence of a measurement performed on the system. Such a rule is connected with interpretational problems of the measurement process in Quantum Mechanics and, therefore, it is postponed to Sect. 5.3.

5.2 Some Comments and Developments

The rules formulated in the previous section have some peculiar aspects with important physical consequences on the quantum description of a system. We schematically summarize such main points:

- the state space of the system is a linear space;
- in general, the self-adjoint operators representing the observables do not commute;
- the state evolution law is linear and deterministic;
- the predictions of the theory are probabilistic (except in some special cases) and the probability is given by a quadratic expression of the state;
- given the state of a system of n particles, in general, the state of a subsystem made of m, with $m < n$, particles is not defined.

In the sequel, we shall give some further comments and developments on these points to better clarify the meaning of the rules and their implications in the physical description of a quantum system.

State

Rule I defines the state as a unit vector ψ in a Hilbert space $\mathscr{H}$. Note, however, that the vectors ψ and $e^{i\alpha}\psi$, with $\alpha \in \mathbb{R}$, give rise to the same experimental predictions for the result of a measurement (see 5.2) and then they are physically indistinguishable. Therefore, if ψ_1 and ψ_2 are two unit vectors such that $\psi_2 = e^{i\alpha}\psi_1$, $\alpha \in \mathbb{R}$, then ψ_1 and ψ_2 define the same state of the system.

The notion of state defined in rule I can be generalized in analogy with the corresponding generalization in Classical Mechanics (see Sect. 1.7).

More precisely, the state as unit vector in a Hilbert space is known as pure state. On the other hand, in many concrete physical situations, it may happen that the pure state of the system is unknown and one only knows that the system is in the state ψ_1 with probability p_1, in the state ψ_2 with probability p_2, and so on, with $\sum_i p_i = 1$. In such cases, one says that the system is in a mixed state described by the operator

$$\rho = \sum_i p_i(\psi_i, \cdot)\psi_i \tag{5.5}$$

called density matrix. Note that ρ is a self-adjoint, positive, trace-class operator with $\operatorname{Tr}\rho = 1$ (for the definition of trace-class operators see, e.g., [8], Sect. VI.6).

As an exercise, the reader can verify that the evolution law for the density matrix is

$$\rho_t := \sum_i p_i(\psi_{i,t}, \cdot)\psi_{i,t} = e^{-i\frac{t}{\hbar}H}\rho_0\, e^{i\frac{t}{\hbar}H}\,, \tag{5.6}$$

where $\psi_{i,t} = e^{-i\frac{t}{\hbar}H}\psi_{i,0}$ and $\rho_0 = \sum_i p_i(\psi_{i,0}, \cdot)\psi_{i,0}$. At least formally, from (5.6), one obtains the evolution equation for the density matrix

$$\dot{\rho}_t = \frac{i}{\hbar}[\rho_t, H]\,, \tag{5.7}$$

called von Neumann equation. Moreover, the probabilistic prediction relative to an observable $\mathscr{A}$ when the system is in the mixed state ρ_t is

$$\mathscr{P}(\mathscr{A} \in I; \rho_t) := \sum_i p_i(\psi_{i,t}, E_A(I)\psi_{i,t}) = \mathrm{Tr}\,(\rho_t E_A(I))\,. \tag{5.8}$$

We also note that if $p_1 = 1$ and $p_i = 0$, $i = 2, 3, \ldots$, then $\rho = (\psi_1, \cdot)\psi_1$ and we reduce to the description given by the pure state ψ_1.

Probabilistic predictions

Let us discuss with more details rule IV, which can be considered as a generalized version of Born's rule for the position observable discussed in Sect. 3.3.

We first observe that, given the observable $\mathscr{A}$, the interval I, and the state ψ_t, in Formula (5.2), the l.h.s. is a quantity that is measured in experiments while the r.h.s. is theoretically computed. Therefore, Formula (5.2) provides theoretical predictions of experimental results.

Moreover, rule IV states that, in general, one can only formulate probabilistic predictions on the possible results of a measurement of an observable. Equivalently, this means that one can only predict the statistical distribution of the detected results of a measurement of a given observable in a large number of experiments realized in identical conditions. We emphasize that this is a peculiar aspect of Quantum Mechanics.

Let us recall the mathematical meaning of the r.h.s. of (5.2). Given the observable $\mathscr{A}$ represented by the self-adjoint operator A on the Hilbert space $\mathscr{H}$, by the spectral theorem we have the spectral family $\{E_A(\lambda)\}$. Then, as we discussed in Chap. 4, for any state $\psi \in \mathscr{H}$ we define the distribution function $F_A^\psi(\lambda) = (\psi, E_A(\lambda)\psi)$. Recalling Definition 4.24, the probability that the result of the measurement is a value belonging to the interval $I = (a, b]$ is given by

$$(\psi, E_A(I)\psi) = (\psi, (E_A(b) - E_A(a))\psi) \tag{5.9}$$

and similarly for other type of intervals. When the interval reduces to a point, we obtain

$$(\psi, E_A(\{a\})\psi) = (\psi, (E_A(a) - E_A(a - 0))\psi) \tag{5.10}$$

which represents the probability that the result of the measurement is the real value a. Moreover, the distribution function $F_A^\psi(\lambda)$ defines the associated spectral measure m_A^ψ via Formula 4.117. Recall that m_A^ψ is a probability measure supported on the spectrum $\sigma(A)$ and note that the r.h.s. of (5.2) is the measure of the interval I according to the probability measure m_A^ψ.

As for any probability measure, we define the expectation or mean value

$$\langle A \rangle := \int \lambda \, d(\psi, E_A(\lambda)\psi) \tag{5.11}$$

and the mean square deviation

$$\Delta A^2 := \int (\lambda - \langle A \rangle)^2 \, d(\psi, E_A(\lambda)\psi) \tag{5.12}$$

for the observable $\mathscr{A}$ when the system is described by the state ψ. By the spectral theorem we have $\langle A \rangle = (\psi, A\psi)$ and $\Delta A^2 = \langle A^2 \rangle - \langle A \rangle^2$.

Let us exemplify the above formulas in some typical cases. For the position observable represented by the operator Q, $D(Q)$ in $L^2(\mathbb{R})$, we have (see Example 4.30)

$$F_Q^\psi(\lambda) = \int_{-\infty}^{\lambda} dx \, |\psi(x)|^2 , \tag{5.13}$$

$$(\psi, E_Q(\mathrm{I})\psi) = \int_{\mathrm{I}} dx \, |\psi(x)|^2, \tag{5.14}$$

$$\langle Q \rangle = \int dx \, x \, |\psi(x)|^2 , \qquad \Delta Q^2 = \int dx \, x^2 |\psi(x)|^2 - \langle Q \rangle^2 . \tag{5.15}$$

Note that (5.14) coincides with Born's rule for the position of the particle discussed in Sect. 3.3.

For the momentum observable represented by the operator P, $D(P)$ in $L^2(\mathbb{R})$, we have (see Example 4.31)

$$F_P^\psi(\lambda) = \frac{1}{\hbar} \int_{-\infty}^{\lambda} dp \, |\tilde{\psi}(\hbar^{-1}p)|^2 , \tag{5.16}$$

$$(\psi, E_P(\mathrm{I})\psi) = \frac{1}{\hbar} \int_{\mathrm{I}} dp \, |\tilde{\psi}(\hbar^{-1}p)|^2, \tag{5.17}$$

$$\langle P \rangle = \hbar \int dk \, k \, |\tilde{\psi}(k)|^2 , \qquad \Delta P^2 = \hbar^2 \int dk \, k^2 |\tilde{\psi}(k)|^2 - \langle P \rangle^2 . \tag{5.18}$$

Note that in both cases the spectral measure is absolutely continuous with respect to Lebesgue measure for any $\psi \in L^2(\mathbb{R})$ and therefore $(\psi, E_Q(\{x_0\})\psi) = (\psi, E_P(\{p_0\})\psi) = 0$ for any $x_0, p_0 \in \mathbb{R}$ and for any $\psi \in L^2(\mathbb{R})$, i.e., the prob-

ability that the result of a position or momentum measurement is a definite real value
is always zero.

Let us consider an observable $\mathscr{C}$ represented by a self-adjoint operator C in the
Hilbert space $\mathscr{H}$ with a spectrum composed only by nondegenerate eigenvalues
$c_1, c_2, \ldots$, and let $\{\phi_i\}$ be the corresponding orthonormal basis of eigenvectors (see,
e.g., Example 4.32). Then

$$F_C^{\psi}(\lambda) = \sum_{j : c_j \leq \lambda} |(\phi_j, \psi)|^2, \tag{5.19}$$

$$(\psi, E_C(I)\psi) = \sum_{j : c_j \in I} |(\phi_j, \psi)|^2, \tag{5.20}$$

$$\langle C \rangle = \sum_j c_j |(\phi_j, \psi)|^2, \qquad \Delta C^2 = \sum_j c_j^2 |(\phi_j, \psi)|^2 - \langle C \rangle^2. \tag{5.21}$$

Note that in this case the spectral measure is pure point and it is supported on the
eigenvalues, so that the probability that the result of a measurement of $\mathscr{C}$ is a definite
real value c is

$$\mathscr{P}(\mathscr{C} = c; \psi) = (\psi, E_C(\{c\})\psi) = \begin{cases} 0 & \text{if} \quad c \notin \{c_1, c_2, \ldots\} \\ |(\phi_i, \psi)|^2 & \text{if} \quad c = c_i \text{ for some } i \in \mathbb{N}. \end{cases} \tag{5.22}$$

In particular, this means that

$$\mathscr{P}(\mathscr{C} = c_k; \phi_k) = 1, \qquad \langle C \rangle = c_k, \qquad \Delta C = 0 \tag{5.23}$$

for any $k \in \mathbb{N}$. In other words, if the state is an eigenvector ϕ_k of the operator
C then we can predict with certainty, i.e., with probability one, that the result of
a measurement of the observable $\mathscr{C}$ will be the corresponding eigenvalue c_k. We
stress that, by Proposition 4.27, this is the only case in which the theory provides
a non-probabilistic prediction. In the general case, one can formulate the following
weaker assertion. Let us consider an observable $\mathscr{A}$ represented by the operator A
and let $a \in \sigma(A)$. For any $\varepsilon > 0$ let us fix the interval $(a - \varepsilon, a + \varepsilon)$ and a state
$\psi_\varepsilon \in Ran\,(E_A((a - \varepsilon, a + \varepsilon)))$. Then, by the spectral theorem, we have

$$\mathscr{P}(\mathscr{A} \in (a - \varepsilon, a + \varepsilon); \psi_\varepsilon) = 1, \qquad \langle A \rangle = a + O(\varepsilon), \qquad \Delta A = O(\varepsilon), \tag{5.24}$$

where $\langle A \rangle$ and ΔA are computed on the state ψ_ε.

Non-commuting observables

The fact that the operators representing the observables of a system may not commute has strong physical consequences. Indeed, consider two observables $\mathscr{A}, \mathscr{B}$
represented by the self-adjoint operators A, B in the Hilbert space $\mathscr{H}$. Assume,
for simplicity, that A and B are bounded and satisfy $[A, B] = iC$, where C
is another bounded self-adjoint operator. Then, by Proposition 3.2, we have that

$\Delta A \cdot \Delta B \geq \frac{1}{2} \langle C \rangle$. As we already explained in Sect. 3.4, this means that, given any state $\psi \in \mathcal{H}$, we cannot predict with probability close to one both the result of a measurement of $\mathcal{A}$ and the result of a measurement of $\mathcal{B}$. It is worth to stress the radical difference with Classical Mechanics where, given any pure state of the system, we can predict with probability one the result of a measurement of any observable relative to the system.

Heisenberg representation

According to the formulation of the rules that we have given, the states of the system evolve in time while the observables are represented by time-independent operators. This formulation is usually called Schrödinger representation. There is an alternative equivalent formulation where the operators evolve in time and the states are fixed, called Heisenberg representation, which is close to the original approach of Matrix Mechanics developed by Heisenberg, Born, and Jordan in 1925. More precisely, in the Heisenberg representation, the observable $\mathcal{A}_t$ at time t is represented by the time-dependent operator

$$A_t = e^{i \frac{t}{\hbar} H} A \, e^{-i \frac{t}{\hbar} H}, \tag{5.25}$$

where A is the operator representing the observable at time zero (we neglect technical problems arising when A is unbounded). Moreover, the rule IV for the predictions is $\mathcal{P}(\mathcal{A}_t \in I; \, \psi) = (\psi, E_{A_t}(I)\psi)$, where ψ is the state at time zero. Note that, by Proposition 4.24, we have

$$(\psi, E_{A_t}(I)\psi) = (\psi_t, E_A(I)\psi_t) \tag{5.26}$$

and moreover

$$(\psi, A_t \psi) = (\psi_t, A\psi_t), \tag{5.27}$$

$$(\psi, A_t^2 \psi) - (\psi, A_t \psi)^2 = (\psi_t, A^2 \psi_t) - (\psi_t, A\psi_t)^2. \tag{5.28}$$

Therefore, the predictions of the theory in the Schrödinger and the Heisenberg representations coincide. From (5.25), one also obtains the evolution equation for the operator

$$\dot{A}_t = -\frac{i}{\hbar} [A_t, H], \tag{5.29}$$

called Heisenberg equation. We note that Eq. (5.29) formally reduces to the corresponding evolution equation for a classical observable (see Eq. 1.157) if one replaces the commutator $-\frac{i}{\hbar}[\cdot, H]$ with the Poisson bracket $\{\cdot, H_{cl}\}$, where H_{cl} is the Hamiltonian of the corresponding classical system.

Constants of motion and symmetries

Let us consider a quantum system with state space $\mathcal{H}$ and described by the Hamiltonian H. We say that an observable $\mathcal{A}$ represented by the operator A is a constant of motion if $\dot{A}_t = 0$. Note that, by (5.29), an equivalent definition is $[H, A] = 0$.

Moreover, we say that the one parameter group of unitary operators $e^{-i\lambda G}$, $\lambda \in \mathbb{R}$, where G is self-adjoint, is a symmetry for the Hamiltonian if $e^{i\lambda G} H e^{-i\lambda G} = H$.

Note that

$$\frac{d}{d\lambda} e^{i\lambda G} H e^{-i\lambda G} = i e^{i\lambda G}[G, H] e^{-i\lambda G} \tag{5.30}$$

and then $e^{-i\lambda G}$ is a symmetry if and only if $[G, H] = 0$.

Thus, we obtain Noether's theorem in Quantum Mechanics: $\mathscr{A}$ is a constant of motion if and only if $e^{-i\lambda A}$ is a symmetry.

We recall that a completely analogous result holds in Hamiltonian Mechanics: given a Hamiltonian system with Hamiltonian H_{cl}, $f(q, p)$ is a constant of motion if and only if the Hamiltonian flow $\phi_f^\lambda(q, p)$ generated by f is a symmetry for the Hamiltonian, i.e., $H_{cl}(\phi_f^\lambda(q, p)) = H_{cl}(q, p)$ (see, e.g., [9], Sect. 10.9).

Superposition principle and entanglement

We conclude this section by commenting briefly on two points which can be considered as the main peculiar aspects of the quantum description.

The first one is the superposition principle. From the mathematical point of view, it simply means that if ψ_1 and ψ_2 are two states of the system then also a linear combination (called superposition state)

$$\psi = \alpha\psi_1 + \beta\psi_2, \qquad \alpha, \beta \in \mathbb{C}, \quad \|\psi\| = 1 \tag{5.31}$$

is a possible state of the system. This property is preserved by the (linear) evolution law (5.1), in the sense that $e^{-i\frac{t}{\hbar}H}\psi = \alpha\, e^{-i\frac{t}{\hbar}H}\psi_1 + \beta\, e^{-i\frac{t}{\hbar}H}\psi_2$. These apparently trivial observations have relevant physical consequences, due to the fact that the predictions of the theory are given by a quadratic expression with respect to the state (see 5.2). To exemplify, let us consider a particle in one dimension so that $\psi_1, \psi_2 \in L^2(\mathbb{R})$. Then, the probability density for the position of the particle in the state $\psi = \alpha\psi_1 + \beta\psi_2$ is

$$|\psi(x)|^2 = |\alpha|^2|\psi_1(x)|^2 + |\beta|^2|\psi_2(x)|^2 + 2\,\Re\alpha\overline{\beta}\psi_1(x)\overline{\psi_2}(x). \tag{5.32}$$

Formula (5.32) shows that the physical situation described by the superposition state ψ can not be considered in any sense as the "sum" (with weights $|\alpha|^2$ and $|\beta|^2$) of the situations described by ψ_1 and ψ_2 separately due to the presence of the last term in (5.32), called interference term. In particular, such a term is responsible for the appearance of the interference effects, typical of wave phenomena, in the statistical distribution of the detected positions of the particle in a large number of experiments performed in identical conditions. These effects, which can be directly observed in experiments, are purely quantum effects with no classical counterpart. We refer to [10], Chap. 1, for a more detailed physical discussion. In Sect. 6.4, we shall give a quantitative description of the interference effect in the case of a free particle.

The second crucial aspect of the quantum description is the entanglement. For a system composed of n particles in $\mathbb{R}^d$ the state space is $\otimes_{j=1}^n L^2(\mathbb{R}^d)$, which is isomorphic to $L^2(\mathbb{R}^{nd})$. Hence, the state, or wave function, $\psi = \psi(x_1, \ldots, x_n)$, $\psi \in L^2(\mathbb{R}^{nd})$, is a "wave" in the configuration space $\mathbb{R}^{nd}$ of the corresponding classical system. In particular, it can happen that the wave function is a product state, i.e., a product of one-particle wave functions $\psi = \prod_{j=1}^{n} \psi_j(x_j)$. In such a case, we can associate a state ψ_j, $j = 1, \ldots, n$, to the jth particle of the system. It should be emphasized that this is a very special situation. Indeed, a product state can be realized at a certain instant of time but it is immediately destroyed by the interactions among the particles. This means that for a system of interacting particles the state is, in general, an entangled state, i.e., a state that cannot be written as a product state. When a system is described by an entangled state, we cannot associate a definite (pure) state to each subsystem and this fact has important physical consequences. In particular, it is at the origin of the "nonlocal effects" which can be produced on a subsystem S_1 acting on another spatially separated subsystem S_2 (for more details see, e.g., [11, 12]). It is worth mentioning that in Classical Mechanics the situation is radically different due to the fact that, given a state $(x_1, p_1, \ldots, x_n, p_n)$ of the n particle system, it is automatically well-defined state (x_j, p_j), $j = 1, \ldots, n$, of the jth particle and then of any subsystem composed by $m < n$ particles.

5.3 The Measurement Problem

There is a general and complete agreement in the physics community on the validity of the rules of Quantum Mechanics formulated in Sect. 5.1. They are sufficient to give an extremely accurate description of the microscopic world and they can be considered the basis of the "pragmatic" view of the physicists working with Quantum Mechanics.

On the other hand, one can be tempted to understand better various aspects involved in the above rules, like the nature of the probability arising in the theory, the meaning of physical properties of a system, the role of the measurement process, and so on.

When one attempts to deepen these issues, one enters the field of the so-called interpretational problem. Here, since the birth of quantum theory, many different views have been proposed that have stimulated a long and intense epistemological debate which is still active. A detailed analysis of the debate is out of the scope of these notes and we refer the interested reader to [11–15].

In order to give an idea of the interpretational problems involved, here we briefly discuss some aspects of the so-called Copenhagen interpretation of Quantum Mechanics which, more or less consciously, is the point of view accepted by many physicists. The aim is to highlight a conceptual difficulty arising in such an interpretation when one describes the measurement process or, more generally, the connection between the quantum and the classical description of the world.

We schematically summarize the points to be discussed.

We start with a description of the measurement process of an observable in Quantum Mechanics and the consequent formulation of the wave function collapse rule.

As a second step, we try to explain a relevant point of the Copenhagen interpretation, i.e., the completeness assumption of the wave function, and its implications on the idea of physical property possessed by a system.

Then, we reinterpret the measurement process in the light of the completeness assumption of the wave function and we stress the new meaning acquired by the measurement process within the Copenhagen interpretation.

Finally, we discuss a difficulty arising when one tries to describe the measurement process according to the Copenhagen interpretation using only the Schrödinger equation. We also point out that such a difficulty arises at conceptual level and it has no practical consequences.

Measurement process

The description of the measurement process is a crucial point in Quantum Mechanics. Let us consider a quantum system with state space $\mathscr{H}$ and let $\mathscr{A}$ be an observable. For simplicity, in this section, we assume that the observable is represented by a self-adjoint operator A in $\mathscr{H}$ with a spectrum composed only by nondegenerate eigenvalues $a_1, a_2, \ldots$, so that these eigenvalues are the only possible results of a measurement of $\mathscr{A}$. Moreover, we denote by $\{\phi_i\}$ the corresponding orthonormal basis of eigenvectors.

Let us assume that the system is in a state ψ which is not an eigenvector of A. According to (5.22), this implies that only probabilistic predictions on the result of a measurement of $\mathscr{A}$ are possible.

Let us suppose that a measurement of the observable $\mathscr{A}$ is performed on the system and that the result is the eigenvalue a_1.

Let us clarify the situation determined by the measurement. Suppose that, after a very short time and avoiding any external perturbation on the system, we perform a second measurement of $\mathscr{A}$. It is obvious that the result that we shall find is the same value a_1. In other words, immediately after the first measurement, we can predict with probability one the value a_1 for $\mathscr{A}$. This means that, after the first measurement, the state of the system is the eigenvector ϕ_1 of A (see 5.23). Thus, we arrive at the conclusion that the measurement process caused the instantaneous change $\psi \to \phi_1$ of the state of the system. Note that the transition is stochastic and nonlinear. Such instantaneous and abrupt change of the state determined by the measurement process is known as wave function collapse rule.

Copenhagen interpretation

A basic assumption in the Copenhagen interpretation is the completeness of the wave function. This statement means that the wave function encodes the maximal information available about the physical properties of the quantum system under consideration. The immediate consequence of the completeness assumption is that the probabilistic predictions of the theory have an ontological character, i.e., they are not due to our ignorance about some property of the system but they are a consequence of an intrinsic indeterminism present in nature. To be more specific, let us assume that the system is in a state ψ which is not an eigenvector of A. In this case (see 5.22), we cannot predict with probability one the result of a measurement of $\mathscr{A}$.

Then, according to the completeness assumption, the system in the state ψ does not possess a definite value of $\mathscr{A}$. It is important to stress this point. We are not saying that the system has a value of $\mathscr{A}$ which is simply unknown. We are saying that it makes no physical sense to attribute a value of $\mathscr{A}$ to the system.

On the other hand, if the system is in the eigenstate ϕ_1 of A, then we can predict with probability one that the result of a measurement of $\mathscr{A}$ will be a_1. In this case, we can affirm that the system in the state ϕ_1 has the definite value a_1.

It is worth mentioning that also in Classical Physics one often encounters cases where only probabilistic predictions are available for a given system. In these cases, however, the probabilistic predictions have an epistemic character, i.e., they originate from our ignorance about the precise initial state of the system or the forces acting on it. At least in principle, in Classical Physics one could always restore a deterministic description by a more precise analysis of the properties of the system. We observe that such a possibility is excluded, even in principle, according to the Copenhagen interpretation of Quantum Mechanics.

Meaning of a measurement in the Copenhagen interpretation

Let us analyze the meaning of the measurement process in the light of the Copenhagen interpretation. Assume that, before the measurement, the system is in the state ψ which is not an eigenvector of A. Then, according to the Copenhagen interpretation, the system does not possess a definite value of the observable $\mathscr{A}$. Suppose that a measurement of $\mathscr{A}$ is performed and that the result is a_1. Due to the wave function collapse rule, immediately after the measurement, the system is in the state ϕ_1 and then we can affirm that it has the definite value a_1 of the observable $\mathscr{A}$.

Thus, according to the Copenhagen interpretation, the measurement process is a stochastic mechanism that, via the wave function collapse, assigns a definite value of the observable to the system (which was not possessed before the measurement).

It is important to stress the difference with Classical Physics. Indeed, in Classical Physics, it makes perfect sense to think that a system possesses a definite, even if possibly unknown, value of all the observables and a measurement of an observable is an innocuous process that simply reveals the value of the observable.

Finally, we note that the measurement process is a concrete physical process where the system to be measured interacts with another system playing the role of measurement apparatus. Therefore, its meaning within the Copenhagen interpretation should be discussed in terms of such an interaction.

A conceptual difficulty

A first attempt was made by Bohr ([16]) who claimed that a measurement apparatus is a macroscopic object described by the laws of Classical Mechanics and therefore it always has a definite value of all its observables. Then, he assumed that the interaction between the classical apparatus and the quantum system determines the wave function collapse, inducing the transition of the quantum system state.

Bohr's explanation has been criticized in many respects. We simply mention the following observations.

It is not clear where the borderline between the measurement apparatus (showing a classical behavior) and the system (showing a quantum behavior) should be fixed. The problem is usually solved pragmatically for each specific situation but, at the conceptual level, there is an ambiguity.

An even more relevant point is the fact that it is not explained why a measurement apparatus, despite being made of atoms, cannot be described by Quantum Mechanics and it is a priori considered as a classical object with well-defined classical properties.

The problem was better clarified by von Neumann ([1]). He discussed the measurement process considering both the system and the apparatus as quantum objects. In this case, the measurement process can be schematically described as follows.

Let us consider a system with state space $\mathscr{H}$ and let $\mathscr{A}$ be an observable to be measured (as usual, we assume that the corresponding operator A has only eigenvalues $a_1, a_2, \ldots$ and eigenvectors $\phi_1, \phi_2, \ldots$). The measurement apparatus can be thought as a system characterized by a sequence of possible states $\xi_0, \xi_1, \xi_2, \ldots$, so that when the apparatus is in the state ξ_j we say that the "index" points toward the "position j". In particular, when the apparatus is in the state ξ_0 we say that the index is at rest. The state space of "system + apparatus" is $\mathscr{H} \otimes \mathscr{H}'$, where $\mathscr{H}'$ is the Hilbert space of the apparatus generated by $\xi_0, \xi_1, \xi_2, \ldots$.

Let us assume that, before the measurement, the apparatus is prepared in the state ξ_0 and the system is in the state ϕ_j. The Hamiltonian describing the interaction between the system and the apparatus, i.e., the measurement process, is designed in such a way to produce the following evolution law:

$$\phi_j \otimes \xi_0 \;\rightarrow\; \phi_j \otimes \xi_j, \tag{5.33}$$

i.e., after the measurement, the state of the system is unchanged and the index of the apparatus points toward the position j. It is now sufficient to "look" at the index of the apparatus to conclude that the result of the measurement is a_j. It is worth observing that such a scheme for the measurement process is surely idealized. Nevertheless, it captures the essence of the conceptual problem, as we try to explain in the following lines.

Let us now consider a superposition state as initial state of the system, e.g.,

$$\psi = c_1 \phi_1 + c_2 \phi_2, \qquad c_1, c_2 \in \mathbb{C}, \qquad |c_1|^2 + |c_2|^2 = 1. \tag{5.34}$$

Using (5.33) and the linearity of the evolution law, we have

$$\psi \otimes \xi_0 = (c_1 \phi_1 + c_2 \phi_2) \otimes \xi_0 \;\rightarrow\; c_1 \phi_1 \otimes \xi_1 + c_2 \phi_2 \otimes \xi_2. \tag{5.35}$$

Thus, after the measurement process, the state of the system + apparatus is a typical entangled state corresponding to the superposition of $\phi_1 \otimes \xi_1$ and $\phi_2 \otimes \xi_2$. In terms of density matrix, the pure state after the measurement is equivalently described by the orthogonal projector on $c_1 \phi_1 \otimes \xi_1 + c_2 \phi_2 \otimes \xi_2$, i.e.,

$$\rho_p = |c_1|^2 (\phi_1 \otimes \xi_1, \cdot)\phi_1 \otimes \xi_1 + |c_2|^2 (\phi_2 \otimes \xi_2, \cdot)\phi_2 \otimes \xi_2$$
$$+ \overline{c_1} c_2 (\phi_1 \otimes \xi_1, \cdot)\phi_2 \otimes \xi_2 + c_1 \overline{c_2} (\phi_2 \otimes \xi_2, \cdot)\phi_1 \otimes \xi_1 , \qquad (5.36)$$

where the last two terms represent the interference terms. It is important to stress that, according to the Copenhagen interpretation, the pure state at the r.h.s. of (5.35), or the density matrix ρ_p, describes a situation in which the index of the apparatus does not have a definite position and some interference effects between the position 1 and the position 2 occur. This is clearly an unsatisfactory situation since we expect to find the index of the apparatus in a well-defined position after the measurement, either position 1 or position 2.

Note that the pure state ρ_p describes a situation different from that described by the mixed state

$$\rho_m = |c_1|^2 (\phi_1 \otimes \xi_1, \cdot)\phi_1 \otimes \xi_1 + |c_2|^2 (\phi_2 \otimes \xi_2, \cdot)\phi_2 \otimes \xi_2 , \qquad (5.37)$$

Indeed, if the system + apparatus was described by (5.37), then we could affirm that either the system is in ϕ_1 and the apparatus is in ξ_1 (with probability $|c_1|^2$) or the system is in ϕ_2 and the apparatus is in ξ_2 (with probability $|c_2|^2$). Therefore, it would be legitimate to assert that the index of the apparatus is in a well-defined position.

A possible way to solve the above difficulty is to invoke the wave function collapse, i.e., to assume that, after the measurement, the superposition state at the r.h.s. of (5.35) instantaneously reduces to one of the two states $\phi_1 \otimes \xi_1$ or $\phi_2 \otimes \xi_2$.

Observe that, at conceptual level, this transition can be thought as the combination of two successive steps: first the reduction to the mixed state (5.37) and then the (random) choice between the states $\phi_1 \otimes \xi_1$ and $\phi_2 \otimes \xi_2$.

Some comments on this solution of the difficulty are in order.

(i) There is no possibility to obtain the transition from the pure state (5.36) to the mixed state (5.37), and then the wave function collapse, by a dynamics governed by the Schrödinger equation.

(ii) As an immediate consequence, one has to admit that there are two dynamical laws: the Schrödinger dynamics when no measurement is performed and the wave function collapse when one measures the system.

(iii) It is not explained who or what determines the collapse and, moreover, it is not clear when and under which circumstances the collapse should occur.

By the above remarks, one can understand the reason why the proposed solution has been considered unsatisfactory by many physicists and philosophers.

We do not enter the delicate conceptual details of the debate. We only point out that from a "pragmatic" point of view the difficulty arising in the measurement process has no practical consequences. In fact, in more realistic models of system + apparatus described by the Schrödinger dynamics, it turns out that the replacement of the pure state (5.36) with the mixed state (5.37) can be done at the cost of such a small error that it would be practically unobservable in a real experiment. In other words, the difference between the situation "the pointer has no definite position" and the situation "the pointer has a definite, even if possibly unknown, position" is so small that it is practically undetectable.

On the other hand, from a conceptual point of view the difficulty remains. Indeed, Quantum Mechanics in the Copenhagen interpretation would be a fundamental theory that acquires a precise meaning only through an approximation procedure, although very accurate, depending on the specific model employed.

To summarize the situation, Bell ([13]) wrote that everything works perfectly well F.A.P.P., an acronym of For All Practical Purposes, but he also insisted that a conceptual inconsistency in the Copenhagen interpretation of the formalism related to the measurement problem undoubtedly exists.

5.4   Mathematical Treatment of a Quantum Mechanical Problem

In this section, we describe the steps that are usually needed to analyze a quantum system from the point of view of Mathematical Physics.

Given a physical system, the preliminary step is the definition of the Hilbert space of states $\mathscr{H}$ and of a Hamiltonian acting on $\mathscr{H}$. In the applications described in these notes, the system is a particle in $\mathbb{R}^d$, $d = 1, 2, 3$, and then we fix $\mathscr{H} = L^2(\mathbb{R}^d)$. As for the Hamiltonian, a standard procedure is to consider the operator $\dot{H}$ obtained from the quantization rule of the classical Hamiltonian of the system. Usually, the operator $\dot{H}$ is initially defined on a domain of smooth functions so that, in general, it is a symmetric but not a self-adjoint operator.

The first mathematical problem is the self-adjointness of the Hamiltonian, i.e., the construction of a self-adjoint extension H of $\dot{H}$ (note that from now on we do not distinguish between an observable and the corresponding self-adjoint operator).

The operator $\dot{H}$ may have more than one self-adjoint extensions and the choice of the "right one" depends on the physical problem under consideration. We also recall that self-adjointness of H implies that the spectrum $\sigma(H)$ is real and it guarantees the existence of the time evolution $e^{-i\frac{t}{\hbar}H}$ of the system.

The next step is the analysis of the spectrum of H. In the general case, one tries to obtain qualitative information on $\sigma(H)$ and on its subsets $\sigma_p(H)$, $\sigma_{ac}(H)$, $\sigma_{sc}(H)$, and on the corresponding (proper and generalized) eigenfunctions. It is worth mentioning that for any "reasonable" Hamiltonian H one has $\sigma_{sc}(H) = \emptyset$.

In some simple cases, i.e., for the so-called solvable models, the different parts of the spectrum and the corresponding eigenfunctions can be explicitly determined.

The study of the eigenvalue problem for H, i.e., the solution of $H\psi = E\psi$, with $\psi \in D(H)$, is of particular physical interest. When the system is described by an eigenfunction ψ of the Hamiltonian with eigenvalue E, then the system has the definite value E of the energy, i.e., we can predict with probability one that the result of a measurement of the energy would be E. Since the time evolution of ψ is simply given by $\psi_t = e^{-i\frac{t}{\hbar}E}\psi$, this property does not change with time. Moreover, taking into account of (5.2), the probabilistic predictions of any observable do not change with time. These states are called stationary states. The stationary state corresponding to the minimum eigenvalue (if it exists) is called ground state and the minimum eigenvalue is the ground-state energy.

A further relevant problem is the analysis of the time evolution $e^{-i\frac{t}{\hbar}H}\psi$ for t large and $\psi \notin \mathcal{H}_p(H)$, i.e., when ψ is not an eigenvector or linear combination of eigenvectors of H. This topic is part of scattering theory where, roughly speaking, the main problem is to compare a given asymptotic behavior for $t \to -\infty$ with the emerging asymptotic behavior for $t \to +\infty$.

The above considerations suggest that the properties of the time evolution $e^{-i\frac{t}{\hbar}H}\psi$ strongly depend on the choice of the state ψ. This fact is better clarified by introducing the notions of bound state and scattering state. We limit ourselves to the case of a particle in $\mathbb{R}^d$ described by the Hamiltonian H.

Definition 5.1 A state $\phi \in L^2(\mathbb{R}^d)$ is called a bound state if

$$\lim_{R\to\infty} \sup_t \int_{|x|>R} dx \, |(e^{-i\frac{t}{\hbar}H}\phi)(x)|^2 = 0. \tag{5.38}$$

The set of all bound states will be denoted by $\mathcal{M}_0(H)$.

Formula (5.38) means that the probability to find the particle outside the ball $B_R = \{x \in \mathbb{R}^d \mid |x| < R\}$ can be made arbitrarily small, uniformly in time, by taking R sufficiently large. In this sense, a particle described by a bound state is essentially localized in position in a bounded region for all times. Note that a stationary state is a bound state.

Definition 5.2 A state $\phi \in L^2(\mathbb{R}^d)$ is called a scattering state if, for each $R > 0$,

$$\lim_{t\to\pm\infty} \int_{|x|<R} dx \, |(e^{-i\frac{t}{\hbar}H}\phi)(x)|^2 = 0. \tag{5.39}$$

The set of all scattering states will be denoted by $\mathcal{M}_\infty(H)$.

Formula (5.39) means that the probability to find the particle in the ball B_R, for each $R > 0$, goes to zero for t going to infinity. In other words, a particle described by a scattering state escapes from any bounded region for sufficiently large times.

Remark 5.1 The above definitions translate in quantum mechanical terms the notions of bound state and scattering state in Classical Mechanics. In fact, let (x, p) be a state of a classical particle in $\mathbb{R}^d$. We say that (x, p) is a bound state if the trajectory starting from (x, p) remains in a compact set of $\mathbb{R}^d$ for all times and it is a scattering state if the trajectory escapes from any given compact set.

We also mention an interesting general result establishing the connection between bound and scattering states with the spectral properties of the Hamiltonian. The result can be roughly formulated as follows: for any reasonable Hamiltonian H such that

$\mathcal{H}_{sc}(H) = \emptyset$ one has $\mathcal{M}_0(H) = \mathcal{H}_p(H)$ and $\mathcal{M}_\infty(H) = \mathcal{H}_{ac}(H)$ (for more details, we refer to [17], Chap. 5).

We conclude by recalling that the kind of analysis outlined in this section will be concretely developed in the next chapters for the cases described by the following Hamiltonians:

(i) $H_0 = -\frac{\hbar^2}{2m}\Delta$ in $L^2(\mathbb{R}^d)$ (free particle),

(ii) $H_\omega = -\frac{\hbar^2}{2m}\Delta + \frac{1}{2}m\omega^2 x^2$ in $L^2(\mathbb{R})$ (harmonic oscillator),

(iii) $H_\alpha = -\frac{\hbar^2}{2m}\Delta + \alpha\delta_0$ in $L^2(\mathbb{R})$ (point interaction),

(iv) $H_e = -\frac{\hbar^2}{2m}\Delta - \frac{e^2}{|x|}$ in $L^2(\mathbb{R}^3)$ (hydrogen atom).

References

1. von Neumann, J.: Mathematical Foundations of Quantum Mechanics. Princeton University Press, Princeton (1955)
2. Dell'Antonio, G.: Lectures on the Mathematics of Quantum Mechanics. Atlantis Press (2015)
3. Dimock, J.: Quantum Mechanics and Quantum Field Theory. Cambridge University Press, Cambridge (2011)
4. Faddeev, L.D., Yakubovskii, O.A.: Lectures on Quantum Mechanics for Mathematics Students. AMS (2009)
5. Hall, B.C.: Quantum Theory for Mathematicians. Springer, New York (2013)
6. Strocchi, F.: An Introduction to the Mathematical Structure of Quantum Mechanics, 2nd edn. World Scientific, Singapore (2008)
7. Thirring, W.: Quantum Mechanics of Atoms and Molecules. Springer, New York (1981)
8. Reed, M., Simon, B.: Methods of Modern Mathematical Physics, I: Functional Analysis. Academic Press, New York (1980)
9. Fasano, A., Marmi, S.: Analytical Mechanics. Oxford University Press, New York (2006)
10. Feynman, R., Leighton, R.B., Sands, M.L.: The Feynman Lectures on Physics, vol. III. Addison Wesley Publ, Com (2005)
11. Ghirardi, G.C.: Sneaking a Look at God's Cards—Unraveling the Mysteries of Quantum Mechanics. Princeton University Press, Princeton (2007)
12. Isham, C.J.: Lectures on Quantum Theory. Imperial College Press (1995)
13. Bell, J.: Speakable and Unspeakable in Quantum Mechanics. Cambridge University Press (1987)
14. Bricmont, J.: Making Sense of Quantum Mechanics. Springer (2016)
15. Dürr, D., Teufel, S.: Bohmian Mechanics. The Physics and Mathematics of Quantum Theory. Springer, Berlin (2009)
16. Bohr, N.: The quantum postulate and the recent development of atomic theory. Nature **121**, 580–590 (1928)
17. Amrein, W.O.: Hilbert Space Methods in Quantum Mechanics. EPFL Press, Lausanne (2009)

Chapter 6
Free Particle

6.1 Free Hamiltonian

Here, we study the quantum dynamics of a free particle with mass m in $\mathbb{R}^d, d = 1, 2, 3$ (see also [1–3]). The corresponding classical problem is trivial while, as we shall see, in the quantum case some mathematical problems must be approached. Moreover, the free particle is the natural context where some peculiar conceptual aspects of quantum dynamics (e.g., the interference effect) can be discussed.

The first step is the construction of the Hamiltonian as a self-adjoint operator in the Hilbert space $L^2(\mathbb{R}^d)$. Following the quantization procedure, we consider the Hamiltonian function of the classical case

$$H_0^{cl} = \frac{p^2}{2m}, \qquad p \in \mathbb{R}^d \tag{6.1}$$

and, by the substitution $p_j \to \frac{\hbar}{i} \frac{\partial}{\partial x_j}$, we obtain the formal operator

$$\dot{H}_0 = -\frac{\hbar^2}{2m} \Delta . \tag{6.2}$$

Let us define $\dot{H}_0$ on a domain made of smooth functions, e.g.,

$$D(\dot{H}_0) = \mathscr{S}(\mathbb{R}^d) . \tag{6.3}$$

The operator $\dot{H}_0$, $D(\dot{H}_0)$ is symmetric but not self-adjoint in $L^2(\mathbb{R}^d)$ (verify). In order to find the (unique) self-adjoint extension, we consider the Fourier transform in $L^2(\mathbb{R}^d)$

$$\mathscr{F} : f \to \tilde{f}, \qquad \tilde{f}(p) = \frac{1}{(2\pi)^{d/2}} \int dx \, f(x) \, e^{-ip \cdot x} . \tag{6.4}$$

© Springer International Publishing AG, part of Springer Nature 2018
A. Teta, *A Mathematical Primer on Quantum Mechanics*,
UNITEXT for Physics, https://doi.org/10.1007/978-3-319-77893-8_6

(In this section, we denote by p the independent variable in the Fourier space). By an integration by parts, one verifies that the unitary operator $\mathscr{F}$ maps the operator (6.2) into a multiplication operator. Indeed,

$$\left(\widetilde{H_0 f}\right)(p) := (\mathscr{F} \dot{H}_0 \mathscr{F}^{-1} \tilde{f})(p) = \frac{1}{(2\pi)^{d/2}} \frac{\hbar^2}{2m} \int dx\, (-\Delta f)(x)\, e^{-ip\cdot x}$$

$$= \frac{1}{(2\pi)^{d/2}} \frac{\hbar^2}{2m} \int dx\, f(x)\, (-\Delta e^{-ip\cdot x}) = \frac{\hbar^2 p^2}{2m} \tilde{f}(p)\,, \qquad (6.5)$$

$$D(\widetilde{H_0}) = \mathscr{S}(\mathbb{R}^d)\,. \qquad (6.6)$$

As we know from Chap. 4, such multiplication operator, defined in the L^2-space of the Fourier transforms, is essentially self-adjoint and its unique self-adjoint extension is

$$\left(\widetilde{H_0 f}\right)(p) = \frac{\hbar^2 p^2}{2m} \tilde{f}(p)\,, \quad D(\widetilde{H_0}) = \left\{ \tilde{f} \in L^2(\mathbb{R}^d) \mid \int dp\, |p^2 \tilde{f}(p)|^2 < \infty \right\}. \tag{6.7}$$

Using Proposition 4.17 of Chap. 4, we conclude that the self-adjoint Hamiltonian in the original L^2-space describing a free particle in dimension d is

$$(H_0 f)(x) = (\mathscr{F}^{-1} \widetilde{H_0} \mathscr{F} f)(x) = \frac{1}{(2\pi)^{d/2}} \int dp\, \frac{\hbar^2 p^2}{2m} \tilde{f}(p)\, e^{ip\cdot x}\,, \qquad (6.8)$$

$$D(H_0) = \left\{ f \in L^2(\mathbb{R}^d) \mid \int dp\, |p^2 \tilde{f}(p)|^2 < \infty \right\} := H^2(\mathbb{R}^d)\,. \qquad (6.9)$$

Remark 6.1 The functions $e^{ip\cdot x}$, $p \in \mathbb{R}^d$, are solutions of the equation $H_0 u = \frac{\hbar^2 p^2}{2m} u$ but they are not eigenfunctions of H_0 since they are not square integrable and, in particular, they cannot represent a state of the system. These functions are called generalized eigenfunctions of H_0, or plane waves. In our context, their role is to define, via (6.4), the Fourier transform, i.e., the unitary operator which diagonalizes the free Hamiltonian.

Remark 6.2 The free Hamiltonian is invariant under translations $x_k \to x_k + \lambda$, for $\lambda \in \mathbb{R}$ and $k = 1, \ldots, d$, and, by Noether's theorem (Sect. 5.2), this means that each component p_k of the momentum is a constant of motion.

Let us characterize the spectrum of $H_0, D(H_0)$. By Proposition 4.24, the resolvent operator $(H_0 - z)^{-1}$, called free resolvent, is given by

$$(H_0 - z)^{-1} = \mathscr{F}^{-1} (\widetilde{H_0} - z)^{-1} \mathscr{F}\,, \qquad (6.10)$$

where $(\widetilde{H_0} - z)^{-1}$ is the multiplication operator

$$\left((\widetilde{H_0} - z)^{-1} \tilde{f}\right)(p) = \frac{1}{\frac{\hbar^2 p^2}{2m} - z} \tilde{f}(p)\,. \qquad (6.11)$$

From (6.11) one sees that, for $z \in \mathbb{R}$, the resolvent is well defined as a bounded operator if and only if $z < 0$. Therefore, the spectrum is $[0, \infty)$ and, in particular, the Hamiltonian is a positive operator. Moreover, proceeding as in the cases of the operators Q and P, one verifies that there are no eigenvalues and then

$$\sigma(H_0) = \sigma_c(H_0) = [0, +\infty). \tag{6.12}$$

Exercise 6.1 Construct a Weyl sequence for H_0 in $d = 1$.
(Hint: for any $E \geq 0$, consider $\psi_n = \mathscr{F}^{-1}\tilde{\psi}_n$ with

$$\tilde{\psi}_n(p) = c_n \chi_{(p_0-1/n, p_0+1/n)}(p), \tag{6.13}$$

where $p_0 = \sqrt{2mE}/\hbar$. Determine c_n in such a way that $\|\psi_n\| = 1$ and verify that $\psi_n \in D(H_0)$, $\|(H_0 - E)\psi_n\| \to 0$, $\psi_n \rightharpoonup 0$.)

Using the Fourier transform, one can explicitly compute the spectral family and the corresponding spectral measure uniquely associated to H_0, $D(H_0)$ via the spectral theorem. Indeed, defining

$$\widetilde{(E_0(\lambda)f)}(p) := \chi_{(-\infty,\lambda)}\left(\frac{\hbar^2 p^2}{2m}\right)\tilde{f}(p) = \begin{cases} \tilde{f}(p) & \text{if } \dfrac{\hbar^2 p^2}{2m} < \lambda \\ 0 & \text{otherwise} \end{cases} \tag{6.14}$$

we obtain for $d = 1$

$$\begin{aligned}
(E_0(\lambda)f, g) &= \int dp \, \chi_{(-\infty,\lambda)}\left(\frac{\hbar^2 p^2}{2m}\right)\overline{\tilde{f}(p)}\tilde{g}(p) \\
&= \chi_{(0,\infty)}(\lambda) \int_0^\infty dp \, \chi_{(0,\lambda)}\left(\frac{\hbar^2 p^2}{2m}\right)\left(\overline{\tilde{f}(p)}\tilde{g}(p) + \overline{\tilde{f}(-p)}\tilde{g}(-p)\right) \\
&= \chi_{(0,\infty)}(\lambda) \int_0^{\frac{\sqrt{2m\lambda}}{\hbar}} dp \left(\overline{\tilde{f}(p)}\tilde{g}(p) + \overline{\tilde{f}(-p)}\tilde{g}(-p)\right)
\end{aligned} \tag{6.15}$$

and, with a slight abuse of notation, for $d = 2, 3$ we have

$$(E_0(\lambda)f, g) = \chi_{(0,\infty)}(\lambda) \int_0^{\frac{\sqrt{2m\lambda}}{\hbar}} d|p| \, |p|^{d-1} \int_{S^{d-1}} d\Omega \, \overline{\tilde{f}(|p|, \Omega)}\tilde{g}(|p|, \Omega). \tag{6.16}$$

Exercise 6.2 Verify that $\{E_0(\lambda)\}_{\lambda \in \mathbb{R}}$ is the spectral family associated to H_0, $D(H_0)$, i.e.,

$$D(H_0) = \left\{ u \in L^2(\mathbb{R}) \mid \int \lambda^2 d(E_0(\lambda)u, u) < \infty \right\}, \tag{6.17}$$

$$(H_0 u, v) = \int_0^\infty \lambda^2 d(E_0(\lambda)u, v), \qquad \forall u \in D(H_0), \qquad \forall v \in L^2(\mathbb{R}^d). \tag{6.18}$$

Note that the distribution function $\lambda \to (E_0(\lambda)f, f)$ is absolutely continuous for any $f \in L^2(\mathbb{R}^d)$ and then the associated spectral measure m_0^f is absolutely continuous with respect to Lebesgue measure for any $f \in L^2(\mathbb{R}^d)$. Therefore, we conclude

$$\sigma(H_0) = \sigma_{ac}(H_0) = [0, +\infty). \tag{6.19}$$

It is useful to derive the explicit expression of the integral kernel of the free resolvent in the x-space. Let us fix $z \in \mathbb{C} \setminus \mathbb{R}^+$ and $f \in \mathscr{S}(\mathbb{R}^d)$. We write

$$z = -k^2, \qquad \text{where} \quad k \in \mathbb{C} \quad \text{with} \quad \Re k > 0, \tag{6.20}$$

so that $\Re z = (\Im k)^2 - (\Re k)^2$, $\Im z = -2 \Im k \Re k$ (z in the upper half complex plane corresponds to $\Im k < 0$). Let us consider the function

$$\phi_k : \mathbb{R}^d \to \mathbb{C}, \qquad \phi_k(p) := \frac{1}{\frac{\hbar^2 p^2}{2m} + k^2} \tag{6.21}$$

and note that for $d = 1$ we have $\phi_k \in L^1(\mathbb{R}) \cap L^2(\mathbb{R})$, while for $d = 2, 3$ we have $\phi_k \in L^2(\mathbb{R}^d)$ but $\phi_k \notin L^1(\mathbb{R})$. Then, using unitarity of the Fourier transform, we find

$$\left[(H_0 - z)^{-1} f \right](x) = \left[(H_0 + k^2)^{-1} f \right](x) = \frac{1}{(2\pi)^{d/2}} \int dp \, \overline{e^{-ipx} \overline{\phi_k}(p)} \, \tilde{f}(p)$$

$$= \frac{1}{(2\pi)^d} \int dy \left(\int dp \, e^{ip(x-y)} \phi_k(p) \right) f(y) \tag{6.22}$$

$$=: \int dy \, (H_0 + k^2)^{-1}(x - y) \, f(y), \tag{6.23}$$

where we have denoted by $(H_0 + k^2)^{-1}(x - y)$ the integral kernel of the operator $(H_0 + k^2)^{-1}$. Let us compute the integral kernel in (6.23), also called Green's function, in the two cases $d = 1$ and $d = 3$.

(a) Integral kernel of the resolvent in $d = 1$.
In this case, the integral in parenthesis in (6.22) exists for any x, y and we have

$$(H_0 + k^2)^{-1}(x) = \frac{\sqrt{2m}}{2\pi \hbar} \int dq \, \frac{e^{iaq}}{(q - ik)(q + ik)}, \qquad a := \frac{\sqrt{2m}}{\hbar} x. \tag{6.24}$$

Exercise 6.3 Using the residue theorem, verify that

$$\int dq \, \frac{e^{iaq}}{(q - ik)(q + ik)} = \pi \, \frac{e^{-|a|k}}{k} \, . \tag{6.25}$$

From (6.24) and (6.25), we find

$$(H_0 - z)^{-1}(x - y) = (H_0 + k^2)^{-1}(x - y) = \frac{\sqrt{2m}}{\hbar} \, \frac{e^{-\frac{\sqrt{2m}}{\hbar} k \, |x-y|}}{2k} \, . \tag{6.26}$$

In particular, for $z = \mu$, with μ real and negative, we have

$$(H_0 - \mu)^{-1}(x - y) = \frac{\sqrt{2m}}{\hbar} \, \frac{e^{-\frac{\sqrt{2m}}{\hbar} \sqrt{-\mu} \, |x-y|}}{2\sqrt{-\mu}} \, , \qquad \mu < 0 \, . \tag{6.27}$$

On the other hand, let us fix $z = \mu \pm i\varepsilon$, with $\mu, \varepsilon > 0$, and consider the limit $\varepsilon \to 0$ (i.e., $\Re k \to 0$ and $\Im k \to \mp\sqrt{\mu}$). In this way, we find the limiting values of the integral kernel of the resolvent when z approaches a point of the spectrum of H_0 from the upper and the lower half plane

$$(H_0 - (\mu \pm i0))^{-1} (x - y) := \lim_{\varepsilon \to 0} (H_0 - (\mu \pm i\varepsilon))^{-1} (x - y)$$

$$= \frac{\sqrt{2m}}{\hbar} \, \frac{e^{\pm i \frac{\sqrt{2m}}{\hbar} \sqrt{\mu} \, |x-y|}}{\mp 2i \sqrt{\mu}} \, , \qquad \mu > 0 \, . \tag{6.28}$$

As it should be expected, the two operators defined by (6.28) are not bounded in $L^2(\mathbb{R})$.

(b) Integral kernel of the resolvent in $d = 3$.

In this case, the computation requires some care (see also [1, 4]). The integral in parenthesis in (6.22) is, by definition, given by

$$(H_0 + k^2)^{-1}(x) = \left(\frac{\sqrt{2m}}{2\pi\hbar}\right)^3 \lim_{R \to \infty} \int_{|q|<R} dq \, \frac{e^{ib \cdot q}}{q^2 + k^2} \, , \qquad b := \frac{\sqrt{2m}}{\hbar} \, x \, , \tag{6.29}$$

where the limit is taken in the L^2-sense. For the evaluation of the integral in (6.29), we use spherical coordinates

$$\int_{|q|<R} dq \, \frac{e^{ib \cdot q}}{q^2 + k^2} = 2\pi \int_0^R dr \, \frac{r^2}{r^2 + k^2} \int_0^\pi d\theta \, \sin\theta \, e^{i|b|r \cos\theta}$$

$$= \frac{2\pi}{i|b|} \int_{-R}^R dr \, \frac{r \, e^{i|b|r}}{r^2 + k^2} \tag{6.30}$$

and then the residue theorem. In particular, we choose the rectangular path in the upper half plane defined as follows for any $R > 0$: from $-R$ to R on the real axis, from R to $R + i\sqrt{R}$ on the line $\gamma_1(t) = R + it$, from $R + i\sqrt{R}$ to $-R + i\sqrt{R}$ on the line $\gamma_2(t) = t + i\sqrt{R}$ and, finally, from $-R + i\sqrt{R}$ to $-R$ on the line $\gamma_3(t) = -R + it$. Note that for R sufficiently large the path contains the pole $z = ik$. We obtain

$$
\int_{|q|<R} dq \, \frac{e^{ib \cdot q}}{q^2 + k^2} = \frac{2\pi}{i|b|} \left(2\pi i \, \text{Res} \, (ik) - \sum_{j=1}^{3} \int_{\gamma_j} dz \, \frac{z \, e^{i|b|z}}{z^2 + k^2} \right)
$$

$$
= 2\pi^2 \frac{e^{-k|b|}}{|b|} + \frac{2\pi i}{|b|} \sum_{j=1}^{3} \int_{\gamma_j} dz \, \frac{z \, e^{i|b|z}}{z^2 + k^2} . \tag{6.31}
$$

It remains to show that the L^2-norm of the last term in (6.31) vanishes for $R \to \infty$.

Exercise 6.4 Verify that

$$
\lim_{R \to \infty} \int dx \, \left| \frac{1}{|x|} \int_{\gamma_j} dz \, \frac{z \, e^{i|x|z}}{z^2 + k^2} \right|^2 = 0 , \qquad j = 1, 2, 3 . \tag{6.32}
$$

(Hint: the result follows from the estimates valid for R sufficiently large (c denotes a positive constant)

$$
\frac{1}{|x|} \left| \int_{\gamma_1} dz \, \frac{z \, e^{i|x|z}}{z^2 + k^2} \right| \leq \frac{1}{|x|} \int_0^{\sqrt{R}} dt \, \frac{|R + it| \, e^{-|x|t}}{|(R + it)^2 + k^2|} = \frac{1}{\sqrt{R}|x|} \int_0^1 ds \, \frac{|1 + \frac{is}{\sqrt{R}}| \, e^{-\sqrt{R}|x|s}}{|(1 + \frac{is}{\sqrt{R}})^2 + \frac{k^2}{R^2}|}
$$

$$
\leq \frac{c}{\sqrt{R}|x|} \int_0^1 ds \, e^{-\sqrt{R}|x|s} = \frac{c}{R|x|^2} \left(1 - e^{-\sqrt{R}|x|} \right) , \tag{6.33}
$$

$$
\frac{1}{|x|} \left| \int_{\gamma_2} dz \, \frac{z \, e^{i|x|z}}{z^2 + k^2} \right| \leq \frac{e^{-\sqrt{R}|x|}}{|x|} \int_{-R}^{R} dt \, \frac{|t + i\sqrt{R}|}{|(t + i\sqrt{R}) + k^2|} = \frac{e^{-\sqrt{R}|x|}}{|x|} \int_{-1}^{1} ds \, \frac{|s + \frac{i}{\sqrt{R}}|}{|(s + \frac{i}{\sqrt{R}})^2 + \frac{k^2}{R^2}|}
$$

$$
\leq c \frac{e^{-\sqrt{R}|x|}}{|x|} \int_0^1 ds \, \frac{1}{\sqrt{s^2 + R^{-1}}} = c \frac{e^{-\sqrt{R}|x|}}{|x|} \log \left(\sqrt{R} + \sqrt{R+1} \right) \tag{6.34}
$$

and analogously for the integral on γ_3).

Taking into account (6.29), (6.31), and (6.32), we find

$$
(H_0 - z)^{-1}(x - y) = (H_0 + k^2)^{-1}(x - y) = \frac{2m}{\hbar^2} \frac{e^{-\frac{\sqrt{2m}}{\hbar} k \, |x-y|}}{4\pi |x - y|} \tag{6.35}
$$

and then

$$(H_0 - \mu)^{-1}(x - y) = \frac{2m}{\hbar^2} \frac{e^{-\frac{\sqrt{2m}}{\hbar}\sqrt{-\mu}|x-y|}}{4\pi|x - y|}, \qquad \mu < 0, \quad (6.36)$$

$$(H_0 - (\mu \pm i0))^{-1}(x - y) = \frac{2m}{\hbar^2} \frac{e^{\pm i \frac{\sqrt{2m}}{\hbar}\sqrt{\mu}|x-y|}}{4\pi|x - y|}, \qquad \mu > 0, \quad (6.37)$$

where the two operators defined by (6.37) are not bounded in $L^2(\mathbb{R}^3)$.

6.2 Free Propagator

In this section, we discuss some properties of the unitary group $U_0(t) := e^{-i\frac{t}{\hbar}H_0}$ generated by H_0, $D(H_0)$, also called free propagator. Note that in the Fourier space the unitary group is simply given by the multiplication operator by $e^{-i\frac{\hbar}{2m}tk^2}$ (see Proposition 4.24). Then, we have

$$(U_0(t)f)(x) = \frac{1}{(2\pi)^{d/2}} \int dk \, e^{ik \cdot x - i\frac{\hbar}{2m}tk^2} \, \tilde{f}(k). \qquad (6.38)$$

We also recall that $U_0(t)f$, for $f \in D(H_0)$, is the (unique) solution of the Cauchy problem

$$i\hbar \frac{\partial \psi(t)}{\partial t} = H_0 \psi(t), \qquad \psi(0) = f. \qquad (6.39)$$

In the next proposition, we derive the explicit expression of the unitary group, from now on denoted by $e^{-i\frac{t}{\hbar}H_0}$, as an integral operator in the x-space.

Proposition 6.1 *Let $f \in L^2(\mathbb{R}^d) \cap L^1(\mathbb{R}^d)$. Then*

$$\left(e^{-i\frac{t}{\hbar}H_0}f\right)(x) = \left(\frac{m}{2\pi i\hbar t}\right)^{d/2} \int dy \, f(y) \, e^{i\frac{m}{2\hbar t}(x-y)^2}. \qquad (6.40)$$

Moreover, if $f \in L^2(\mathbb{R}^d)$ then

$$\left(e^{-i\frac{t}{\hbar}H_0}f\right)(x) = \left(\frac{m}{2\pi i\hbar t}\right)^{d/2} \lim_{n \to \infty} \int_{|y|<n} dy \, f(y) \, e^{i\frac{m}{2\hbar t}(x-y)^2}, \qquad (6.41)$$

where the limit is in the L^2-sense.

Proof To simplify the notation, we set $e^{-i\frac{t}{\hbar}H_0} = e^{-i\tau(-\Delta)}$, $\tau := \frac{\hbar}{2m}t$ and fix $\tau > 0$. For any $f, g \in \mathscr{S}(\mathbb{R}^d)$ we have

$$
\begin{aligned}
(g, e^{-i\frac{t}{\hbar}H_0} f) &= (\tilde{g}, \widetilde{e^{-i\frac{t}{\hbar}H_0} f}) = \int dk \, \overline{\tilde{g}(k)} \, \tilde{f}(k) \, e^{-i\tau k^2} = \lim_{\varepsilon \to 0} \int dk \, \overline{\tilde{g}(k)} \, \tilde{f}(k) \, e^{-(\varepsilon+i\tau)k^2} \\
&= \frac{1}{(2\pi)^d} \lim_{\varepsilon \to 0} \int dx\, dy \, \overline{g(x)} \, f(y) \int dk \, e^{-(\varepsilon+i\tau)k^2 + ik\cdot(x-y)} \\
&= \frac{1}{(2\pi)^d} \lim_{\varepsilon \to 0} \int dx\, dy \, \overline{g(x)} \, f(y) \prod_{j=1}^{d} \int dk_j \, e^{-(\varepsilon+i\tau)k_j^2 + ik_j(x-y)_j}, \quad (6.42)
\end{aligned}
$$

where we have denoted by q_j the jth component of the vector $q \in \mathbb{R}^d$. The last integral in (6.42) can be explicitly computed (see Exercise 6.5) and we obtain

$$
\begin{aligned}
(g, e^{-i\frac{t}{\hbar}H_0} f) &= \frac{(\sqrt{\pi})^d}{(2\pi)^d} \lim_{\varepsilon \to 0} \int dx\, dy \, \overline{g(x)} \, f(y) \prod_{j=1}^{d} \frac{1}{(\tau^2 + \varepsilon^2)^{1/4}} \, e^{-\frac{i}{2}\tan^{-1}\left(\frac{\tau}{\varepsilon}\right)} \, e^{-\frac{(x-y)_j^2}{4(i\tau+\varepsilon)}} \\
&= \left(\frac{1}{4\pi}\right)^{d/2} \lim_{\varepsilon \to 0} \frac{e^{-\frac{i}{2}d\tan^{-1}\left(\frac{\tau}{\varepsilon}\right)}}{(\tau^2 + \varepsilon^2)^{d/4}} \int dx\, dy \, \overline{g(x)} \, f(y) \, e^{-\frac{(x-y)^2}{4(i\tau+\varepsilon)}} \\
&= \left(\frac{1}{4\pi i\tau}\right)^{d/2} \int dx\, dy \, \overline{g(x)} \, f(y) \, e^{i\frac{(x-y)^2}{4\tau}}, \quad (6.43)
\end{aligned}
$$

where, in the last step, we have used the dominated convergence theorem.

Let us fix $f \in L^2(\mathbb{R}^d) \cap L^1(\mathbb{R}^d)$. Then, using a density argument, (6.40) is proved.

Let us fix $f \in L^2(\mathbb{R}^d)$ and let us define $f_n(x) := \chi_{|x|<n}(x) f(x)$. We have $f_n \in L^2(\mathbb{R}^d) \cap L^1(\mathbb{R}^d)$, $\|f - f_n\| \to 0$ for $n \to \infty$ and

$$
\left(e^{-i\frac{t}{\hbar}H_0} f_n\right)(x) = \left(\frac{m}{2\pi i\hbar t}\right)^{d/2} \int_{|y|<n} dy \, f(y) \, e^{i\frac{m}{2\hbar t}(x-y)^2}. \quad (6.44)
$$

Then, formula (6.41) is proved if we observe that

$$
\lim_{n \to \infty} \|e^{-i\frac{t}{\hbar}H_0} f - e^{-i\frac{t}{\hbar}H_0} f_n\| = \lim_{n \to \infty} \|f - f_n\| = 0. \quad (6.45)
$$

Exercise 6.5 Verify that

$$
\int_{-\infty}^{+\infty} dx \, e^{-ax^2 + i\lambda x} = \sqrt{\frac{\pi}{|a|}} \, e^{-\frac{i}{2}\phi} \, e^{-\frac{\lambda^2}{4a}}, \quad (6.46)
$$

where $\lambda \in \mathbb{R}$, $a = |a|e^{i\phi}$, with $\phi \in (0, \pi/2]$.
(Hint: consider

$$
I(\lambda) := \int_{-\infty}^{+\infty} dx \, e^{-ax^2 + i\lambda x} \quad (6.47)
$$

and verify that $I'(\lambda) = -(\lambda/2a) I(\lambda)$, so $I(\lambda) = e^{-\frac{\lambda^2}{4a}} I(0)$. In order to compute $I(0)$, one can integrate the function of complex variable $f(z) = e^{-|a|z^2}$ on the path:

from 0 to $R > 0$ on the real axis, from R to $Re^{i\frac{\phi}{2}}$ on $\hat{\gamma}_1(t) = Re^{it}$, and from $Re^{i\frac{\phi}{2}}$ to the origin on $\hat{\gamma}_2(t) = te^{i\frac{\phi}{2}}$. Then, take the limit $R \to \infty$.)

A first useful application of the explicit formula for the unitary group is the characterization of the asymptotic behavior for large t of the solution of the free Schrödinger equation.

Proposition 6.2 *For any $\phi \in L^2(\mathbb{R}^d)$, we have*

$$\lim_{t \to \pm\infty} \|e^{-i\frac{t}{\hbar}H_0}\phi - \phi_t^a\| = 0, \tag{6.48}$$

where

$$\phi_t^a(x) := \left(\frac{m}{i\hbar t}\right)^{d/2} e^{i\frac{m}{2\hbar t}x^2} \tilde{\phi}\left(\frac{mx}{\hbar t}\right). \tag{6.49}$$

Proof Let us fix $\phi \in L^2(\mathbb{R}^d) \cap L^1(\mathbb{R}^d)$. Then

$$
\begin{aligned}
(e^{-i\frac{t}{\hbar}H_0}\phi)(x) &= \left(\frac{m}{2\pi i\hbar t}\right)^{d/2} \int dy\, \phi(y)\, e^{i\frac{m}{2\hbar t}(x-y)^2} \\
&= \left(\frac{m}{2\pi i\hbar t}\right)^{d/2} e^{i\frac{m}{2\hbar t}x^2} \int dy\, \phi(y)\, e^{i\frac{m}{2\hbar t}y^2 - i\frac{m}{\hbar t}x \cdot y} \\
&:= \phi_t^a(x) + \mathscr{R}_t(x),
\end{aligned}
\tag{6.50}
$$

where

$$\mathscr{R}_t(x) = \left(\frac{m}{i\hbar t}\right)^{d/2} e^{i\frac{m}{2\hbar t}x^2} \frac{1}{(2\pi)^{d/2}} \int dy\, \phi(y) \left(e^{i\frac{m}{2\hbar t}y^2} - 1\right) e^{-i\frac{m}{\hbar t}x \cdot y}. \tag{6.51}$$

Note that $\mathscr{R}_t(x)$ is proportional to the Fourier transform of the function $\phi(y)(e^{i\frac{m}{2\hbar t}y^2} - 1)$ evaluated at $mx/\hbar t$. Then, applying Plancherel theorem, we have

$$
\begin{aligned}
\|\mathscr{R}_t\|^2 &= \left(\frac{m}{2\pi\hbar|t|}\right)^d \int dx \left|\int dy\, \phi(y) \left(e^{i\frac{m}{2\hbar t}y^2} - 1\right) e^{-i\frac{mx}{\hbar t} \cdot y}\right|^2 \\
&= \frac{1}{(2\pi)^d} \int dz \left|\int dy\, \phi(y) \left(e^{i\frac{m}{2\hbar t}y^2} - 1\right) e^{-iz \cdot y}\right|^2 \\
&= \int dy\, |\phi(y)|^2 \left|e^{i\frac{m}{2\hbar t}y^2} - 1\right|^2.
\end{aligned}
\tag{6.52}
$$

By dominated convergence theorem, we obtain (6.48) for $\phi \in L^2(\mathbb{R}^d) \cap L^1(\mathbb{R}^d)$. For $\phi \in L^2(\mathbb{R}^d)$, the result follows from a density argument. $\qed$

Remark 6.3 It is possible to estimate the rate of convergence in the limit (6.48) if the initial datum decays sufficiently fast at infinity. For example, from (6.52), one obtains

$$\|e^{-i\frac{t}{\hbar}H_0}\phi - \phi_t^a\|^2 \le \frac{m^2}{4\hbar^2 t^2}\int dy\, y^4|\phi(y)|^2\,. \tag{6.53}$$

Remark 6.4 From the explicit expression (6.40), one also easily obtains the estimate

$$\sup_{x\in\mathbb{R}^d}\left|\left(e^{-i\frac{t}{\hbar}H_0}f\right)(x)\right| \le \left(\frac{m}{2\pi\hbar}\right)^{d/2}\frac{\|f\|_{L^1(\mathbb{R}^d)}}{|t|^{d/2}}, \tag{6.54}$$

which shows the "dispersive" character of the free evolution.

In conclusion, we note that the above proposition can be used to show that for any state $\phi \in L^2(\mathbb{R}^d)$ one has $\phi \in \mathcal{M}_\infty(H_0)$ (see Definition 5.39). In other words, any state is a scattering state with respect to the free dynamics, i.e.,

$$\lim_{t\to\pm\infty}\int_{|x|<R}dx\,|(e^{-i\frac{t}{\hbar}H_0}\phi)(x)|^2 = 0, \tag{6.55}$$

for any fixed $R > 0$. This means that the probability to find the particle in any fixed ball $\{x \in \mathbb{R}^d \mid |x| < R\}$ goes to zero for $t \to \pm\infty$. For the proof of (6.55), we observe that

$$\int_{|x|<R}dx\,|(e^{-i\frac{t}{\hbar}H_0}\phi)(x)|^2 \le 2\int_{|x|<R}dx\,|\phi_t^a(x)|^2 + 2\int_{|x|<R}dx\,|(e^{-i\frac{t}{\hbar}H_0}\phi)(x) - \phi_t^a(x)|^2$$

$$\le 2\int dx\,\chi_{\{|x|<R\}}(x)\left(\frac{m}{\hbar t}\right)^d|\tilde\phi\left(\frac{mx}{\hbar t}\right)|^2 + 2\|e^{-i\frac{t}{\hbar}H_0}\phi - \phi_t^a\|^2$$

$$= 2\int dk\,\chi_{\{|k|<\frac{mR}{\hbar t}\}}(k)|\tilde\phi(k)|^2 + 2\|e^{-i\frac{t}{\hbar}H_0}\phi - \phi_t^a\|^2. \tag{6.56}$$

Then, taking the limit for $t \to \pm\infty$, we obtain (6.55).

6.3   Evolution of a Wave Packet

The properties of the free evolution are well illustrated if we choose an initial state having the form of a wave packet, i.e., a state "well concentrated" both in position and in momentum. We consider for simplicity the one-dimensional case. More precisely, we consider a function $f \in \mathscr{S}(\mathbb{R})$, real, even and such that $\|f\| = 1$. Then, we define the wave packet as

$$\psi_0^{x_0,p_0,\sigma}(x) = \frac{1}{\sqrt\sigma}f\left(\sigma^{-1}(x+x_0)\right)e^{i\frac{p_0}{\hbar}x} \tag{6.57}$$

where $x_0, p_0, \sigma > 0$ are given parameters. For the sake of brevity, in the following we shall omit the dependence of the state on the parameters x_0, p_0, σ.

Let us compute mean value and variance of the position for a particle described by (6.57)

$$\langle x \rangle(0) = \int dx\, x\, |\psi_0(x)|^2 = \frac{1}{\sigma} \int dx\, x\, |f(\sigma^{-1}(x + x_0))|^2$$

$$= \int dy\, (\sigma y - x_0)|f(y)|^2 = -x_0\,, \tag{6.58}$$

$$\Delta x(0)^2 = \langle x^2 \rangle(0) - \langle x \rangle(0)^2 = \frac{1}{\sigma} \int dx\, x^2 |f(\sigma^{-1}(x + x_0))|^2 - x_0^2$$

$$= \int dy\, (\sigma y - x_0)^2 |f(y)|^2 - x_0^2 = \sigma^2 \|\tilde{f}'\|^2\,. \tag{6.59}$$

Concerning the momentum, we first observe that

$$\tilde{\psi}_0(k) = \frac{1}{\sqrt{2\pi\sigma}} \int dx\, f(\sigma^{-1}(x + x_0)) e^{i\frac{p_0}{\hbar} x} e^{-ikx} = \sqrt{\sigma}\, \tilde{f}(\sigma(k - k_0)) e^{ix_0 k - ik_0 x_0} \tag{6.60}$$

where we have denoted $k_0 = \hbar^{-1} p_0$. Then, we have

$$\langle p \rangle(0) = \hbar \int dk\, k\, |\tilde{\psi}_0(k)|^2 = \hbar\sigma \int dk\, k\, |\tilde{f}(\sigma(k - k_0))|^2$$

$$= \hbar \int dq\, (\sigma^{-1} q + k_0)|\tilde{f}(q)|^2 = p_0\,, \tag{6.61}$$

$$\Delta p(0)^2 = \langle p^2 \rangle(0) - \langle p \rangle(0)^2 = \hbar^2 \sigma \int dk\, k^2 |\tilde{f}(\sigma(k - k_0))|^2 - p_0^2$$

$$= \hbar^2 \int dq\, (\sigma^{-1} q + k_0)^2 |\tilde{f}(q)|^2 - p_0^2 = \frac{\hbar^2}{\sigma^2} \|f'\|^2\,. \tag{6.62}$$

The above formulas show that the wave packet (6.57) is concentrated in position around $-x_0$ and in momentum around p_0, with dispersions satisfying

$$\Delta x(0)\Delta p(0) = \hbar \,\|\tilde{f}'\| \|f'\|\,. \tag{6.63}$$

Note that for "σ of order $\sqrt{\hbar}$" we obtain that the dispersions in position and in momentum are both "small and of the same order $\sqrt{\hbar}$".

Remark 6.5 The situation of minimal uncertainty is realized if we choose a Gaussian function, i.e., the state (6.57) with f given by

$$f_g(x) = \pi^{-1/4} e^{-\frac{x^2}{2}}\,. \tag{6.64}$$

Recall that $\tilde{f}_g(k) = \pi^{-1/4} e^{-\frac{k^2}{2}}$. In this case, we have

$$\| f'_g \| = \| \tilde{f}'_g \| = \frac{1}{\sqrt{2}} \tag{6.65}$$

and therefore the right-hand side of (6.63) reduces to $\hbar/2$, which is the minimum allowed by the Heisenberg uncertainty relations. An initial state of this type is called coherent state.

Let us discuss the time evolution $\psi_t = e^{-i \frac{t}{\hbar} H_0} \psi_0$. Taking into account that

$$\left(\widetilde{e^{-i \frac{t}{\hbar} H_0} \psi_0} \right) (k) := \tilde{\psi}_t(k) = e^{-i \frac{\hbar t}{2m} k^2} \tilde{\psi}_0(k), \tag{6.66}$$

we immediately have

$$\langle p \rangle(t) = \langle p \rangle(0), \qquad \Delta p(t) = \Delta p(0) . \tag{6.67}$$

The above formulas are a consequence of the fact that the momentum is a constant of motion for the free particle. Moreover, we know that the multiplication operator by x becomes the derivation operator $i \frac{d}{dk}$ in the Fourier space. Then

$$\langle x \rangle(t) = \int dk \, \overline{\tilde{\psi}_t(k)} i \tilde{\psi}'_t(k) = \int dk \, \overline{\tilde{\psi}_0(k)} \left(\frac{\hbar t}{m} k \tilde{\psi}_0(k) + i \tilde{\psi}'_0(k) \right) = \frac{p_0}{m} t - x_0 . \tag{6.68}$$

Note that (6.68) can also be derived from Ehrenfest theorem. Let us compute $\langle x^2 \rangle(t)$

$$\langle x^2 \rangle(t) = \int dk \, |\tilde{\psi}'_t(k)|^2 = \left(\frac{\hbar t}{m} \right)^2 \int dk \, k^2 \, |\tilde{\psi}_0(k)|^2 + \int dk \, |\tilde{\psi}'_0(k)|^2$$
$$+ \frac{t}{m} \left(\int dk \, \overline{i \tilde{\psi}'_0(k)} \, \hbar k \, \tilde{\psi}_0(k) + \int dk \, \overline{\hbar k \, \tilde{\psi}_0(k)} \, i \tilde{\psi}'_0(k) \right)$$
$$= \left(\frac{t}{m} \right)^2 \langle p^2 \rangle(0) + \langle x^2 \rangle(0) + \frac{t}{m} \langle xp + px \rangle(0) . \tag{6.69}$$

Therefore

$$\Delta x(t)^2 = \left(\frac{t}{m} \right)^2 \Delta p(0)^2 + \Delta x(0)^2 + \frac{t}{m} \left(\langle xp + px \rangle(0) + 2 p_0 x_0 \right) . \tag{6.70}$$

From (6.68), we see that the mean value of the position, i.e., the center of the wave packet, moves with constant velocity, like the corresponding classical particle. On the other hand, from (6.70) we have that the dispersion of the position increases with time. Therefore, the wave packet, even if it is well localized at initial time, inevitably spreads in space as time goes by. The consequence is that the evolution of the wave packet can be assimilated to the motion of a (classical) point particle only for a time interval "not too long" (see, e.g., the next exercise) (Fig. 6.1).

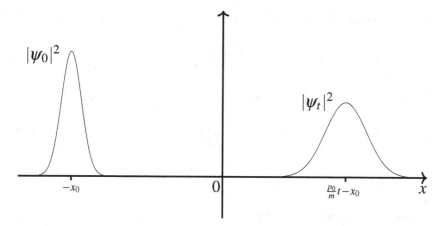

Fig. 6.1 Spreading of the wave packet

Exercise 6.6 Verify that

$$\left(e^{-i\frac{t}{\hbar}H_0}\psi_0\right)(x) = \frac{e^{-i\frac{p_0^2}{2m}\frac{t}{\hbar}+i\frac{p_0}{\hbar}x}}{\sqrt{\sigma}}\left(e^{-i\nu(-\Delta)}f\right)\left(\frac{x-\frac{p_0}{m}t+x_0}{\sigma}\right), \qquad \nu := \frac{\hbar t}{2m\sigma^2}.$$
(6.71)

Note that for $\nu \ll 1$ the dispersion of the position is negligible.

In some special cases, it is also possible to compute the time evolution in explicit form. In the next exercise, it is shown the result in the case of a Gaussian initial state.

Exercise 6.7 Verify that if $f(x) = f_g(x) = \pi^{-1/4}e^{-\frac{x^2}{2}}$ then

$$\left(e^{-i\frac{t}{\hbar}H_0}\psi_0\right)(x) = e^{-i\frac{p_0^2}{2m}\frac{t}{\hbar}+i\frac{p_0}{\hbar}x}\frac{1}{\pi^{1/4}\sqrt{\sigma}\sqrt{1+i\frac{\hbar t}{m\sigma^2}}}e^{-\frac{\left(x-\frac{p_0}{m}t+x_0\right)^2}{2\sigma^2\left(1+i\frac{\hbar t}{m\sigma^2}\right)}}.$$
(6.72)

(Hint: use the results of Exercises 6.6, 6.5).

From (6.72), we derive the probability density to find the particle at time t in x

$$|\psi_t(x)|^2 = \frac{1}{\sqrt{\pi}\sigma\sqrt{1+\frac{\hbar^2 t^2}{m^2\sigma^4}}}e^{-\frac{\left(x+x_0-\frac{p_0}{m}t\right)^2}{\sigma^2\left(1+\frac{\hbar^2 t^2}{m^2\sigma^4}\right)}}.$$
(6.73)

Such a Gaussian distribution is centered in

$$\langle x\rangle(t) = \frac{p_0}{m}t - x_0$$
(6.74)

and the mean square deviation is

$$\Delta x(t)^2 = \frac{\sigma^2}{2}\left(1 + \frac{\hbar^2 t^2}{m^2 \sigma^4}\right). \tag{6.75}$$

Exercise 6.8 Consider the initial state

$$\psi_0(x) = \frac{1}{\pi^{1/4}\sqrt{\sigma}}e^{-\frac{x^2}{2\sigma^2}+i\frac{p_0}{\hbar}x} \tag{6.76}$$

and let ψ_t be its evolution at time t. Compute the two limits: $\lim_{t\to 0} t^{-2}\big(1 - |(\psi_0, \psi_t)|^2\big)$ and $\lim_{t\to\infty} t|(\psi_0, \psi_t)|^2$.

Exercise 6.9 Consider the initial state

$$\psi_0(x) = \sqrt{\frac{2}{a}}\,\chi_{(0,a)}(x)\sin\left(\frac{\pi}{a}x\right), \tag{6.77}$$

where $a > 0$. Let ψ_t be the state at time t and $\mathscr{P}_{t,a}$ the probability to find the particle at time t in $(0, a)$. Determine the asymptotic behavior of $\mathscr{P}_{t,a}$ for $t \to \infty$ and for $t \to 0$. Repeat the computation when the initial state is

$$\psi_1(x) = \sqrt{\frac{2}{a}}\,\chi_{(0,a)}(x)\sin\left(\frac{2\pi}{a}x\right). \tag{6.78}$$

6.4 Interference of Two Wave Packets

In this section, we give a quantitative description of the interference phenomenon that takes place when the initial state of the particle is chosen in the form of a "coherent superposition" of two wave packets. We stress that this phenomenon is of fundamental physical importance. It can be considered as the main aspect that characterizes the quantum description, having no counterpart in Classical Physics.

Let us consider an initial state of the form

$$\begin{aligned}
\psi_0(x) &= C\left(\psi^+(x) + \psi^-(x)\right)\\
&= C\left[\frac{1}{\sqrt{\sigma}}f\left(\sigma^{-1}(x+x_0)\right)e^{i\frac{p_0}{\hbar}x} + \frac{1}{\sqrt{\sigma}}f\left(\sigma^{-1}(x-x_0)\right)e^{-i\frac{p_0}{\hbar}x}\right]. \tag{6.79}
\end{aligned}$$

Note that ψ^+ is the wave packet considered in the previous section, while ψ^- is a wave packet of the same form but localized in position around x_0 and in momentum around $-p_0$. The constant C is fixed in such a way that $\|\psi_0\| = 1$.

From (6.79), the probability density for the position of the particle at $t = 0$ is

$$|\psi_0(x)|^2 = C^2 \left[\frac{1}{\sigma} |f\left(\sigma^{-1}(x+x_0)\right)|^2 + \frac{1}{\sigma} |f\left(\sigma^{-1}(x-x_0)\right)|^2 \right.$$

$$\left. + \frac{2}{\sigma} f\left(\sigma^{-1}(x+x_0)\right) f\left(\sigma^{-1}(x-x_0)\right) \cos \frac{2p_0}{\hbar} x \right]. \quad (6.80)$$

The last term in (6.80) is called interference term. Its presence implies that the probability density for the position of the particle is not the sum of the probability densities associated with each wave packet, as it would be the case for a classical particle.

Let us consider the time evolution $\psi_t = e^{-i\frac{t}{\hbar}H_0}\psi_0$. Since the time evolution is linear, we have

$$\psi_t(x) = C\left(e^{-i\frac{t}{\hbar}H_0}\psi^+\right)(x) + C\left(e^{-i\frac{t}{\hbar}H_0}\psi^-\right)(x) := C\left(\psi_t^+(x) + \psi_t^-(x)\right). \quad (6.81)$$

From the analysis developed in the previous section, we know that the centers of the two wave packets $\psi_t^{\pm}$ move with opposite velocities p_0/m and $-p_0/m$, starting from the initial positions $-x_0$ and x_0, respectively. Then, at the time

$$\tau = \frac{mx_0}{p_0} \quad (6.82)$$

the two centers reach the origin and we are in the condition of maximal overlapping of the two wave packets. Using (6.71), the wave function at time $t = \tau$ is

$$\psi_\tau(x) = e^{-i\frac{p_0^2}{2m}\frac{\tau}{\hbar}} C\left[\frac{1}{\sqrt{\sigma}}\left(e^{-i\alpha(-\frac{1}{2}\Delta)}f\right)\left(\frac{x}{\sigma}\right)e^{i\frac{p_0}{\hbar}x} + \frac{1}{\sqrt{\sigma}}\left(e^{-i\alpha(-\frac{1}{2}\Delta)}f\right)\left(\frac{x}{\sigma}\right)e^{-i\frac{p_0}{\hbar}x}\right]$$

$$= e^{-i\frac{p_0^2}{2m}\frac{\tau}{\hbar}} \frac{2C}{\sqrt{\sigma}}\left(e^{-i\alpha(-\frac{1}{2}\Delta)}f\right)\left(\frac{x}{\sigma}\right)\cos\frac{p_0}{\hbar}x, \qquad \alpha := \frac{\hbar x_0}{\sigma^2 p_0} \quad (6.83)$$

and the corresponding probability density for the position of the particle is

$$|\psi_\tau(x)|^2 = C^2\left[\frac{1}{\sigma}\left|\left(e^{-i\alpha(-\frac{1}{2}\Delta)}f\right)\left(\frac{x}{\sigma}\right)\right|^2 + \frac{1}{\sigma}\left|\left(e^{-i\alpha(-\frac{1}{2}\Delta)}f\right)\left(\frac{x}{\sigma}\right)\right|^2 \right.$$

$$\left. + \frac{2}{\sigma}\left|\left(e^{-i\alpha(-\frac{1}{2}\Delta)}f\right)\left(\frac{x}{\sigma}\right)\right|^2 \cos 2\frac{p_0}{\hbar}x\right]$$

$$= \frac{4C^2}{\sigma}\left|\left(e^{-i\alpha(-\frac{1}{2}\Delta)}f\right)\left(\frac{x}{\sigma}\right)\right|^2 \cos^2\frac{p_0}{\hbar}x. \quad (6.84)$$

Formula (6.84) is particularly relevant. In order to discuss its meaning more clearly, it is convenient to introduce the following assumptions on the parameters:

$$\frac{\sigma}{x_0} \ll \frac{\hbar}{\sigma p_0} \ll 1. \quad (6.85)$$

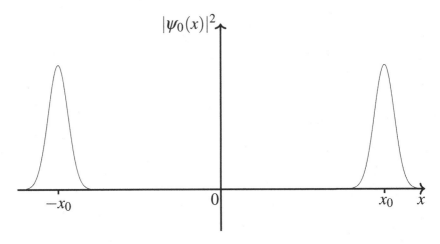

Fig. 6.2 Probability distribution of the position at $t = 0$

From (6.85), we also have

$$\alpha \gg 1 . \tag{6.86}$$

Note that the condition $\sigma/x_0 \ll 1$ guarantees that at $t = 0$ the region where the two wave packets overlap is small see (6.80) and therefore the effect of the interference term is negligible (Fig. 6.2).

The situation is different at $t = \tau$, since the effect of the interference term is crucial. In fact, using (6.86) and proposition 6.2, we have

$$|\psi_\tau(x)|^2 \simeq \frac{4C^2}{\sigma\alpha} \left| \tilde{f}\left(\frac{x}{\sigma\alpha}\right) \right|^2 \cos^2 \frac{p_0}{\hbar} x . \tag{6.87}$$

Let us analyze formula (6.87) more closely. The term $\left| \tilde{f}\left(x/\sigma\alpha\right) \right|^2$ is significantly different from zero only for x in the interval

$$(-\sigma\alpha, \sigma\alpha) \tag{6.88}$$

(for instance, in the Gaussian case we have $\left| \tilde{f}\left(x/\sigma\alpha\right) \right|^2 = (\pi)^{-1/2} e^{-x^2/\sigma^2\alpha^2}$).

On the other hand, the oscillating term $\cos^2 p_0 x/\hbar$ has its zeros at the points

$$x_k^0 = (2k+1)\frac{\pi}{2}\frac{\hbar}{p_0}, \qquad k \in \mathbb{Z} . \tag{6.89}$$

These points x_k^0 belong to the interval (6.88) when the index k satisfies the condition

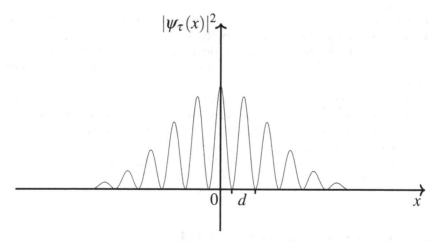

Fig. 6.3 Probability distribution of the position at $t = \tau$ $(d = \frac{\pi \hbar}{2p_0})$

$$|2k + 1| \frac{\pi}{2} \frac{\hbar}{p_0} < \sigma \alpha \qquad (6.90)$$

or

$$-\frac{1}{\pi} \frac{x_0}{\sigma} - \frac{1}{2} < k < \frac{1}{\pi} \frac{x_0}{\sigma} - \frac{1}{2}. \qquad (6.91)$$

By the assumptions (6.85), we conclude that there are many integers k satisfying (6.91). Therefore, the function (6.87) has a strongly oscillating behavior in the interval (6.88), with the amplitude of the oscillations modulated by the function $\left|\tilde{f}(x/\sigma\alpha)\right|^2$ and the distance between two consecutive zeros very small compared with the length of the interval (6.88). In other terms, the function exhibits a sequence of many zeros and maxima very close to each other (Fig. 6.3).

The result is the typical interference effect, characteristic of any classical field whose evolution is governed by a wave equation, like in Acoustics or in Electromagnetism. In these cases, however, the interference effect concerns a physical quantity, the intensity of the wave, which is directly measured in experiments.

Instead, in our quantum case, the interference effect is produced by the dynamical evolution of a single particle and it concerns the probability density for the position of the particle itself. As a consequence, if we perform a position measurement when the state is (6.83), we have a negligible probability to find the particle around the zeros of (6.84) and a high probability to find the particle around the maxima of (6.84).

According to Born's statistical interpretation, this means that if we repeat the same position measurement many times in identical conditions, we find that the statistical distribution of the detected positions of the particles is concentrated around the maxima of (6.84), while very few particles are detected around the zeros of (6.84).

We could say that the particles "avoid" the regions around the zeros of (6.84) and "prefer" the regions around the maxima of (6.84).

It is worth to emphasize that such a behavior is highly non-classical and it is surely hard to understand at intuitive level. In any case, it must be considered as the main distinctive characteristic of a quantum system. As Feynman wrote in his Lectures: "*We choose to examine a phenomenon*" (i.e., the interference) "*which is impossible, absolutely impossible, to explain in any classical way, and which has in it the heart of quantum mechanics. In reality, it contains the* only *mystery. We cannot make the mystery go away by 'explaining' how it works. We will just tell you how it works. In telling you how it works we will have told you about the basic peculiarities of all quantum mechanics.*" ([5], Chap. 1).

6.5 Particle in a Constant Force Field

The methods used to construct the Hamiltonian and the unitary propagator for a free particle can be generalized to the case of a particle in a constant force field (see also [3, 6]). Let us consider a particle in one dimension subject to a constant force F. We recall that the classical Hamiltonian is

$$H_F^{cl} = \frac{p^2}{2m} - Fx, \qquad F > 0 \tag{6.92}$$

and therefore the solutions of Hamilton's equations are

$$x(t) = x(0) + \frac{p(0)}{m}t + \frac{F}{2m}t^2, \qquad p(t) = p(0) + Ft. \tag{6.93}$$

By the quantization rule, we find the following symmetric operator in $L^2(\mathbb{R})$ which, for convenience, we define on a domain of smooth functions

$$\dot{H}_F = -\frac{\hbar^2}{2m}\Delta - Fx, \qquad D(\dot{H}_F) = \mathscr{S}(\mathbb{R}). \tag{6.94}$$

In order to construct the (unique) self-adjoint extension of the above operator, for $f \in L^2(\mathbb{R})$ we define

$$\tilde{f}_F(k) := \sqrt{F}\,\tilde{f}(-Fk) = \sqrt{\frac{F}{2\pi}} \int dx\, f(x)\, e^{iFkx} \tag{6.95}$$

and note that

$$\widetilde{(\dot{H}_F f)}_F(k) = \left(\frac{\hbar^2 F^2}{2m}k^2 + i\frac{d}{dk}\right)\tilde{f}_F(k). \tag{6.96}$$

Moreover, the following identity holds:

$$i\frac{d}{dk}\left(e^{-ia\hbar^3k^3}\,\tilde{f}_F(k)\right) = e^{-ia\hbar^3k^3}\left(\frac{\hbar^2 F^2}{2m}k^2 + i\frac{d}{dk}\right)\tilde{f}_F(k)\,, \qquad a := \frac{F^2}{6m\hbar}\,. \tag{6.97}$$

Making use of (6.96) and (6.97), one can verify that the unitary operator W_F : $L^2(\mathbb{R}) \to L^2(\mathbb{R})$ defined by

$$(W_F f)(\xi) \equiv f^\sharp(\xi) := \frac{1}{\sqrt{2\pi}}\int dk\, e^{ik\xi - ia\hbar^3k^3}\,\tilde{f}_F(k) \tag{6.98}$$

diagonalizes the operator $\dot{H}_F$, i.e.,

$$\left(W_F \dot{H}_F W_F^{-1} f^\sharp\right)(\xi) = \xi\, f^\sharp(\xi)\,. \tag{6.99}$$

Proceeding as in the case of the free particle, we conclude that the self-adjoint Hamiltonian describing a particle in $\mathbb{R}$ subject to a constant force field is

$$(H_F f)(x) = \frac{\sqrt{F}}{2\pi}\int dk\, e^{-iFkx + ia\hbar^3k^3}\int d\xi\, \xi\, f^\sharp(\xi)\, e^{-ik\xi} \tag{6.100}$$

$$D(H_F) = \left\{f \in L^2(\mathbb{R}) \mid \int d\xi\, |\xi f^\sharp(\xi)|^2 < \infty\right\}\,. \tag{6.101}$$

Moreover, one has $\sigma(H_F) = \sigma_{ac}(H_F) = \mathbb{R}$.

Remark 6.6 The unitary operator W_F can be written as

$$(W_F f)(\xi) = \int dx\, f(x)\,\overline{\phi_F(x,\xi)}\,, \qquad \phi_F(x,\xi) := \frac{\sqrt{F}}{2\pi}\int dk\, e^{-ik(Fx+\xi) + ia\hbar^3k^3}\,, \tag{6.102}$$

where the functions $\phi_F(\cdot,\xi)$, $\xi \in \mathbb{R}$, are distributional solutions of the equation $H_F u = \xi u$. Moreover,

$$\phi_F(x,\xi) = \left(\frac{2m}{\hbar^2 F^{1/2}}\right)^{1/3} Ai\left(-\left(\frac{2mF}{\hbar^2}\right)^{1/3}(x + F^{-1}\xi)\right)\,, \tag{6.103}$$

where

$$Ai(x) := \frac{1}{\pi}\int_0^\infty dt\, \cos\left(\frac{t^3}{3} + xt\right) \tag{6.104}$$

denotes the Airy function and it is a solution of the equation $u'' - xu = 0$. The asymptotic behavior of the Airy function is given by ([6, 7])

$$Ai(x) \simeq \frac{1}{2\sqrt{\pi}\,x^{1/4}}\, e^{-\frac{2}{3}x^{3/2}} \qquad\qquad \text{for } x \to +\infty \tag{6.105}$$

$$Ai(x) \simeq \frac{1}{\sqrt{\pi}\,|x|^{1/4}}\, \cos\left(\frac{2}{3}|x|^{3/2} - \frac{\pi}{4}\right) \qquad \text{for } x \to -\infty\,. \tag{6.106}$$

Such a behavior implies that $\phi_F(\cdot, \xi)$ does not belong to $L^2(\mathbb{R})$. Therefore, it is not a possible state of the system and, also, it is not an eigenfunction of the Hamiltonian $H_F, D(H_F)$. The function $\phi(\cdot, \xi)$ is called generalized eigenfunction. Its importance lies in the fact that it defines, via (6.102), the unitary operator which diagonalizes the Hamiltonian, exactly as the function $e^{ik\cdot x}$ defines, via Fourier transform, the unitary operator which diagonalizes the free Hamiltonian.

Exploiting again the unitary operator W_F, one can also construct the unitary propagator generated by $H_F, D(H_F)$. Indeed, for any $f \in L^2(\mathbb{R})$ we have

$$\left(e^{-i\frac{t}{\hbar}H_F}f\right)(x) = \left(W_F^{-1}e^{-i\frac{t}{\hbar}(\cdot)}W_F f\right)(x)$$

$$= e^{i\frac{Ft}{\hbar}x - i\frac{F^2t^3}{6m\hbar}}\left(e^{-i\frac{t}{\hbar}H_0}f\right)\left(x - \frac{Ft^2}{2m}\right). \qquad (6.107)$$

The proof of the last equality in (6.107) is left as an exercise.

Exercise 6.10 Using formula (6.107), compute $\langle x \rangle(t)$, $\langle p \rangle(t)$, $\Delta x(t)$, $\Delta p(t)$.

Exercise 6.11 Given the initial state

$$\psi_0(x) = \frac{1}{\pi^{1/4}\sqrt{\sigma}}e^{-\frac{x^2}{2\sigma^2}}, \qquad (6.108)$$

determine the probability $\mathscr{P}_-(t)$ to find the particle at time t in the "classically forbidden region" $(-\infty, 0)$ and verify that

$$\mathscr{P}_-(t) = \frac{1}{2} - \frac{F}{2\sqrt{\pi}m\sigma}t^2 + O(t^4) \qquad \text{for } t \to 0. \qquad (6.109)$$

References

1. Reed, M., Simon, B.: Methods of Modern Mathematical Physics, II: Fourier Analysis, Self-Adjointness. Academic Press, New York (1975)
2. Teschl, G.: Mathematical Methods in Quantum Mechanics. American Mathematical Society, Providence (2009)
3. Thaller, B.: Visual Quantum Mechanics. Springer, New York (2000)
4. de Oliveira, C.R.: Intermediate Spectral Theory and Quantum Dynamics. Birkhäuser, Basel (2009)
5. Feynman, R., Leighton, R.B., Sands, M.L.: The Feynman Lectures on Physics, vol. III. Addison Wesley Publ. Com (2005)
6. Landau, L.D., Lifshitz, E.M.: Quantum Mechanics, 3th edn. Pergamon Press, Oxford (1977)
7. Lebedev, N.N.: Special Functions and Their Applications. Dover Publ, New York (1972)

Chapter 7
Harmonic Oscillator

7.1 Hamiltonian

The one-dimensional harmonic oscillator is the simplest model describing a particle subject to a confining potential, i.e., to a potential $V(x)$ such that $\lim_{|x| \to \infty} V(x) = \infty$. It consists of a point particle of mass m moving on a line (the x-axis) and subject to a conservative force $F(x) = -V'(x)$, with potential $V(x) = \frac{1}{2}m\omega^2 x^2$, where $\omega > 0$ denotes the frequency of the oscillator. The potential has a proper minimum at the origin, corresponding to a stable equilibrium position.

In Classical Mechanics, the dynamics of the harmonic oscillator is described by the quadratic Hamiltonian

$$H_\omega^{cl} = \frac{p^2}{2m} + \frac{1}{2}m\omega^2 q^2, \qquad (q, p) \in \mathbb{R}^2. \tag{7.1}$$

For any initial datum (q_0, p_0), the solution of the corresponding Hamilton's equations

$$\dot{q} = \frac{p}{m}, \qquad \dot{p} = -m\omega^2 q \tag{7.2}$$

is

$$q(t) = q_0 \cos \omega t + \frac{p_0}{m\omega} \sin \omega t, \qquad p(t) = -m\omega q_0 \sin \omega t + p_0 \cos \omega t. \tag{7.3}$$

Note that all the orbits are bounded. For $(q_0, p_0) = (0, 0)$, the solution coincides with the stable equilibrium position while for $(q_0, p_0) \neq (0, 0)$ the solution is periodic with period $T = \frac{2\pi}{\omega}$ which does not depend on the initial condition.

In the quantum case (see also [1–3]), using the quantization procedure, we start considering the operator

$$\dot{H}_\omega = -\frac{\hbar^2}{2m}\frac{d^2}{dx^2} + \frac{1}{2}m\omega^2 x^2, \qquad D(\dot{H}_\omega) = \mathscr{S}(\mathbb{R}) \tag{7.4}$$

© Springer International Publishing AG, part of Springer Nature 2018
A. Teta, *A Mathematical Primer on Quantum Mechanics*,
UNITEXT for Physics, https://doi.org/10.1007/978-3-319-77893-8_7

acting in the Hilbert space $L^2(\mathbb{R})$. In this section, we shall prove that the above symmetric operator has an infinite number of positive eigenvalues and, correspondingly, an orthonormal and complete set of eigenfunctions. This fact implies that the operator is essentially self-adjoint and its unique self-adjoint extension defines the Hamiltonian of the harmonic oscillator.

To simplify the notation, we introduce the dimensionless variable

$$y = \sqrt{\frac{m\omega}{\hbar}}x \tag{7.5}$$

and the unitary operator Y in $L^2(\mathbb{R})$

$$(Yu)(y) = \left(\frac{\hbar}{m\omega}\right)^{1/4} u\left(\sqrt{\frac{\hbar}{m\omega}}y\right), \qquad (Y^{-1}f)(x) = \left(\frac{m\omega}{\hbar}\right)^{1/4} f\left(\sqrt{\frac{m\omega}{\hbar}}x\right). \tag{7.6}$$

Then

$$(Y\dot{H}_\omega Y^{-1}f)(y) = \hbar\omega(\dot{H}f)(y), \qquad \dot{H} := -\frac{1}{2}\frac{d^2}{dy^2} + \frac{1}{2}y^2. \tag{7.7}$$

Remark 7.1 The solution of the eigenvalue problem for $\dot{H}_\omega$ is reduced to the solution of the eigenvalue problem for $\dot{H}$, i.e., we have if

$$\dot{H}_\omega u = Eu, \qquad u \in \mathscr{S}(\mathbb{R}), \qquad \|u\| = 1 \tag{7.8}$$

then $\dot{H}(Yu) = \frac{E}{\hbar\omega}(Yu)$. Conversely, if

$$\dot{H}f = \lambda f, \qquad f \in \mathscr{S}(\mathbb{R}), \qquad \|f\| = 1 \tag{7.9}$$

then $\dot{H}_\omega(Y^{-1}f) = \hbar\omega\lambda(Y^{-1}f)$.

We shall solve problem (7.9) using an algebraic method based on the factorization property of the operator $\dot{H}$. Let us denote

$$Q = y, \qquad P = -i\frac{d}{dy} \tag{7.10}$$

and note that $[Q, P]f = if$, $f \in \mathscr{S}(\mathbb{R})$, and

$$\dot{H} = \left(\frac{1}{2}P^2 + \frac{1}{2}Q^2\right). \tag{7.11}$$

Let us also introduce the following operators defined on $\mathscr{S}(\mathbb{R})$:

$$A := \frac{Q+iP}{\sqrt{2}} = \frac{1}{\sqrt{2}}\left(\frac{d}{dy}+y\right), \qquad A^* := \frac{Q-iP}{\sqrt{2}} = \frac{1}{\sqrt{2}}\left(-\frac{d}{dy}+y\right).$$

$$(7.12)$$

The operators A and A^* are called annihilation operator and creation operator for reasons that will be clear later on. By a direct computation, for any $f, g \in \mathscr{S}(\mathbb{R})$ one verifies that

$$(g, Af) = (A^*g, f), \tag{7.13}$$

$$[A, A^*]f = f, \tag{7.14}$$

$$\dot{H}f = A^*Af + \frac{1}{2}f = AA^*f - \frac{1}{2}f, \tag{7.15}$$

$$[\dot{H}, A]f = -Af, \qquad [\dot{H}, A^*]f = A^*f. \tag{7.16}$$

Observe that (7.15) expresses the factorization property of $\dot{H}$ in terms of A and A^*. Using the properties of A and A^*, we prove the following result.

Proposition 7.1 Let f_λ be solution of problem (7.9). We have

(i) $\lambda \geq \frac{1}{2}$;
(ii) for any $\lambda \geq \frac{1}{2}$ the vector

$$\frac{A^* f_\lambda}{\sqrt{\lambda + \frac{1}{2}}} \tag{7.17}$$

is an eigenvector of $\dot{H}$ with eigenvalue $\lambda' = \lambda + 1$;
(iii) for any $\lambda \geq \frac{3}{2}$ the vector

$$\frac{A f_\lambda}{\sqrt{\lambda - \frac{1}{2}}} \tag{7.18}$$

is an eigenvector of $\dot{H}$ with eigenvalue $\lambda' = \lambda - 1$.

Proof Using (7.15), we have

$$\lambda = (f_\lambda, \dot{H}f_\lambda) = \left(f_\lambda, \left(A^*A + \frac{1}{2}\right)f_\lambda\right) = \|Af_\lambda\|^2 + \frac{1}{2} \tag{7.19}$$

and then point (i) is proved. Let us consider $A^* f_\lambda$. Using again (7.15), we have

$$\dot{H}A^* f_\lambda = \left(A^*A + \frac{1}{2}\right)A^* f_\lambda = A^*\left(AA^* + \frac{1}{2}\right)f_\lambda = A^*(\dot{H}+1)f_\lambda$$

$$= (\lambda + 1)A^* f_\lambda. \tag{7.20}$$

Moreover

$$\|A^* f_\lambda\|^2 = (f_\lambda, A A^* f_\lambda) = \left(f_\lambda, \left(\dot{H} + \frac{1}{2} \right) f_\lambda \right) = \lambda + \frac{1}{2}. \qquad (7.21)$$

Taking into account of (7.20) and (7.21), also point (ii) is proved. Analogously, point (iii) follows from

$$\dot{H} A f_\lambda = \left(A A^* - \frac{1}{2} \right) A f_\lambda = A \left(A^* A - \frac{1}{2} \right) f_\lambda = A(\dot{H} - 1) f_\lambda$$
$$= (\lambda - 1) A f_\lambda \qquad (7.22)$$

and

$$\|A f_\lambda\|^2 = (f_\lambda, A^* A f_\lambda) = \left(f_\lambda, \left(\dot{H} - \frac{1}{2} \right) f_\lambda \right) = \lambda - \frac{1}{2}. \qquad (7.23)$$

□

Note that point (ii) of the above proposition states that the application of A^* to an eigenvector with eigenvalue λ produces an eigenvector with eigenvalue $\lambda + 1$ and this fact motivates the name of creation operator for A^*. Analogously, point (iii) motivates the name of annihilation operator for A.

In the next proposition, we solve problem (7.9) using the properties of A and A^*.

Proposition 7.2 *Problem* (7.9) *admits a solution if and only if* $\lambda = n + \frac{1}{2}$, *with n non negative integer. If* $\lambda = n + \frac{1}{2}$ *then the unique solution is*

$$f_n(y) = \frac{1}{\sqrt{n!}} \left[(A^*)^n f_0 \right] (y), \qquad (7.24)$$

$$f_0(y) = \frac{1}{\pi^{1/4}} e^{-\frac{y^2}{2}}. \qquad (7.25)$$

Moreover, the system of eigenvectors $\{f_n\}$ *is orthonormal and complete in* $L^2(\mathbb{R})$.

Proof Let us consider $\lambda = n + \frac{1}{2}$, with n nonnegative integer. For $n = 0$, i.e., $\lambda = \frac{1}{2}$, by (7.19), we have that problem (7.9) has a solution f_0 if and only if

$$A f_0 = 0. \qquad (7.26)$$

Taking into account of the explicit expression of A in (7.12), Eq. (7.26) reads

$$\frac{df_0}{dy} + y f_0 = 0. \qquad (7.27)$$

Solving the equation, we find that the unique (apart for an irrelevant constant phase factor) solution of problem (7.9) for $\lambda = \frac{1}{2}$ is given by (7.25). Let us fix $n = 1$, i.e., $\lambda = \frac{3}{2}$. By Proposition 7.1, point (ii), we know that

$$f_1 = A^* f_0 \tag{7.28}$$

is solution of the problem with $\lambda = \frac{3}{2}$. Let us verify that the solution is unique. Let g_1 be a solution for $\lambda = \frac{3}{2}$. Then, by Proposition 7.1, point (iii), we have that Ag_1 is solution for $\lambda = \frac{1}{2}$, so that

$$Ag_1 = f_0. \tag{7.29}$$

Applying A^* and using (7.15), we have

$$A^* f_0 = A^* A g_1 = \left(\dot{H} - \frac{1}{2} \right) g_1 = g_1. \tag{7.30}$$

By (7.28), (7.30), we conclude that g_1 coincides with f_1. Repeating the argument for $n = 2$, one shows that

$$f_2 = \frac{1}{\sqrt{2}} A^* f_1 = \frac{1}{\sqrt{2}} (A^*)^2 f_0 \tag{7.31}$$

is the unique solution of the problem for $\lambda = \frac{5}{2}$, and so on. After n steps, we find that (7.24) is the unique solution of the problem for $\lambda = n + \frac{1}{2}$.

Let us observe that two vectors f_n and $f_{n'}$, with $n \neq n'$, are orthogonal since they are eigenvectors of a symmetric operator. The proof that the system $\{f_n\}$ is complete is postponed in Sect. 7.5.

Finally, let us assume that problem (7.9) has a solution $h_{\lambda'}$ for $\lambda' \neq n + \frac{1}{2}$. Then $(h_{\lambda'}, f_n) = 0$ for any n and the completeness of $\{f_n\}$ implies that $h_{\lambda'} = 0$, concluding the proof of the proposition. $\qquad\square$

The eigenfunctions f_n found in the previous proposition can be written in terms of the Hermite polynomials [4]. We recall that the Hermite polynomial of degree n is defined as

$$H_n(x) = (-1)^n e^{x^2} \frac{d^n}{dx^n} e^{-x^2}, \qquad x \in \mathbb{R}. \tag{7.32}$$

Exercise 7.1 *Verify the properties*

$$H_n(-x) = (-1)^n H_n(x), \qquad H_{n+1}(x) = -H_n'(x) + 2x\, H_n(x). \tag{7.33}$$

We have

Proposition 7.3

$$f_n(y) = \frac{1}{\sqrt{2^n n!}\, \pi^{1/4}} H_n(y)\, e^{-\frac{y^2}{2}}. \tag{7.34}$$

Proof From (7.24), (7.25) and (7.12), we obtain

$$f_n(y) = \frac{1}{\sqrt{2^n \, n! \, \pi^{1/4}}} \left(-\frac{d}{dy} + y \right)^n e^{-\frac{y^2}{2}}. \tag{7.35}$$

The proposition is proved if we show that the following formula

$$\left(-\frac{d}{dy} + y \right)^n e^{-\frac{y^2}{2}} = (-1)^n e^{\frac{y^2}{2}} \frac{d^n}{dy^n} e^{-y^2} \tag{7.36}$$

holds for any n. Let us proceed by induction. For $n = 1$ (7.36) is easily verified. Let us assume that (7.36) is true for $n = m$. We have

$$
\begin{aligned}
\left(-\frac{d}{dy} + y \right)^{m+1} e^{-\frac{y^2}{2}} &= \left(-\frac{d}{dy} + y \right) (-1)^m e^{\frac{y^2}{2}} \frac{d^m}{dy^m} e^{-y^2} \\
&= (-1)^m \left(-y \, e^{\frac{y^2}{2}} \frac{d^m}{dy^m} e^{-y^2} - e^{\frac{y^2}{2}} \frac{d^{m+1}}{dy^{m+1}} e^{-y^2} \right) \\
&\quad + (-1)^m y \, e^{\frac{y^2}{2}} \frac{d^m}{dy^m} e^{-y^2} \\
&= (-1)^{m+1} e^{\frac{y^2}{2}} \frac{d^{m+1}}{dy^{m+1}} e^{-y^2}. \tag{7.37}
\end{aligned}
$$

Then (7.36) holds and the proposition is proved.   □

Using Proposition 7.2, formula (7.34), and Remark 7.1, we conclude that the eigenvalues of the symmetric operator (7.4) are

$$E_n = \hbar\omega \left(n + \frac{1}{2} \right), \quad n = 0, 1, 2 \ldots \tag{7.38}$$

and the unique eigenfunction associated to E_n is $\phi_n(x) = (Y^{-1} f_n)(x)$, i.e.,

$$\phi_n(x) = \left(\frac{m\omega}{\hbar} \right)^{1/4} \frac{1}{\sqrt{2^n \, n! \, \pi^{1/4}}} H_n \left(\sqrt{\frac{m\omega}{\hbar}} x \right) e^{-\frac{m\omega}{2\hbar} x^2}. \tag{7.39}$$

Moreover, the system of eigenfunctions $\{\phi_n\}$ is orthonormal and complete and the unitary operator

$$U_\omega : L^2(\mathbb{R}) \to l^2, \quad (U_\omega f)_n = (\phi_n, f) \tag{7.40}$$

reduces the operator $\dot{H}_\omega$ to a multiplication operator by E_n. It follows that $\dot{H}_\omega$ is essentially self-adjoint and its unique self-adjoint extension H_ω is the Hamiltonian of the harmonic oscillator. Thus, we have proved the following proposition.

Proposition 7.4 *The unique self-adjoint extension of (7.4) is*

$$(H_\omega u)(x) = \sum_{n=0}^{\infty} E_n u_n \, \phi_n(x), \qquad D(H_\omega) = \left\{ u \in L^2(\mathbb{R}) \mid \sum_{n=0}^{\infty} E_n^2 \, |u_n|^2 < \infty \right\},$$
(7.41)

where $u_n = (\phi_n, u)$ *and* E_n *is given in (7.38). Moreover,* $\sigma(H_\omega) = \sigma_p(H_\omega) = \{E_n, \, n = 0, 1, 2, \ldots\}.$

Remark 7.2

(i) The vectors ϕ_n are stationary states. The stationary state corresponding to $n = 0$

$$\phi_0(x) = \left(\frac{m\omega}{\pi\hbar}\right)^{1/4} e^{-\frac{m\omega}{2\hbar}x^2}$$
(7.42)

is the ground state and satisfies the minimal uncertainty condition $\Delta x \, \Delta p = \frac{\hbar}{2}$ (see also Exercise 3.2). For $n > 0$, ϕ_n is called excited state.

(ii) The eigenvalues E_n are called energy levels. Apart from the term $\frac{\hbar\omega}{2}$, independent of n, they coincide with the discrete values of the energy introduced by Planck in 1900. The minimum eigenvalue E_0 is the ground-state energy.

(iii) The fact that $\inf \sigma(H_\omega) = \frac{\hbar\omega}{2} > 0$ can also be understood as a direct consequence of the uncertainty principle. Indeed,

$$(\psi, H_\omega \psi) = \frac{1}{2m} \langle p^2 \rangle + \frac{1}{2} m\omega^2 \langle x^2 \rangle \geq \omega \sqrt{\langle p^2 \rangle} \sqrt{\langle x^2 \rangle} \geq \frac{\hbar\omega}{2}, \quad (7.43)$$

where we have used the inequality $a^2 + b^2 \geq 2ab$ and the uncertainty principle. Moreover, the minimum value $\frac{\hbar\omega}{2}$ is reached if we choose the Gaussian state (3.47) (with $x_0 = p_0 = 0$) that minimizes the uncertainty relations and we impose the condition $\frac{1}{2m}\langle p^2 \rangle = \frac{1}{2}m\omega^2 \langle x^2 \rangle$. The result of the computation shows that the minimum is reached if ψ is the ground state (7.42).

We conclude writing the spectral measure associated to H_ω

$$m_{H_\omega}^f(B) = (f, E_{H_\omega}(B)f) = \sum_{n:E_n \in B} |(\phi_n, f)|^2, \quad (7.44)$$

where $f \in L^2(\mathbb{R})$ is a generic state and B is a Borel set in $\mathbb{R}$. The measure is pure point and it is concentrated on the eigenvalues E_n.

Exercise 7.2 *Consider a harmonic oscillator in the first excited state* ϕ_1. *Compute the probability to find the particle in the "classically forbidden region".*

(Hint: for a particle with energy E subject to a potential $V(x)$, the classically forbidden region is $\{x \in \mathbb{R} \mid V(x) > E\}$).

7.2 Properties of the Unitary Group

The expression for the unitary group generated by H_ω can be explicitly written using the fact that the Hamiltonian is diagonalized by the unitary operator (7.40). Indeed, for $f \in L^2(\mathbb{R})$ we have

$$\psi(x,t) = \left(e^{-i\frac{t}{\hbar}H_\omega} f\right)(x) = \sum_{n=0}^{\infty} e^{-i\frac{t}{\hbar}E_n} f_n \phi_n(x), \qquad f_n = (f, \phi_n). \qquad (7.45)$$

Starting from the above expression, the unitary group can also be written as an integral operator whose integral kernel can be explicitly computed (see, e.g., [3], Chap. 7).

A relevant property of the time evolution defined by (7.45) is that any state is a bound state (see Definition 5.38). We recall that the corresponding property in the classical case is that any orbit is bounded.

Proposition 7.5 Let $f \in L^2(\mathbb{R})$. Then, for any $\varepsilon > 0$ there exists $R_0 > 0$ such that

$$\sup_t \int_{|x|>R_0} dx \left|\left(e^{-i\frac{t}{\hbar}H_\omega} f\right)(x)\right|^2 < \varepsilon. \qquad (7.46)$$

Proof Let us fix $\varepsilon > 0$. The completeness of $\{\phi_n\}$ implies that there is n_0 such that

$$\left\| f - \sum_{n=0}^{n_0} f_n \phi_n \right\| < \frac{\sqrt{\varepsilon}}{2}. \qquad (7.47)$$

Denoting by $\chi_{>R}$ the characteristic function of the set $\{x \in \mathbb{R} \mid |x| > R\}$ and using (7.47), we have

$$\|\chi_{>R}\, e^{-i\frac{t}{\hbar}H_\omega} f\| \leq \left\| \chi_{>R}\, e^{-i\frac{t}{\hbar}H_\omega}\left(f - \sum_{n=0}^{n_0} f_n\phi_n\right) \right\| + \left\| \chi_{>R}\, e^{-i\frac{t}{\hbar}H_\omega} \sum_{n=0}^{n_0} f_n\phi_n \right\|$$

$$\leq \frac{\sqrt{\varepsilon}}{2} + \left\| \chi_{>R} \sum_{n=0}^{n_0} e^{-i\frac{t}{\hbar}E_n} f_n\phi_n \right\| \leq \frac{\sqrt{\varepsilon}}{2} + \sum_{n=0}^{n_0} |f_n| \|\chi_{>R}\,\phi_n\|. \qquad (7.48)$$

Since $\phi_n \in L^2(\mathbb{R})$, there exists $R_0 > 0$ such that

$$\|\chi_{>R_0}\,\phi_n\| < \frac{\sqrt{\varepsilon}}{2(n_0 + 1)\,\max_{\{n \leq n_0\}} |f_n|}. \qquad (7.49)$$

Using this estimate in (7.48), we conclude the proof.   □

Remark 7.3 In the above proof, we have not used the explicit form of the eigenfunctions ϕ_n. It follows that the proposition holds for any Hamiltonian with a purely discrete spectrum.

Let us verify a specific property of the harmonic oscillator, i.e., the fact that the mean values of position $\hat{x}$ and momentum $\hat{p}$ observables evolve in time following the classical law (7.3). The property is an immediate consequence of the Ehrenfest theorem, but it is useful a direct check. Given $f \in \mathscr{S}(\mathbb{R})$, we have

$$\frac{d}{dt}\langle x \rangle(t) = \frac{i}{\hbar}\left(e^{-i\frac{t}{\hbar}H_\omega}f, [H_\omega, \hat{x}]e^{-i\frac{t}{\hbar}H_\omega}f\right), \tag{7.50}$$

$$\frac{d}{dt}\langle p \rangle(t) = \frac{i}{\hbar}\left(e^{-i\frac{t}{\hbar}H_\omega}f, [H_\omega, \hat{p}]e^{-i\frac{t}{\hbar}H_\omega}f\right). \tag{7.51}$$

Since

$$[H_\omega, \hat{x}] = -\frac{i\hbar}{m}\hat{p}, \qquad [H_\omega, \hat{p}] = i\hbar m\omega^2\hat{x}, \tag{7.52}$$

we obtain

$$\frac{d}{dt}\langle x \rangle(t) = \frac{1}{m}\langle p \rangle(t), \qquad \frac{d}{dt}\langle p \rangle(t) = -m\,\omega^2\langle x \rangle(t), \tag{7.53}$$

i.e., the equation for the mean values of $\hat{x}$ and $\hat{p}$ coincide with Hamilton's equation and therefore

$$\langle x \rangle(t) = \langle x \rangle(0)\cos\omega t + \frac{\langle p \rangle(0)}{m\omega}\sin\omega t, \tag{7.54}$$

$$\langle p \rangle(t) = -m\omega\langle x \rangle(0)\sin\omega t + \langle p \rangle(0)\cos\omega t. \tag{7.55}$$

7.3 Time Evolution of a Gaussian State

In this section, we consider the Schrödinger equation for the Hamiltonian of the harmonic oscillator with a Gaussian initial state. By a direct computation, we shall see that the Gaussian form is preserved by the evolution. However, the relevant aspect is that the mean square deviation is a bounded and periodic function of the time. We remark that this is a peculiar property of the harmonic oscillator.

Let us consider a Gaussian initial state written in the form

$$\psi_0(x) = \frac{1}{(\pi\hbar)^{1/4}\sqrt{a_0}}\,e^{-b_0\,a_0^{-1}\frac{(x-q_0)^2}{2\hbar}+i\frac{p_0}{\hbar}(x-q_0)}, \tag{7.56}$$

where $q_0, p_0 \in \mathbb{R}$, and a_0, b_0 are two complex numbers such that $\Re(\bar{a}_0 b_0) = 1$. It is easy to verify that

$$\Re(b_0 a_0^{-1}) = \frac{1}{|a_0|^2}, \qquad \Re(a_0 b_0^{-1}) = \frac{1}{|b_0|^2}. \tag{7.57}$$

We underline that such a Gaussian form of the initial state is particularly useful in the study of the semiclassical limit of Quantum Mechanics (see [5] and Appendix A).

Note that if we fix $a_0 = \frac{\sigma}{\sqrt{\hbar}}$ and $b_0 = a_0^{-1}, \sigma > 0$, then (7.56) reduces to the Gaussian state introduced in Sect. 6.3. By an explicit computation, one finds the mean value and the mean square deviation of the position

$$\langle x \rangle (0) = q_0, \qquad \Delta x(0) = \sqrt{\frac{\hbar}{2}} |a_0|. \tag{7.58}$$

Moreover, the Fourier transform of (7.56) is

$$\tilde{\psi}_0(k) = \frac{\hbar^{1/4}}{\pi^{1/4}\sqrt{b_0}} e^{-a_0 b_0^{-1} \hbar \frac{(k-p_0/\hbar)^2}{2} - ikx_0} \tag{7.59}$$

and then

$$\langle p \rangle (0) = p_0, \qquad \Delta p(0) = \sqrt{\frac{\hbar}{2}} |b_0|. \tag{7.60}$$

We have

$$\Delta x(0) \Delta p(0) = \frac{\hbar}{2} |a_0||b_0|, \tag{7.61}$$

so that for $|a_0||b_0| = 1$ the initial state (7.56) is a coherent state, i.e., the condition of minimal uncertainty is realized. The time evolution of the initial state (7.56) can be explicitly computed.

Proposition 7.6 *The solution of the Schrödinger equation for the harmonic oscillator with initial datum (7.56) is*

$$\psi(x,t) = e^{\frac{i}{\hbar}S(t)} \frac{1}{(\pi\hbar)^{1/4}\sqrt{a(t)}} e^{-b(t)a(t)^{-1}\frac{(x-q(t))^2}{2\hbar} + i\frac{p(t)}{\hbar}(x-q(t))}, \tag{7.62}$$

where $S(t), a(t), b(t), q(t),$ and $p(t)$ are functions satisfying the following equations:

$$\dot{q}(t) = \frac{p(t)}{m}, \qquad \dot{p}(t) = -m\omega^2 q(t), \tag{7.63}$$

$$\dot{a}(t) = i\frac{b(t)}{m}, \qquad \dot{b}(t) = i m\omega^2 a(t), \tag{7.64}$$

$$\dot{S}(t) = \frac{p^2(t)}{2m} - \frac{1}{2}m\omega^2 q^2(t) := L(q(t), p(t)) \tag{7.65}$$

with initial conditions $S(0) = 0, a(0) = a_0, b(0) = b_0, q(0) = q_0,$ and $p(0) = p_0$.

Proof We set

$$k(t) := e^{\frac{i}{\hbar}S(t)}, \qquad g(x,t) := \frac{1}{(\pi\hbar)^{1/4}\sqrt{a(t)}} e^{-b(t)a(t)^{-1}\frac{(x-q(t))^2}{2\hbar} + i\frac{p(t)}{\hbar}(x-q(t))} \tag{7.66}$$

and compute the time derivative of (7.62)

$$\dot{\psi} = \dot{k}g + k\left[-\frac{\dot{a}}{2a}g + g\left(-\frac{\dot{b}}{2\hbar a}(x-q)^2 + \frac{b\dot{a}}{2\hbar a^2}(x-q)^2 + \frac{b}{\hbar a}\dot{q}\,(x-q)\right.\right.$$
$$\left.\left. - \frac{i}{\hbar}p\dot{q} + \frac{i}{\hbar}(x-q)\dot{p}\right)\right]$$
$$= \psi\left[\frac{\dot{k}}{k} - \frac{\dot{a}}{2a} - \frac{i}{\hbar}p\dot{q} + \left(\frac{b}{\hbar a}\dot{q} + \frac{i}{\hbar}\dot{p}\right)(x-q) + \frac{1}{2}\left(\frac{b\dot{a}}{\hbar a^2} - \frac{\dot{b}}{\hbar a}\right)(x-q)^2\right].$$

$$(7.67)$$

For the derivatives with respect to x, we have

$$\psi' = \psi\left(-\frac{b}{\hbar a}(x-q) + \frac{i}{\hbar}p\right), \tag{7.68}$$

$$\psi'' = \psi\left[-\frac{b}{\hbar a} - \frac{p^2}{\hbar^2} - \frac{2i}{\hbar^2}\frac{b}{a}p\,(x-q) + \frac{b^2}{\hbar^2 a^2}(x-q)^2\right]. \tag{7.69}$$

Moreover,

$$x^2\psi = \psi\left[q^2 + 2q(x-q) + (x-q)^2\right]. \tag{7.70}$$

By (7.67), (7.69), (7.70), we obtain

$$i\hbar\dot{\psi} - H_\omega\psi = \psi\left\{-\left(\dot{S} + \frac{p^2}{2m} - p\dot{q} + \frac{1}{2}m\omega^2 q^2\right) - i\frac{\hbar}{2a}\left(\dot{a} - i\frac{b}{m}\right)\right.$$
$$+ \left[i\frac{b}{a}\left(\dot{q} - \frac{p}{m}\right) - (\dot{p} + m\omega^2 q)\right](x-q)$$
$$\left. + \frac{1}{2}\left[i\frac{b}{a^2}\left(\dot{a} - i\frac{b}{m}\right) - \frac{i}{a}(\dot{b} - i\,m\omega^2 a)\right](x-q)^2\right\}. \tag{7.71}$$

It is now easy to check that if (7.63), (7.64), (7.65) hold then the right-hand side of (7.71) vanishes and the proposition is proved. □

Remark 7.4

(i) Note that (7.63) is the classical equation of motion of the harmonic oscillator, so that the solution is a Gaussian centered on the classical motion, in agreement with the result of the previous section.

(ii) By (7.65), we have that $S(t) = L(q(0), p(0))\frac{\sin 2\omega t}{2\omega}$ and it coincides with the classical action computed along the classical motion (7.63).

(iii) The solution of Eq. (7.64) is

$$a(t) = a_0 \cos\omega t + i\frac{b_0}{m\omega}\sin\omega t, \qquad b(t) = i\,m\omega a_0 \sin\omega t + b_0 \cos\omega t. \tag{7.72}$$

Since $\Delta x(t) = \sqrt{\frac{\hbar}{2}}|a(t)|$, $\Delta p(t) = \sqrt{\frac{\hbar}{2}}|b(t)|$, we conclude that the mean square deviations of position and momentum are bounded and periodic functions of the time. If we choose an initial state with

$$a_0 = \frac{1}{\sqrt{m\omega}}, \qquad b_0 = \sqrt{m\omega}, \qquad\qquad (7.73)$$

i.e., the shifted ground state (see 7.42), then the solution reduces to

$$\psi(x,t) = e^{\frac{i}{\hbar}S(t)-\frac{i}{2}\omega t} \left(\frac{m\omega}{\pi\hbar}\right)^{1/4} e^{-\frac{m\omega}{2\hbar}(x-q(t))^2 + i\frac{p(t)}{\hbar}(x-q(t))}. \qquad (7.74)$$

Note that the evolution preserves the form of the initial state, the mean square deviations of position and momentum do not depend on time, and the minimal uncertainty condition is realized for any time t.

Exercise 7.3 *Let us consider a harmonic oscillator with frequency ω which is in the ground state at time $t = 0$. Assume that an external perturbation, acting only in the time interval $(0, T)$, $T > 0$, modifies the frequency from ω to ω'. Find the probability that the oscillator is in an excited state at $t > T$.*

7.4 Particle in a Constant Magnetic Field

We shall briefly discuss the Hamiltonian of a particle in a constant magnetic field. As we shall see, the analysis can be essentially reduced to the study of a harmonic oscillator [1, 3]. For simplicity, we consider a particle of mass m and charge e moving in the xy-plane subject to a constant magnetic field B directed along the z-axis of strength $|B| > 0$. In this case, a possible choice for the vector potential is

$$A = (-|B|\, y, 0, 0). \qquad\qquad (7.75)$$

The classical Hamiltonian reads (see the end of Sect. 1.1)

$$H_B^{cl} = \frac{1}{2m}\left(p_x + \frac{e|B|}{c}y\right)^2 + \frac{p_y^2}{2m} = \frac{p_x^2}{2m} + \frac{p_y^2}{2m} + \omega_B p_x y + \frac{1}{2}m\omega_B^2 y^2, \quad (7.76)$$

where ω_B is the frequency

$$\omega_B := \frac{e|B|}{mc}. \qquad\qquad (7.77)$$

The solution of Hamilton's equations is straightforward and one finds that the motion $t \rightarrow (x(t), y(t))$ of the particle is

$$x(t) = x_0 + \frac{p_{y_0}}{m\omega_B} + \left(y_0 + \frac{p_{x_0}}{m\omega_B}\right)\sin\omega_B t - \frac{p_{y_0}}{m\omega_B}\cos\omega_B t, \qquad (7.78)$$

$$y(t) = -\frac{p_{x_0}}{m\omega_B} + \left(y_0 + \frac{p_{x_0}}{m\omega_B}\right)\cos\omega_B t + \frac{p_{y_0}}{m\omega_B}\sin\omega_B t, \qquad (7.79)$$

where $(x_0, y_0, p_{x_0}, p_{y_0})$ are the initial conditions. We see that the motion is periodic with frequency ω_B and the trajectory is the circle with center and radius given by

$$(x_c, y_c) = \left(x_0 + \frac{p_{y_0}}{m\omega_B}, -\frac{p_{x_0}}{m\omega_B} \right), \qquad R = \sqrt{ \left(y_0 + \frac{p_{x_0}}{m\omega_B} \right)^2 + \left(\frac{p_{y_0}}{m\omega_B} \right)^2 }.$$

(7.80)

Let us consider the quantum case. We start with the symmetric operator in $L^2(\mathbb{R}^2)$

$$\dot{H}_B = \frac{1}{2m} \left(\frac{\hbar}{i} \frac{\partial}{\partial x} + \frac{e|B|}{c} y \right)^2 + \frac{1}{2m} \left(\frac{\hbar}{i} \frac{\partial}{\partial y} \right)^2, \qquad D(\dot{H}_B) = \mathscr{S}(\mathbb{R}^2). \quad (7.81)$$

We shall see that, by successive unitary transformations, the above operator is reduced to a multiplication operator. We first consider the Fourier transform with respect to the x variable

$$\mathscr{F}_x : L^2(\mathbb{R}^2) \to L^2(\mathbb{R}^2), \qquad (\mathscr{F}_x f)(k, y) = \frac{1}{\sqrt{2\pi}} \int dx \, e^{-ikx} f(x, y) \quad (7.82)$$

and define

$$\dot{H}_{B,1} := \mathscr{F}_x \dot{H}_B \mathscr{F}_x^{-1} = -\frac{\hbar^2}{2m} \frac{\partial^2}{\partial y^2} + \frac{1}{2} m \omega_B^2 \left(y + l^2 k \right)^2, \quad (7.83)$$

where l is the length

$$l := \sqrt{\frac{\hbar c}{e|B|}}. \quad (7.84)$$

Next, we consider the translation operator

$$T_k : L^2(\mathbb{R}^2) \to L^2(\mathbb{R}^2), \qquad (T_k f)(k, \xi) = f(k, \xi - l^2 k) \quad (7.85)$$

and define

$$\dot{H}_{B,2} := T_k \dot{H}_{B,1} T_k^{-1} = I \otimes \left(-\frac{\hbar^2}{2m} \frac{d^2}{d\xi^2} + \frac{1}{2} m \omega_B^2 \xi^2 \right), \quad (7.86)$$

i.e., $\dot{H}_{B,2}$ acts as the identity on the variable k and as the harmonic oscillator Hamiltonian on the variable ξ. Finally, we consider the unitary operator which diagonalizes the Hamiltonian of the harmonic oscillator (see 7.40)

$$U_{\omega_B} : L^2(\mathbb{R}^2) \to L^2(\mathbb{R}) \otimes l^2, \qquad (U_{\omega_B} f)(k, n) = \int d\xi \, \phi_n^B(\xi) f(k, \xi), \quad (7.87)$$

where n is a nonnegative integer and ϕ_n^B are the eigenfunction (7.39) which we rewrite for convenience

$$\phi_n^B(\xi) = \frac{1}{\sqrt{l}} \frac{1}{\sqrt{2^n \, n! \, \pi^{1/4}}} \, H_n(l^{-1}\xi) \, e^{-\frac{\xi^2}{2l^2}}. \tag{7.88}$$

So, we arrive at the multiplication operator

$$\dot{H}_{B,3} := U_{\omega_B} \dot{H}_{B,2} U_{\omega_B}^{-1} = I \otimes \hbar \omega_B \left(n + \frac{1}{2}\right). \tag{7.89}$$

Then, we construct the unique self-adjoint extension of $\dot{H}_{B,3}$

$$(H_{B,3} f)(k, n) = \hbar \omega_B \left(n + \frac{1}{2}\right) f(k, n), \tag{7.90}$$

$$D(H_{B,3}) = \left\{ f \in L^2(\mathbb{R}) \otimes l^2 \mid \int dk \sum_n n^2 |f(k, n)|^2 < \infty \right\}. \tag{7.91}$$

Using the unitary operator $V_B := U_{\omega_B} T_k \mathscr{F}_x$, where

$$V_B : L^2(\mathbb{R}^2) \to L^2(\mathbb{R}) \otimes l^2, \qquad (V_B f)(k, n) = \int dy \, \phi_n^B(y + l^2 k) \int dx \, \frac{e^{-ikx}}{\sqrt{2\pi}} \, f(x, y),$$
$$\tag{7.92}$$

we conclude that the Hamiltonian in the original Hilbert space describing the particle in a constant magnetic field is the self-adjoint operator

$$(H_B f)(x, y) = \left(V_B^{-1} H_{B,3} V_B f\right)(x, y), \tag{7.93}$$

$$D(H_B) = \left\{ f \in L^2(\mathbb{R}^2) \mid \int dk \sum_n n^2 |(V_B f)(k, n)|^2 < \infty \right\}. \tag{7.94}$$

Moreover, we have

$$\sigma(H_B) = \sigma_p(H_B) = \left\{ E_n^B = \hbar \omega_B \left(n + \frac{1}{2}\right), \ n = 0, 1, 2, \dots \right\}. \tag{7.95}$$

Observe that the spectrum is made of eigenvalues, called Landau levels, and it has the same structure of the spectrum of the harmonic oscillator. However, there is an important difference due to the fact that each Landau level has infinite degeneracy. Indeed, for a fixed Landau level $E_{n_0}^B$ one can easily check that

$$\Phi_{n_0}(x, y) = \int dk \, \frac{e^{ikx}}{\sqrt{2\pi}} \, \phi_{n_0}^B(y + l^2 k) g(k) \tag{7.96}$$

is an eigenvector for any choice of $g \in L^2(\mathbb{R})$. Thus, we have $\sigma(H_B) = \sigma_{ess}(H_B)$, $\sigma_d(H_B) = \emptyset$.

Exercise 7.4 *Verify that*

$$\psi_0(x, y) = \frac{1}{\sqrt{2\pi}\, l}\, e^{-i\frac{xy}{2l^2}}\, e^{-\frac{x^2+y^2}{4l^2}} \tag{7.97}$$

is a ground state for H_B, i.e., an eigenvector corresponding to the minimum eigenvalue E_0^B.

Remark 7.5 We recall that the vector potential A associated to a given magnetic field B is defined by the condition $B = \nabla \wedge A$ and, therefore, it is not uniquely defined. More precisely, if

$$A' = A + \nabla\gamma, \tag{7.98}$$

where γ is a smooth scalar function, then A and A' define the same magnetic field (gauge invariance). As a consequence, for the corresponding Hamiltonians we have

$$e^{i\frac{e}{\hbar c}\gamma} H(A)\, e^{-i\frac{e}{\hbar c}\gamma} = H(A'), \qquad H(A) := \frac{1}{2m}\left(\frac{\hbar}{i}\nabla - \frac{e}{c}A\right)^2, \tag{7.99}$$

i.e., the two Hamiltonians are unitarily equivalent under a multiplication operator. In particular, this means that they have the same spectrum and if ψ is eigenvector of $H(A)$ then $e^{i\frac{e}{\hbar c}\gamma}\psi$ is eigenvector of $H(A')$ corresponding to the same eigenvalue.

In the case of a constant magnetic field, we have chosen the vector potential (7.75). Another possible choice (see, e.g., [3]) is $A' = (-\frac{|B|}{2}y, \frac{|B|}{2}x, 0)$ and $A' = A + \nabla\gamma$, with $\gamma = \frac{|B|}{2}xy$. Note that $e^{-i\frac{e}{\hbar c}\gamma} = e^{-i\frac{xy}{2l^2}}$ is the phase factor in (7.97).

7.5   Completeness of the Hermite Polynomials

Here, we show that the orthonormal system $\{\phi_n\}$ defined in (7.39) is complete in $L^2(\mathbb{R})$. It is sufficient to prove that if $h \in L^2(\mathbb{R})$ satisfies

$$\int dx\, h(x)\, H_n(x)\, e^{-\frac{x^2}{2}} = 0, \qquad n = 0, 1, 2, \ldots \tag{7.100}$$

then h is the null vector. We observe that (7.100) implies

$$\int dx\, h(x)\, x^n\, e^{-\frac{x^2}{2}} = 0, \qquad n = 0, 1, 2, \ldots . \tag{7.101}$$

Let us consider the Fourier transform of $g(x) := h(x)\, e^{-\frac{x^2}{2}}$, where $g \in L^1(\mathbb{R}) \cap L^2(\mathbb{R})$,

$$\tilde{g}(k) = \frac{1}{\sqrt{2\pi}}\int dx\, h(x)\, e^{-\frac{x^2}{2}} e^{-ikx}. \tag{7.102}$$

Since

$$e^{-ikx} = \lim_N \sum_{n=0}^{N} \frac{(-ikx)^n}{n!} \tag{7.103}$$

and

$$\left| h(x) e^{-\frac{x^2}{2}} \sum_{n=0}^{N} \frac{(-ikx)^n}{n!} \right| \le |h(x)| e^{-\frac{x^2}{2}} \sum_{n=0}^{N} \frac{(|k||x|)^n}{n!} \le |h(x)| e^{-\frac{x^2}{2}} e^{|k||x|}, \tag{7.104}$$

we can apply the dominated convergence theorem and we obtain

$$\tilde{g}(k) = \frac{1}{\sqrt{2\pi}} \lim_N \sum_{n=0}^{N} \frac{(-ik)^n}{n!} \int dx \, h(x) \, x^n e^{-\frac{x^2}{2}}. \tag{7.105}$$

By (7.101), we find that $\tilde{g}$ is the null vector and this implies that also h is the null vector.

References

1. Landau, L.D., Lifshitz, E.M.: Quantum Mechanics, 3rd edn. Pergamon Press, Oxford (1977)
2. Teschl, G.: Mathematical Methods in Quantum Mechanics. American Mathematical Society, Providence (2009)
3. Thaller, B.: Visual Quantum Mechanics. Springer, New York (2000)
4. Lebedev, N.N.: Special Functions and Their Applications. Dover Publications, New York (1972)
5. Hagedorn, G.: Raising and lowering operators for semiclassical wave packets. Ann. Phys. **269**, 77–104 (1998)

Chapter 8
Point Interaction

8.1 Hamiltonian and Spectrum

In this chapter, we shall study the dynamics of a particle in dimension one subject to a point interaction (also called zero-range or delta interaction), i.e., to a potential having the form of a Dirac distribution.

The reason for this choice is that the corresponding Hamiltonian, apart for some difficulties arising in its definition as a self-adjoint operator, is relatively easy to treat and, on the other hand, it generates a nontrivial dynamics. More specifically, such a dynamics reproduces some relevant typical features of that generated by generic Hamiltonians with an interaction potential $V(x)$ such that $\lim_{|x| \to \infty} V(x) = 0$.

At a formal level, the Hamiltonian with a point interaction has the form

$$H = -\frac{\hbar^2}{2m}\Delta + \alpha\delta_0 , \tag{8.1}$$

where δ_0 is the Dirac distribution placed, for simplicity, in the origin and $\alpha \in \mathbb{R}$ is a coupling constant which represents the strength of the interaction.

From the physical point of view, such kind of Hamiltonians are introduced to describe situations where a particle is subject to a very intense force with a range much shorter than the wavelength associated to the particle. A typical example, studied by Fermi in 1936, is that of slow neutrons interacting with the nuclei of a system of atoms (see, e.g., [1], p. 640).

From the mathematical point of view, we note that δ_0 is not an operator in $L^2(\mathbb{R})$ and, therefore, the first problem is to give a meaning to the formal expression (8.1) as a self-adjoint operator in $L^2(\mathbb{R})$. The problem can be solved by using the theory of self-adjoint extensions of symmetric operators (see also [2] where a comprehensive mathematical treatment of Hamiltonians with point interactions is developed).

The first step is to give a reasonable definition of a Hamiltonian with a point interaction at the origin. We note that such a Hamiltonian should act as the free Hamiltonian on smooth functions vanishing at the origin. This fact suggests to consider the operator A_0, $D(A_0)$ in $L^2(\mathbb{R})$ defined as

© Springer International Publishing AG, part of Springer Nature 2018
A. Teta, *A Mathematical Primer on Quantum Mechanics*,
UNITEXT for Physics, https://doi.org/10.1007/978-3-319-77893-8_8

$$A_0 = -\frac{\hbar^2}{2m}\Delta, \qquad D(A_0) = \{u \in L^2(\mathbb{R}) \mid u \in H^2(\mathbb{R}),\ u(0) = 0\}. \qquad (8.2)$$

It is easy to verify that (8.2) is symmetric but not self-adjoint. Moreover, a self-adjoint extension of (8.2) is the free Hamiltonian H_0, $D(H_0)$. It is natural to think that any other possible self-adjoint extension of (8.2) defines a Hamiltonian with an interaction supported only at the origin. Thus, we arrive at the following.

Definition 8.1 A Hamiltonian with a point interaction at the origin in $L^2(\mathbb{R})$ is a self-adjoint extension of (8.2) different from H_0, $D(H_0)$.

The second step is the explicit construction of such self-adjoint extensions. The idea is the following: we compute the adjoint of (8.2), which is not symmetric, and we characterize all the self-adjoint restrictions of the adjoint. Each of such restrictions different from the free Hamiltonian is, by definition, a Hamiltonian with a point interaction.

In order to formulate the result, we denote by G^λ, $\lambda > 0$, the integral kernel of the free resolvent $R_0(-\lambda)$ see (6.27), i.e., the Green's function

$$G^\lambda(x) = \frac{\sqrt{2m}}{\hbar} \frac{e^{-\frac{\sqrt{2m}}{\hbar}\sqrt{\lambda}|x|}}{2\sqrt{\lambda}} \qquad (8.3)$$

and by $\tilde{G}^\lambda$ its Fourier transform

$$\tilde{G}^\lambda(k) = \frac{1}{\sqrt{2\pi}} \frac{1}{\frac{\hbar^2 k^2}{2m} + \lambda}. \qquad (8.4)$$

It is easy to verify that $G^\lambda \in H^1(\mathbb{R})$ and $G^\lambda \notin H^2(\mathbb{R})$. Then, we have the following proposition.

Proposition 8.1 *For any $\alpha \in \mathbb{R} \setminus \{0\}$, the Hamiltonian with point interaction at the origin with strength α is the self-adjoint operator*

$$D(H_\alpha) = \Big\{u \in L^2(\mathbb{R}) \mid u = w^\lambda + q\, G^\lambda,\ w^\lambda \in H^2(\mathbb{R}),\ q \in \mathbb{C},$$

$$w^\lambda(0) = -\big(\alpha^{-1} + G^\lambda(0)\big)q\Big\}, \qquad (8.5)$$

$$(H_\alpha + \lambda)u = (H_0 + \lambda)w^\lambda. \qquad (8.6)$$

Moreover, the resolvent $R_\alpha(z) = (H_\alpha - z)^{-1}$, with $\Im z \neq 0$, is

$$R_\alpha(z) = R_0(z) - \frac{\alpha}{1 + \alpha\, G^{-z}(0)}\left(\overline{G^{-z}},\ \cdot\ \right)G^{-z}, \qquad (8.7)$$

where G^{-z} is the integral kernel of $R_0(z) = (H_0 - z)^{-1}$ see (6.26).

The proof is postponed to Sect. 8.6. Here we only add some comments.

Remark 8.1 (i) The generic element of the domain (8.5) is written as the sum of a regular term $w^\lambda \in H^2(\mathbb{R})$ plus a singular one $q\, G^\lambda \in H^1(\mathbb{R})$. The last equation in (8.5) connects q and $w^\lambda(0)$ and it represents the boundary condition satisfied by an element u of the domain at the origin. Note that for any $u \in D(H_\alpha)$, one has

$$u(0) = w^\lambda(0) + q\, G^\lambda(0), \qquad u'(0^+) - u'(0^-) = -\frac{2m}{\hbar^2}\, q. \qquad (8.8)$$

Therefore, the boundary condition at the origin can be rewritten as

$$u'(0^+) - u'(0^-) = \frac{2m\alpha}{\hbar^2}\, u(0). \qquad (8.9)$$

(ii) For $u \in D(H_\alpha)$ with $u(0) = 0$, we have $q = 0$ and, consequently, $H_\alpha u = H_0 u$, i.e., $H_\alpha, D(H_\alpha)$ is a self-adjoint extension of (8.2).

(iii) For $\alpha = 0$, the operator $H_\alpha, D(H_\alpha)$ reduces to the free Hamiltonian.

(iv) If $u \in D(H_\alpha)$ then $u \in H^2(\mathbb{R} \setminus \{0\})$. Moreover, taking into account of (8.6) and the fact that $(H_0 + \lambda)G^\lambda = 0$ in $\mathbb{R} \setminus \{0\}$, we find

$$\int dx\, \phi(x)(H_\alpha u)(x) = \int dx\, \phi(x)(H_0 u)(x) \qquad \text{for any} \quad \phi \in C_0^\infty(\mathbb{R} \setminus \{0\}),$$

$$(8.10)$$

i.e., $H_\alpha u = H_0 u$ in $\mathbb{R} \setminus \{0\}$.

(v) Using point (iv) and an integration by parts, we have

$$(u, H_\alpha u) = \lim_{\varepsilon \to 0} \int_{|x| > \varepsilon} dx\, \overline{u(x)}(H_0 u)(x)$$

$$= \frac{\hbar^2}{2m} \int dx\, |u'(x)|^2 + \frac{\hbar^2}{2m} \lim_{\varepsilon \to 0} \left(\overline{u(\varepsilon)}\, u'(\varepsilon) - \overline{u(-\varepsilon)}\, u'(-\varepsilon) \right)$$

$$= \frac{\hbar^2}{2m} \int dx\, |u'(x)|^2 + \alpha |u(0)|^2. \qquad (8.11)$$

In the following remark, we mention a different approach to construct the Hamiltonian with a point interaction.

Remark 8.2 Another natural way to obtain the Hamiltonian $H_\alpha, D(H_\alpha)$ is based on an approximation procedure. The idea is to consider a sequence of Hamiltonians with a smooth interaction depending on a parameter $\varepsilon > 0$ such that for $\varepsilon \to 0$ the smooth interaction approximates the delta interaction in H_α. If the interaction has the form $V_\varepsilon(x) = \varepsilon^{-1} V(\varepsilon^{-1} x)$, one can prove that the Hamiltonian $H_0 + V_\varepsilon$ in $L^2(\mathbb{R})$ converges to H_α for $\varepsilon \to 0$, where $\alpha = \int dx\, V(x)$, in the resolvent sense, i.e.,

$$\lim_{\varepsilon \to 0} \| (H_0 + V_\varepsilon - z)^{-1} - (H_\alpha - z)^{-1} \| = 0, \qquad (8.12)$$

for some $z \in \rho(H_0 + V_\varepsilon) \cap \rho(H_\alpha)$ (for the proof we refer to [2], where also the more delicate extension to the case in dimension $d = 2, 3$ is discussed).

A simpler way to approximate a δ-interaction is provided by the so called separable potential. In this case, one considers the Hamiltonian in $L^2(\mathbb{R}^d)$, $d = 1, 2, 3$, i.e.,

$$H_{\alpha,\varepsilon} = H_0 + \alpha(g_\varepsilon, \cdot)g_\varepsilon, \qquad \alpha \in \mathbb{R}, \ \varepsilon > 0, \tag{8.13}$$

where $g_\varepsilon(x) = \varepsilon^{-d}g(\varepsilon^{-1}x)$ and $g \in L^2(\mathbb{R}^d) \cap L^1(\mathbb{R}^d)$ is a real, positive function with $\int dx\, g(x) = 1$. The reader can easily verify that $H_{\alpha,\varepsilon}$ is self-adjoint on $D(H_0)$.

Note that the separable potential $\psi \to \alpha(g_\varepsilon, \psi)g_\varepsilon$ defines a nonlocal interaction, in the sense that the value of the interaction in a point x depends on the values of the wave function ψ in the whole space.

Moreover, the parameter ε gives a measure of the range of the interaction and one has $\lim_{\varepsilon \to 0} g_\varepsilon = \delta_0$ in distributional sense. Therefore, it is reasonable to expect that $H_{\alpha,\varepsilon}$ reduces to H_α for $\varepsilon \to 0$.

Exercise 8.1 Compute the resolvent of the operator (8.13).

Exercise 8.2 Let us consider the case $d = 1$ and take $\lambda > 0$ sufficiently large. Verify that $\lim_{\varepsilon \to 0} \|(H_{\alpha,\varepsilon} + \lambda)^{-1} - (H_\alpha + \lambda)^{-1}\| = 0$. In the case $d = 3$, show that $\lim_{\varepsilon \to 0} \|(H_{\alpha,\varepsilon} + \lambda)^{-1} - (H_0 + \lambda)^{-1}\| = 0$.

In the next proposition, we characterize the spectrum of the Hamiltonian H_α, $D(H_\alpha)$.

Proposition 8.2 *For $\alpha > 0$ ("repulsive" point interaction)*

$$\sigma(H_\alpha) = \sigma_c(H_\alpha) = [0, +\infty) \tag{8.14}$$

and for $\alpha < 0$ ("attractive" point interaction)

$$\sigma(H_\alpha) = \sigma_p(H_\alpha) \cup \sigma_c(H_\alpha) = \{E_0\} \cup [0, +\infty), \tag{8.15}$$

where

$$E_0 = -\frac{m\alpha^2}{2\hbar^2} \tag{8.16}$$

and the only normalized eigenvector corresponding to the eigenvalue $E_0 = -\lambda_0$ is

$$\Xi_0(x) = \frac{\sqrt{m|\alpha|}}{\hbar} e^{-\frac{m|\alpha|}{\hbar^2}|x|} = q_0\, G^{\lambda_0}(x), \qquad q_0 = \frac{\sqrt{m|\alpha|^3}}{\hbar}. \tag{8.17}$$

Proof By (8.7) one sees that the difference between the resolvent of H_α and the free resolvent is a rank one operator. Then, using Weyl's Theorem 4.30, we find

$$\sigma_{ess}(H_\alpha) = \sigma_{ess}(H_0) = [0, +\infty). \tag{8.18}$$

Let us show that the Hamiltonian does not have nonpositive eigenvalues. Let us suppose that there exist $E > 0$ and $\psi \in D(H_\alpha)$, with $\psi \neq 0$, such that $H_\alpha \psi = E \psi$. Then ψ satisfies

$$- \psi''(x) = \frac{2mE}{\hbar^2} \, \psi(x), \qquad \text{for } x \neq 0. \tag{8.19}$$

On the other hand, any solution of Eq. (8.19) for $x > 0$ and for $x < 0$ is a linear combination of the exponentials $e^{i \frac{\sqrt{2mE}}{\hbar} x}$ and $e^{-i \frac{\sqrt{2mE}}{\hbar} x}$ and, therefore, ψ cannot belong to $L^2(\mathbb{R})$, which is absurd. Analogously, one shows the $E = 0$ is not an eigenvalue. Thus, we conclude that there are no eigenvalues in $[0, +\infty)$.

Let us consider $\alpha > 0$. From (8.11), it follows that there are no negative eigenvalues, so (8.14) is proved.

Let us consider the eigenvalue problem for $\alpha < 0$

$$H_\alpha \psi = E \psi, \qquad E < 0, \qquad \psi \in D(H_\alpha), \qquad \|\psi\| = 1. \tag{8.20}$$

Denoted $|E| = \lambda$, we write $\psi = w^\lambda + q G^\lambda$ and we have $0 = (H_\alpha + \lambda)\psi = (H_0 + \lambda)w^\lambda$, which means $w^\lambda = 0$. Then, the solution has the form $\psi = q G^\lambda$, with $0 = -(\alpha^{-1} + G^\lambda(0))q$. A nonzero solution is obtained if and only if $G^\lambda(0) = -\alpha^{-1}$, i.e., for $\lambda = m\alpha^2/2\hbar^2$. Finally, the constant q is determined imposing the normalization condition.

$\square$

Exercise 8.3 Characterize the spectrum of the Hamiltonian (8.13). For $d = 1$ and $\alpha < 0$, verify the convergence of the eigenvalue and the eigenfunction for $\varepsilon \to 0$.

Exercise 8.4 Let us consider the Hamiltonian in $L^2(\mathbb{R})$

$$H = H_0 - V_0 \, \chi_{(0,a)}, \tag{8.21}$$

where $V_0, a > 0$ and $\chi_{(0,a)}$ is the characteristic function of the interval $(0, a)$. Compute the negative eigenvalues and the corresponding eigenfunctions. Study the limit $a \to 0$, $V_0 \to \infty$, with aV_0 constant.

8.2 Eigenfunction Expansion

In the previous section, we have seen that the equation

$$H_\alpha \phi = E \phi, \qquad E \geq 0 \tag{8.22}$$

does not have solutions in $L^2(\mathbb{R})$. On the other hand, the equation has a family of bounded solutions, called generalized eigenfunctions, which play an important role in the analysis of the dynamics. We encountered a similar problem in the study of the free Hamiltonian H_0 in Chap. 6. In particular, in Remark 6.1, we explained the role

of the plane waves $e^{ik \cdot x}$ as generalized eigenfunctions of H_0. Here, we generalize the
result to the case of the Hamiltonian H_α. For the construction of the solutions, we
proceed heuristically, starting from the formal Hamiltonian (8.1). We represent the
solution as a perturbation of the plane waves, i.e., $\phi(x, k) = e^{ikx} + \eta(x, k)$, so that
the equation for η is

$$(-\Delta - k^2)\eta + \alpha_0 \delta_0 \eta = -\alpha_0 \delta_0 e^{ikx}, \qquad (8.23)$$

where

$$\alpha_0 := \frac{2m\alpha}{\hbar^2}, \qquad k^2 := \frac{2mE}{\hbar^2}. \qquad (8.24)$$

It is convenient to rewrite (8.23) as an integral equation applying the inverse of
$(-\Delta - k^2)$. As we know, this is a delicate operation since k^2 is a point of the spectrum
of $-\Delta$. Taking into account of the computation done in Sect. 6.1, we replace k^2 by
$k^2 \pm i\varepsilon$, $\varepsilon > 0$, and we have

$$\eta + \left(-\Delta - (k^2 \pm i\varepsilon)\right)^{-1} (\alpha_0 \delta_0 \eta) = -\left(-\Delta - (k^2 \pm i\varepsilon)\right)^{-1} (\alpha_0 \delta_0 e^{ik \cdot}). \qquad (8.25)$$

Now we take the limit $\varepsilon \to 0$ and use (6.28). We obtain two equations

$$\eta_\mp(x, k) \mp \alpha_0 \frac{e^{\pm i|k||x|}}{2i|k|} \eta_\mp(0, k) = \pm \alpha_0 \frac{e^{\pm i|k||x|}}{2i|k|}. \qquad (8.26)$$

Evaluating (8.26) in $x = 0$ we find $\eta_\mp(0, k) = \pm \alpha_0 (2i|k| \mp \alpha_0)^{-1}$ and then, by
(8.26), we also compute $\eta_\pm(x, k)$. We thus obtain the following generalized eigenfunctions:

$$\phi_+(x, k) = e^{ikx} + \mathscr{R}(k)e^{-i|k||x|}, \qquad \phi_-(x, k) = e^{ikx} + \overline{\mathscr{R}(k)}e^{i|k||x|}, \qquad (8.27)$$

where

$$\mathscr{R}(k) = -\frac{\alpha_0}{\alpha_0 + 2i|k|}. \qquad (8.28)$$

Remark 8.3 Note that $\phi_-(x, k) = \overline{\phi_+(x, -k)}$. Moreover, for any $k \in \mathbb{R}$ the functions (8.27) are bounded and continuous. For $x \neq 0$, they are also differentiable
infinitely many times and satisfy

$$-\Delta\phi_\pm(x, k) = k^2 \phi_\pm(x, k), \qquad x \neq 0, \qquad (8.29)$$

$$\phi'_\pm(0^+, k) - \phi'_\pm(0^-, k) = \frac{2m\alpha}{\hbar^2} \phi_\pm(0, k). \qquad (8.30)$$

Using the generalized eigenfunctions (8.27), we can formulate the eigenfunction
expansion theorem for the Hamiltonian H_α, in the two cases $\alpha > 0$ and $\alpha < 0$.

Proposition 8.3

(i) *For any $f \in L^2(\mathbb{R})$, the generalized transforms of f*

$$\hat{f}_{\pm}(k) := \frac{1}{\sqrt{2\pi}} \lim_{N\to\infty} \int_{|x|<N} dx \, \overline{\phi_{\pm}(x,k)} f(x) \qquad (8.31)$$

exist and belong to $L^2(\mathbb{R})$, where the limits in (8.31) (and the limits below) hold in the L^2-norm. Thus, we define

$$W_{\pm} : L^2(\mathbb{R}) \to L^2(\mathbb{R}), \qquad W_{\pm}f = \hat{f}_{\pm}. \qquad (8.32)$$

(ii) *For $\alpha > 0$ the operators $W_{\pm}$ are unitary with inverse*

$$\left(W_{\pm}^{-1}\hat{g}\right)(x) = \frac{1}{\sqrt{2\pi}} \lim_{N\to\infty} \int_{|k|<N} dk \, \hat{g}(k)\phi_{\pm}(x,k), \qquad \hat{g} \in L^2(\mathbb{R}). \quad (8.33)$$

(iii) *For $\alpha < 0$ the operators*

$$U_{\pm} : L^2(\mathbb{R}) \to L^2(\mathbb{R}) \oplus \mathbb{C}, \qquad U_{\pm}f = \{\hat{f}_{\pm}, \hat{f}_0\}, \qquad \hat{f}_0 = (\Xi_0, f) \quad (8.34)$$

are unitary with inverse

$$(U_{\pm}^{-1}\{\hat{g}, \hat{g}_0\})(x) = \frac{1}{\sqrt{2\pi}} \lim_{N\to\infty} \int_{|k|<N} dk \, \hat{g}(k)\phi_{\pm}(x,k) + \hat{g}_0 \Xi_0(x), \qquad \hat{g} \in L^2(\mathbb{R}), \quad \hat{g}_0 \in \mathbb{C}.$$
$$(8.35)$$

(iv) *if $u \in D(H_{\alpha})$ then $\int dk \, |k^2\hat{u}_{\pm}(k)|^2 < \infty$ and*

$$(H_{\alpha}u)(x) = \frac{1}{\sqrt{2\pi}} \lim_{N\to\infty} \int_{|k|<N} dk \, \frac{\hbar^2 k^2}{2m} \hat{u}_{\pm}(k) \, \phi_{\pm}(x,k) \qquad \text{if } \alpha > 0, \qquad (8.36)$$

$$(H_{\alpha}u)(x) = \frac{1}{\sqrt{2\pi}} \lim_{N\to\infty} \int_{|k|<N} dk \, \frac{\hbar^2 k^2}{2m} \hat{u}_{\pm}(k) \, \phi_{\pm}(x,k) + E_0 \, \hat{u}_0 \Xi_0(x) \qquad \text{if } \alpha < 0.$$
$$(8.37)$$

The proof is discussed in Sect. 8.7.

Remark 8.4 For $\alpha = 0$, formula (8.31) reduces to the definition of the Fourier transform and this is the reason why $\hat{f}_{\pm}$ are called generalized transforms of f. Moreover, by unitarity, we have

$$\|f\|^2 = \|W_{\pm}f\|^2 = \int dk \, |\hat{f}_{\pm}(k)|^2 \qquad \text{if } \alpha > 0, \qquad (8.38)$$

$$\|f\|^2 = \|U_{\pm}f\|^2 = \int dk \, |\hat{f}_{\pm}(k)|^2 + |\hat{f}_0|^2 \qquad \text{if } \alpha < 0. \qquad (8.39)$$

Note that for $\alpha < 0$, by (8.39), we have $W_\pm \varXi_0 = 0$.

Exercise 8.5 Construct the spectral family and the spectral measure associated to H_α. Verify that $\sigma_{sc}(H_\alpha) = \emptyset$ and therefore

$$\mathscr{H}_{ac}(H_\alpha) = L^2(\mathbb{R}) \qquad\qquad \text{if } \alpha > 0, \qquad (8.40)$$

$$\mathscr{H}_p(H_\alpha) = [\varXi_0], \qquad \mathscr{H}_{ac}(H_\alpha) = [\varXi_0]^\perp \qquad \text{if } \alpha < 0, \qquad (8.41)$$

where $[\varXi_0]$ is the one-dimensional subspace generated by $\varXi_0$.

Exercise 8.6 Compute the generalized eigenfunctions of the Hamiltonian (8.13) in $L^2(\mathbb{R}^d)$, $d = 1, 2, 3$ and formulate the corresponding eigenfunction expansion theorem.

Remark 8.5 The validity of the results of this section can be extended to a generic Hamiltonian $H = H_0 + V$ in $L^2(\mathbb{R}^d)$, $d = 1, 2, 3$, for a large class of interaction potential V such that $V(x) \to 0$ for $|x| \to \infty$. In particular, it can be proved that the existence of the generalized eigenfunctions and the corresponding eigenfunction expansion theorem. The proof requires a nontrivial mathematical work and we refer to [3–5] for the details.

8.3 Asymptotic Evolution for $|t| \to \infty$

The eigenfunction expansion theorem allows to give an explicit representation of the time evolution $e^{-i\frac{t}{\hbar}H_\alpha}\psi_0$ of an arbitrary initial state $\psi_0 \in L^2(\mathbb{R})$. Indeed, for $\alpha > 0$, one has

$$\psi_t(x) = \left(e^{-i\frac{t}{\hbar}H_\alpha}\psi_0 \right)(x) = \frac{1}{\sqrt{2\pi}} \int dk\, e^{-it\frac{\hbar k^2}{2m}} \hat{\psi}_{0,+}(k)\phi_+(x,k), \quad (8.42)$$

where $\hat{\psi}_{0,+} = W_+\psi_0$. In this section, we characterize the asymptotic behavior for t large of ψ_t. Using the expression for ϕ_+ and the representation of the free evolution, we have

$$\psi_t(x) = \frac{1}{\sqrt{2\pi}} \int dk\, e^{-it\frac{\hbar k^2}{2m}+ikx} \hat{\psi}_{0,+}(k) + \frac{1}{\sqrt{2\pi}} \int dk\, e^{-it\frac{\hbar k^2}{2m}-i|k||x|} \mathscr{R}(k)\hat{\psi}_{0,+}(k)$$

$$= \left(e^{-i\frac{t}{\hbar}H_0}\mathscr{F}^{-1}W_+\psi_0 \right)(x) + \frac{1}{\sqrt{2\pi}} \int dk\, e^{-it\frac{\hbar k^2}{2m}-i|k||x|} \mathscr{R}(k)\hat{\psi}_{0,+}(k). (8.43)$$

In the next proposition, we prove that the last term in (8.43) is negligible for $t \to +\infty$.

Proposition 8.4 *Let $\alpha > 0$ and $\psi_0 \in L^2(\mathbb{R})$. Then*

$$\lim_{t\to+\infty} \left\| e^{-i\frac{t}{\hbar}H_\alpha}\psi_0 - e^{-i\frac{t}{\hbar}H_0}\mathscr{F}^{-1}W_+\psi_0 \right\| = 0. \qquad (8.44)$$

Proof Let us consider a smooth initial datum $f \in \mathscr{S}(\mathbb{R})$ and let us denote

$$B_f(\cdot, t) := e^{-i\frac{t}{\hbar}H_\alpha} f - e^{-i\frac{t}{\hbar}H_0} \mathscr{F}^{-1} W_+ f. \tag{8.45}$$

As a first step, we prove that $\|B_f(\cdot, t)\|$ converges to zero for $t \to +\infty$. Using (8.43), we write

$$
\begin{aligned}
B_f(x, t) &= \frac{1}{\sqrt{2\pi}} \int_0^\infty dk \, e^{-i\left(t\frac{\hbar k^2}{2m} + k|x|\right)} \mathscr{R}(k) \left(\hat{f}_+(k) + \hat{f}_+(-k)\right) \\
&= \int_0^\infty dk \, e^{-i\left(t\frac{\hbar k^2}{2m} + k|x|\right)} g(k),
\end{aligned}
\tag{8.46}
$$

where $g : [0, \infty) \to \mathbb{C}$ is

$$g(k) := \frac{1}{\pi} \mathscr{R}(k) \int dy \, f(y) \cos ky + \frac{1}{\pi} |\mathscr{R}(k)|^2 \int dy \, f(y) e^{ik|y|}. \tag{8.47}$$

As an exercise, the reader should verify that the following properties hold:

(p_1) g is (at least) twice differentiable;
(p_2) $g(0) = 0$;
(p_3) $\lim_{k\to\infty} g(k) = 0$;
(p_4) g and g' belong to $L^1(\mathbb{R})$.

Making use of these properties, we shall obtain an estimate of $\|B_f(\cdot, t)\|$. Note that $B_f(x, t)$ is represented by an integral containing an exponential term with a rapidly oscillating phase for $t \to \infty$. The standard method to take into account of these oscillations and to show that $|B_f(x, t)|$ goes to zero for $t \to +\infty$ is based on an integration by parts. Following this line, we write

$$
\begin{aligned}
B_f(x, t) &= \int_0^\infty dk \left(\frac{d}{dk} e^{-i\left(\frac{t\hbar}{2m}k^2 + k|x|\right)}\right) \frac{g(k)}{-i\left(\frac{t\hbar}{m}k + |x|\right)} \\
&= -i \int_0^\infty dk \, e^{-i\left(\frac{t\hbar}{2m}k^2 + k|x|\right)} \left(\frac{d}{dk} \frac{g(k)}{\frac{t\hbar}{m}k + |x|}\right),
\end{aligned}
\tag{8.48}
$$

where the boundary terms of the integration by parts are zero due to (p_2), (p_3). Therefore,

$$
\begin{aligned}
\int dx \, |B_f(x, t)|^2 &\leq \int dx \left(\int_0^\infty dk \left|\frac{d}{dk} \frac{g(k)}{\frac{t\hbar}{m}k + |x|}\right|\right)^2 = \frac{2m}{\hbar t} \int_0^\infty dz \left(\int_0^\infty dk \left|\frac{d}{dk} \frac{g(k)}{k + z}\right|\right)^2 \\
&= \frac{2m}{\hbar t} \int_0^\infty dz \frac{1}{(1+z)^2} \left(\int_0^\infty dk \, (1 + z) \left|\frac{d}{dk} \frac{g(k)}{k + z}\right|\right)^2 \\
&= \frac{2m}{\hbar t} \int_0^\infty dz \frac{1}{(1+z)^2} \left(\int_0^\infty dk \frac{1+z}{(k+z)^2} |(k + z)g'(k) - g(k)|\right)^2. \tag{8.49}
\end{aligned}
$$

Let us show that the last integral in the variable k in (B.31) is estimated by a constant independent of z. We separate the integration region in the two regions:
(a) $0 \leq k \leq 1$,   (b) $k > 1$.
In case (a), taking into account of (p_1), (p_2) and using Taylor's formula, we have

$$|(k+z)g'(k) - g(k)| \leq z|g'(0)| + k(z+2k) \sup_{k \in [0,1]} |g''(k)|. \tag{8.50}$$

Using the inequality $\frac{1+z}{k+z} \leq \frac{1}{k}$, for $0 < k \leq 1$, we find

$$\int_0^1 dk \, \frac{1+z}{(k+z)^2} |(k+z)g'(k) - g(k)|$$

$$\leq z(1+z)|g'(0)| \int_0^1 dk \, \frac{1}{(z+k)^2} + \sup_{k \in [0,1]} |g''(k)| \int_0^1 dk \, \frac{z+2k}{z+k}$$

$$\leq |g'(0)| + 2 \sup_{k \in [0,1]} |g''(k)|. \tag{8.51}$$

In case (b), using the inequality $\frac{1+z}{k+z} \leq 1$ for $k \geq 1$, we obtain

$$\int_1^\infty dk \, \frac{1+z}{(z+k)^2} |g'(k)(k+z) - g(k)| \leq \int_1^\infty dk \left(|g'(k)| + |g(k)| \right), \tag{8.52}$$

where the last integral is finite due to (p_4). The estimates for the cases (a) and (b) and (B.31) imply

$$\|B_f(\cdot, t)\| \leq \frac{c_f}{\sqrt{t}}, \tag{8.53}$$

where c_f is a positive constant depending on f. Now we fix an initial datum $\psi_0 \in L^2(\mathbb{R})$. For any $\varepsilon > 0$, we choose $f \in \mathscr{S}(\mathbb{R})$ such that $\|\psi_0 - f\| \leq \frac{\varepsilon}{3}$. Moreover, we denote $t_0 = 9 \, c_f^2 \, \varepsilon^{-2}$. Using (8.53), for $t > t_0$, we have

$$\left\| e^{-i\frac{t}{\hbar} H_\alpha} \psi_0 - e^{-i\frac{t}{\hbar} H_0} \mathscr{F}^{-1} W_+ \psi_0 \right\|$$

$$\leq \left\| e^{-i\frac{t}{\hbar} H_\alpha} (\psi_0 - f) \right\| + \left\| e^{-i\frac{t}{\hbar} H_0} \mathscr{F}^{-1} W_+ (f - \psi_0) \right\| + \|B_f(\cdot, t)\|$$

$$\leq 2\|\psi_0 - f\| + \frac{c_f}{\sqrt{t}} \leq \varepsilon \tag{8.54}$$

and the proposition is proved.

$\square$

Exercise 8.7 Verify that

$$\lim_{t \to -\infty} \left\| e^{-i\frac{t}{\hbar} H_\alpha} \psi_0 - e^{-i\frac{t}{\hbar} H_0} \mathscr{F}^{-1} W_- \psi_0 \right\| = 0. \tag{8.55}$$

An immediate consequence of (8.44), (8.55) (and of 6.55) is that all the states are scattering states for H_α, with $\alpha > 0$, i.e., for any $\psi_0 \in L^2(\mathbb{R})$ and for any $R > 0$

$$\lim_{t \to \pm\infty} \int_{|x|<R} dx \left| \left(e^{-i\frac{t}{\hbar}H_\alpha} \psi_0 \right)(x) \right|^2 = 0. \tag{8.56}$$

Therefore, according to definition 5.39, for $\alpha > 0$, we have $\mathcal{M}_\infty(H_\alpha) = L^2(\mathbb{R})$. Le us consider the case $\alpha < 0$. For any initial state $\psi_0 \in L^2(\mathbb{R})$, the solution of the Schrödinger equation reads

$$\left(e^{-i\frac{t}{\hbar}H_\alpha} \psi_0 \right)(x) = \frac{1}{\sqrt{2\pi}} \int dk\, e^{-it\frac{\hbar k^2}{2m}} \hat{\psi}_{0,+}(k) \phi_+(x,k) + e^{-i\frac{t}{\hbar}E_0} \hat{\psi}_{0,0} \Xi_0(x),$$
$$\tag{8.57}$$

where $\hat{\psi}_{0,+} = W_+\psi_0$ and $\hat{\psi}_{0,0} = (\Xi_0, \psi_0)$. Note that, due to the presence of the last term in (8.57), we cannot expect the asymptotic behavior expressed in Proposition 8.4 for any initial state. More precisely, the result holds only for initial states belonging to $[\Xi_0]^\perp$. Thus, according to Definitions 5.38, 5.39 for $\alpha < 0$, we have $\mathcal{M}_0(H_\alpha) = [\Xi_0]$, $\mathcal{M}_\infty(H_\alpha) = [\Xi_0]^\perp$.

Remark 8.6 The above results on the asymptotic behavior of the time evolution $e^{-i\frac{t}{\hbar}H_\alpha}$ for $|t|$ large are essentially based on the eigenfunction expansion theorem. Therefore, they can be extended to the time evolution generated by the generic Hamiltonians mentioned in Remark 8.5 (see [3–5]).

Exercise 8.8 Let us consider a particle described by the Hamiltonian H_α, $\alpha < 0$, which is in the bound state Ξ_0 at time $t = 0$. Assume that an external perturbation has the effect to cancel the interaction in the time interval $(0, T)$, $T > 0$, so that the Hamiltonian is H_0 for $(0, T)$ and H_α for $t > T$. Denoted by ψ_t the state of the particle at time t, compute the ionization probability $\mathcal{P}_i(t) := 1 - |(\Xi_0, \psi_t)|^2$ for $t > T$. Characterize the behavior of $\mathcal{P}_i(t)$ for $T \to 0$ and for $T \to \infty$.

8.4   A One-Dimensional Scattering Problem

In its simplest form, a quantum scattering problem in dimension one can be formulated is as follows (see also [6]). We consider a particle described by a Hamiltonian $H = H_0 + V$, with $V(x) \to 0$ for $|x| \to \infty$. In this context, the interaction potential V modelizes the effect of a target fixed in the laboratory around the origin. We assume that the initial state ψ_0 of the particle is well localized at a large distance on the left with a positive mean momentum. For t small, the particle moves toward the target in a region where V is negligible and therefore the evolution of the state ψ_t is approximately described by the free Hamiltonian H_0. As time goes by, the particle approaches the target and "feels" the potential V. Therefore, the evolution of ψ_t is described by the Hamiltonian H and this fact remains true for all the times in which the particle moves in the vicinity of the target. On the other hand, for t sufficiently large we expect that the particle moves away from the target and therefore the evolution of the state is again approximately described by a free evolution. Our scattering problem consists in the following: given the initial state ψ_0 of the type described above, compute the asymptotic free evolution for $t \to +\infty$ of the state ψ_t. One typically finds that ψ_t, for t large, is the sum of a transmitted wave (i.e., a wave moving to the right with positive mean momentum) and of a reflected wave (i.e., a wave moving to the left with negative mean momentum). As a consequence, one obtains a transmission probability and a reflection probability for the particle which are both different from zero. Of course, the case of an initial state well localized on the right with negative mean momentum is treated in a completely analogous way.

We underline that the above quantum description is in contrast with the corresponding classical scattering problem in dimension one, where the particle can only be reflected with probability one or transmitted with probability one.

Here, we describe in detail the quantum scattering problem in the case of an interaction potential given by $\alpha\delta_0$, $\alpha > 0$. In this case, the analysis is simplified due to the knowledge of the explicit form of the generalized eigenfunctions. Nevertheless, the result that we will find reproduces the qualitative behavior of the solution of the scattering problem in the case of a generic interaction potential (for an introduction to the general formulation of scattering theory see Appendix B).

Let us fix the initial state

$$\psi_0(x) = \frac{1}{\sqrt{\sigma}\,\pi^{1/4}}\, e^{-\frac{(x+x_0)^2}{2\sigma^2} + i\frac{p_0}{\hbar}x}\,, \qquad \sigma, x_0, p_0 > 0\,, \qquad (8.58)$$

with Fourier transform

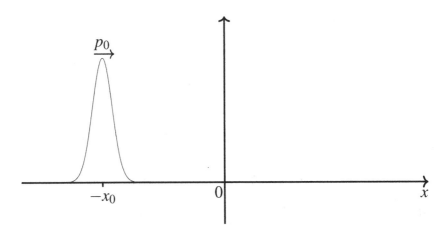

Fig. 8.1 The incident wave

$$\tilde{\psi}_0(k) = \frac{\sqrt{\sigma}}{\pi^{1/4}} e^{ix_0(k-k_0)} e^{-\frac{\sigma^2}{2}(k-k_0)^2}, \qquad k_0 := \frac{p_0}{\hbar}. \qquad (8.59)$$

The initial state ψ_0 represents our incident wave and, following the formulation of the scattering problem, it must describe a particle localized in position on the left of the origin with a positive momentum (Fig. 8.1). In other words, the probability that the particle has a position in $(0, +\infty)$ and a momentum in $(-\infty, 0)$ must be negligible. According to Born's rule, this means that the two quantities

$$\int_0^\infty dx\, |\psi_0(x)|^2 = \frac{1}{\sqrt{\pi}} \int_{\frac{x_0}{\sigma}}^\infty dy\, e^{-y^2}, \qquad \int_{-\infty}^0 dk\, |\tilde{\psi}_0(k)|^2 = \frac{1}{\sqrt{\pi}} \int_{\sigma k_0}^\infty dq\, e^{-q^2} (8.60)$$

must be small. Therefore, we introduce the following further assumptions on the initial state:

$$\delta := \frac{x_0}{\sigma} \gg 1, \qquad \eta := \frac{\sigma p_0}{\hbar} \gg 1. \qquad (8.61)$$

Note that, integrating by parts, we easily see that the integral

$$\frac{1}{\sqrt{\pi}} \int_u^\infty dz\, e^{-z^2} = \frac{e^{-u^2}}{2\sqrt{\pi}\, u} - \frac{1}{2\sqrt{\pi}} \int_u^\infty dz\, \frac{e^{-z^2}}{z^2} < \frac{e^{-u^2}}{2\sqrt{\pi}\, u} \qquad (8.62)$$

is exponentially small for $u \to \infty$. So, the quantities in (8.60) are exponentially small for $\delta, \eta \to \infty$.

Our goal is to characterize the behavior of $e^{-i\frac{t}{\hbar}H_\alpha}\psi_0$ for $t \to +\infty$ by neglecting exponentially small quantities for δ, η large.

As a first step, we use propositions 8.4 and 6.2 and we obtain

$$\left(e^{-i\frac{t}{\hbar}H_\alpha}\psi_0\right)(x) = \left(\frac{m}{i\hbar t}\right)^{1/2} e^{i\frac{m}{2\hbar t}x^2} \left(W_+\psi_0\right)\left(\frac{mx}{\hbar t}\right) + \mathscr{E}_t(x), \qquad (8.63)$$

where $\|\mathscr{E}_t\| \to 0$ for $t \to +\infty$. The second step is to characterize the asymptotic behavior of the generalized transform $W_+\psi_0$ for δ, η large. Let us denote by $\chi_{<0}$ (resp. $\chi_{>0}$) the characteristic function of the interval $(-\infty, 0)$ (resp. $(0, +\infty)$) and define

$$\mathscr{T}(k) := 1 + \mathscr{R}(k) = \frac{2i|k|}{\alpha_0 + 2i|k|}. \qquad (8.64)$$

Then we have

Proposition 8.5

$$\left(W_+\psi_0\right)(k) = \left(W_+^a\psi_0\right)(k) + \mathscr{E}_\delta(k) + \mathscr{E}_\eta(k), \qquad (8.65)$$

where

$$\left(W_+^a\psi_0\right)(k) = \overline{\mathscr{T}(k)}\, \tilde{\psi}_0(k)\, \chi_{>0}(k) + \overline{\mathscr{R}(k)}\, \tilde{\psi}_0(-k)\, \chi_{<0}(k) \qquad (8.66)$$

and the error terms in (8.65) *satisfy the estimates*

$$\|\mathscr{E}_\delta\| < \frac{\sqrt{\sigma\alpha_0}}{\pi^{1/4}} \frac{e^{-\delta^2/2}}{\delta}, \qquad \|\mathscr{E}_\eta\| < \frac{1}{\sqrt{2}\pi^{1/4}} \frac{e^{-\eta^2/2}}{\sqrt{\eta}}. \tag{8.67}$$

Proof By definition of generalized transform, we have

$$\left(W_+\psi_0\right)(k) = \tilde{\psi}_0(k) + \frac{\overline{\mathscr{R}(k)}}{\sqrt{2\pi}} \int dy\, \psi_0(y)\, e^{i|k||y|}. \tag{8.68}$$

We write the Fourier transform of ψ_0 as

$$\tilde{\psi}_0(k) = \tilde{\psi}_0(k)\chi_{>0}(k) + \mathscr{E}_\eta(k) \tag{8.69}$$

and, using (8.62), we have

$$\|\mathscr{E}_\eta\|^2 = \frac{\sigma}{\sqrt{\pi}} \int_{-\infty}^{0} dk\, e^{-\sigma^2(k-k_0)^2} = \frac{1}{\sqrt{\pi}} \int_{\eta}^{\infty} dz\, e^{-z^2} < \frac{1}{2\sqrt{\pi}} \frac{e^{-\eta^2}}{\eta}. \tag{8.70}$$

Let us write the last integral in (8.68) as the sum $J_1 + J_2 + J_3$, where

$$J_1(k) = \sqrt{2\pi}\, \tilde{\psi}_0(|k|), \tag{8.71}$$

$$J_2(k) = \frac{\sqrt{\sigma}}{\pi^{1/4}} e^{-ix_0(|k|+k_0)} \int_{\delta}^{\infty} dz\, e^{-\frac{z^2}{2}+i\sigma(|k|+k_0)z}, \tag{8.72}$$

$$J_3(k) = -\frac{\sqrt{\sigma}}{\pi^{1/4}} e^{ix_0(|k|-k_0)} \int_{\delta}^{\infty} dz\, e^{-\frac{z^2}{2}-i\sigma(|k|-k_0)z}. \tag{8.73}$$

Note that J_2 and J_3 can be estimated as in (8.70). Therefore, we find

$$\frac{\overline{\mathscr{R}(k)}}{\sqrt{2\pi}} \int dy\, \psi_0(y)\, e^{i|k||y|} = \overline{\mathscr{R}(k)}\, \tilde{\psi}_0(|k|) + \mathscr{E}_\delta(k). \tag{8.74}$$

Using (8.69) and (8.74) in (8.68), we conclude the proof.

$\square$

We are now in position to characterize the asymptotic behavior of $e^{-i\frac{t}{\hbar}H_\alpha}\psi_0$. Neglecting small error terms for δ, η large, from (8.63) and Proposition 8.5 we have

$$e^{-i\frac{t}{\hbar}H_\alpha}\psi_0 \simeq \psi_t^{\mathscr{T}} + \psi_t^{\mathscr{R}} \qquad \text{for } t \to +\infty, \tag{8.75}$$

where $\psi_t^{\mathscr{T}}$ and $\psi_t^{\mathscr{R}}$ are the transmitted and reflected waves, given by

$$\psi_t^{\mathcal{T}}(x) = C_t\, e^{i\varphi_t(x)}\, \overline{\mathcal{T}\left(\frac{mx}{\hbar t}\right)} e^{-\frac{m^2\sigma^2}{2\hbar^2 t^2}\left(x-\frac{p_0}{m}t\right)^2} \chi_{>0}(x)\,, \tag{8.76}$$

$$\psi_t^{\mathcal{R}}(x) = C_t\, e^{i\varphi_t(-x)}\, \overline{\mathcal{R}\left(\frac{mx}{\hbar t}\right)} e^{-\frac{m^2\sigma^2}{2\hbar^2 t^2}\left(x+\frac{p_0}{m}t\right)^2} \chi_{<0}(x)\,, \tag{8.77}$$

$$C_t := \frac{\sqrt{\sigma}}{\pi^{1/4}}\left(\frac{m}{i\hbar t}\right)^{1/2}, \qquad \varphi_t(x) := \frac{mx_0}{\hbar t}\left(\frac{x^2}{2x_0} + x - \frac{p_0}{m}t\right). \tag{8.78}$$

Let us observe that $\psi_t^{\mathcal{T}}$ is a wave packet concentrated in position around $x \simeq \frac{p_0}{m}t$. Therefore, for t large, it is localized very far on the right of the origin and, as time increases, it moves to the right with positive momentum p_0. Analogously, $\psi_t^{\mathcal{R}}$ is a wave packet concentrated in position around $x \simeq -\frac{p_0}{m}t$. Then, for t large, it is localized very far on the left of the origin and, as time increases, it moves to the left with negative momentum $-p_0$. These observations motivate the names of transmitted wave and reflected wave, respectively. The two coefficients $\mathcal{T}$ and $\mathcal{R}$, called transmission and reflection coefficients, determine the fractions of the incident wave that are transmitted and reflected (Fig. 8.2).

To summarize, we have obtained a complete qualitative description of the scattering process. We assign an incident wave coming from the left with a positive momentum p_0. Then, after a long enough time, the resulting wave can be explicitly computed and it is the sum of a transmitted wave, moving to the right with the same momentum p_0, and a reflected wave, moving to the left with opposite momentum $-p_0$.

It is also possible to give a quantitative prediction of the probability that the particle is transmitted or reflected. More precisely, according to Born's rule, we define transmission and reflection probabilities as follows:

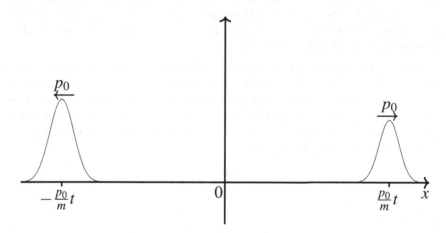

Fig. 8.2 The transmitted and reflected waves

$$\mathscr{P}^{\mathscr{T}}(\psi_0) := \lim_{t \to +\infty} \int_0^\infty dx \; \left| \left(e^{-i\frac{t}{\hbar}H_\alpha} \psi_0 \right)(x) \right|^2, \tag{8.79}$$

$$\mathscr{P}^{\mathscr{R}}(\psi_0) := \lim_{t \to +\infty} \int_{-\infty}^0 dx \; \left| \left(e^{-i\frac{t}{\hbar}H_\alpha} \psi_0 \right)(x) \right|^2. \tag{8.80}$$

As a consequence of (8.63) and Proposition 8.5, we find

$$\mathscr{P}^{\mathscr{T}}(\psi_0) = \int dk \, |\mathscr{T}(k)|^2 |\tilde{\psi}_0(k)|^2 + \mathscr{E}, \tag{8.81}$$

$$\mathscr{P}^{\mathscr{R}}(\psi_0) = \int dk \, |\mathscr{R}(k)|^2 |\tilde{\psi}_0(k)|^2 + \mathscr{E}. \tag{8.82}$$

where the error $\mathscr{E}$ is exponentially small for $\delta, \eta \to \infty$. Note that the incident wave is totally transmitted for $\alpha \to 0$ and totally reflected for $\alpha \to \infty$.

Remark 8.7 Taking into account of the explicit form of $\tilde{\psi}_0$, for $\eta \to \infty$ we have

$$\int dk \, |\mathscr{T}(k)|^2 |\tilde{\psi}_0(k)|^2 = \frac{1}{\sqrt{\pi}} \int dq \, \frac{4k_0^2(1+\eta^{-1}q)^2}{\alpha_0^2 + 4k_0^2(1+\eta^{-1}q)^2} e^{-q^2} = |\mathscr{T}(k_0)|^2 + O(\eta^{-1}) \tag{8.83}$$

and analogously for the integral in (8.82). This implies that if we neglect error terms of order η^{-1} then transmission and reflection probabilities are

$$\mathscr{P}^{\mathscr{T}}(\psi_0) \simeq |\mathscr{T}(k_0)|^2, \qquad\qquad \mathscr{P}^{\mathscr{R}}(\psi_0) \simeq |\mathscr{R}(k_0)|^2. \tag{8.84}$$

These are the expressions usually found in the physics textbooks (see, e.g., [1, 7]). Note that in (8.84), the dependence on the initial state survives only through the mean momentum p_0. We also observe that (8.84) are formally obtained from (8.81), (8.82) by replacing $|\tilde{\psi}_0(k)|^2$ with $\delta(k - k_0)$ and, therefore, formulas (8.84) correspond to an unphysical choice of the initial state.

Remark 8.8 It is interesting to characterize the behavior of transmission and reflection probabilities in the classical limit, i.e., when the typical action of the system is large compared to $\hbar$. Note that the physical dimension of the coupling constant α is energy times length (see, e.g., (8.11)). It follows that the quantity $m\alpha/p_0$ is an action and it can be considered as the typical action of the system in the state ψ_0. Therefore, by classical limit in this case we mean

$$\theta := \frac{m\alpha}{\hbar p_0} \gg 1 \tag{8.85}$$

By (8.81), we have

$$\mathscr{P}^{\mathscr{T}}(\psi_0) \simeq \frac{\sigma}{\sqrt{\pi}} \int dk \, \frac{k^2}{\frac{m^2\alpha^2}{\hbar^4} + k^2} e^{-\sigma^2(k-k_0)^2} = \frac{\sigma k_0}{\sqrt{\pi}} \int dz \, \frac{z^2}{\theta^2 + z^2} e^{-\sigma^2 k_0^2(z-1)^2}$$
$$\to 0 \quad \text{for } \theta \to \infty. \tag{8.86}$$

Thus, we conclude that in the classical limit a point interaction behaves as an impenetrable barrier.

Exercise 8.9 Discuss the scattering problem in the case of the Hamiltonian (8.13) with $d = 1$.

8.5 Two Point Interactions

The construction of the Hamiltonian with $N > 1$ point interactions is similar to the case $N = 1$. In this section, we discuss some properties of the Hamiltonian with two point interactions placed at $y_1 = 0$ and $y_2 = l$, both with strength α, with $\alpha \neq 0$. For $\lambda > 0$, we introduce the 2×2 matrix

$$\Gamma_{ij}(\lambda) = -\frac{1 + \alpha\, G^\lambda(0)}{\alpha} \delta_{ij} + (\delta_{ij} - 1)\, G^\lambda(l) \tag{8.87}$$

and define the operator

$$D(H_{\alpha,l}) = \Big\{ u \in L^2(\mathbb{R}) \mid u = w^\lambda + q_1 G^\lambda + q_2 G^\lambda(\cdot - l),\ w^\lambda \in H^2(\mathbb{R}),\ q_i \in \mathbb{C},$$

$$w^\lambda(y_i) = \sum_{j=1,2} \Gamma_{ij}(\lambda) q_j,\ i = 1, 2 \Big\}, \tag{8.88}$$

$$(H_{\alpha,l} + \lambda)u = (H_0 + \lambda)w^\lambda. \tag{8.89}$$

Exercise 8.10 Verify that the above operator is self-adjoint. Moreover, show that any $u \in D(H_{\alpha,l})$ satisfies the boundary conditions

$$u'(y_i^+) - u'(y_i^-) = \frac{2m\alpha}{\hbar^2} u(y_i), \qquad i = 1, 2. \tag{8.90}$$

By definition, $H_{\alpha,l}$, $D(H_{\alpha,l})$ is the Hamiltonian with point interactions in $y_1 = 0$ and $y_2 = l$ with strength α ([2]). Moreover, one can compute the resolvent and verify that the essential spectrum is $[0, \infty)$. Finally, the absence of nonnegative eigenvalues is shown as in the case of one point interaction. Here, we want to study the negative eigenvalues and the corresponding eigenvectors. For $E < 0$, we have to solve the problem

$$H_{\alpha,l}\psi = -|E|\,\psi, \qquad \psi \in D(H_{\alpha,l}), \qquad \|\psi\| = 1. \tag{8.91}$$

Taking into account of (8.88) and (8.89), ψ has the form

$$\psi = q_1 G^\lambda + q_2 G^\lambda(\cdot - l), \tag{8.92}$$

where q_1, q_2 are solutions of the linear system

$$\sum_{j=1,2} \Gamma_{ij}(|E|)q_j = 0. \tag{8.93}$$

Such a system admits nontrivial solution if and only if $\det \Gamma_{ij}(|E|) = 0$, i.e.,

$$\left(1 + \frac{2\sqrt{|E|}\hbar}{\sqrt{2m}\,\alpha}\right)^2 - e^{-2\frac{\sqrt{2m}}{\hbar}\sqrt{|E|}\,l} = 0. \tag{8.94}$$

For $\alpha > 0$ (repulsive case), equation (8.94) does not have solutions and so there are no negative eigenvalues. Let us consider the attractive case $\alpha < 0$. It is convenient to introduce the dimensionless variables

$$\xi = \frac{\sqrt{2m|E|}\,l}{\hbar}, \qquad \gamma = \frac{m|\alpha|l}{\hbar^2}, \tag{8.95}$$

so that the equation becomes

$$\left[\left(1 - \frac{\xi}{\gamma}\right) + e^{-\xi}\right]\left[\left(1 - \frac{\xi}{\gamma}\right) - e^{-\xi}\right] := F_0(\xi)F_1(\xi) = 0. \tag{8.96}$$

The equation $F_0(\xi) = 0$ always admits a unique positive solution ξ_0, with $\xi_0 > \gamma$, while $F_1(\xi) = 0$ admits a unique positive solution ξ_1 only if $\gamma > 1$ and one has $0 < \xi_1 < \gamma$. For $\alpha < 0$, we conclude that

(i) if $\gamma \leq 1$ then there exists one negative eigenvalue

$$\hat{E}_0 = -\frac{\hbar^2}{2ml^2}\xi_0^2, \qquad \text{where} \quad \xi_0 = \gamma\,(1 + e^{-\xi_0}), \tag{8.97}$$

(ii) if $\gamma > 1$ then there exist two negative eigenvalues $\hat{E}_0 < \hat{E}_1$, with $\hat{E}_1$ given by

$$\hat{E}_1 = -\frac{\hbar^2}{2ml^2}\xi_1^2, \qquad \text{where} \quad \xi_1 = \gamma\,(1 - e^{-\xi_1}). \tag{8.98}$$

Let us compute the eigenvector associated to $\hat{E}_0$. By (8.93), we have

$$\left(\frac{1}{\alpha} + \frac{\sqrt{2m}}{2\hbar\sqrt{|E_0|}}\right)q_1 + \frac{\sqrt{2m}}{2\hbar\sqrt{|E_0|}}e^{-\frac{\sqrt{2m}}{\hbar}\sqrt{|E_0|}\,l}q_2 = 0 \tag{8.99}$$

which can be rewritten as

$$\left(1 - \frac{\xi_0}{\gamma}\right)q_1 + e^{-\xi_0}q_2 = 0. \tag{8.100}$$

Taking into account that $\xi_0 = \gamma(1 + e^{-\xi_0})$, we find $q_1 = q_2$. Then, the eigenvector associated to $\hat{E}_0$ is

$$\Phi_0(x) = c_0\left(e^{-\frac{\xi_0}{l}|x|} + e^{-\frac{\xi_0}{l}|x-l|}\right). \tag{8.101}$$

For $\gamma > 1$, in a similar way, one finds that the eigenvector associated to $\hat{E}_1$ is

$$\Phi_1(x) = c_1\left(e^{-\frac{\xi_1}{l}|x|} - e^{-\frac{\xi_1}{l}|x-l|}\right). \tag{8.102}$$

The normalization constants c_0, c_1 can be explicitly computed and one finds

$$c_j = \left(\frac{\xi_j}{2l}\right)^{1/2}\left(1 + (-1)^j(1+\xi_j)e^{-\xi_j}\right)^{-1/2}, \qquad j = 0, 1. \tag{8.103}$$

Note that if a particle is in the state Φ_0 (or Φ_1) then the probability density of the position has two absolute maxima in $x = 0$ and $x = l$. In this case, one says that the particle is "delocalized".

Exercise 8.11 Consider a particle in the state Φ_0 (or Φ_1) and compute the probability to find the position of the particle in the intervals $(-l/2, l/2)$, $(l/2, 3l/2)$.

Remark 8.9 From Eq. (8.94), we also obtain the behavior of the eigenvalues for $\alpha < 0$ in the limiting cases $l \to \infty$ and $l \to 0$. For $l \to \infty$ we have $\hat{E}_0, \hat{E}_1 \to -m\alpha^2/2\hbar^2$, which is the eigenvalue E_0 in the case of one point interaction with strength $\alpha < 0$ (see (8.16)). For l sufficiently small, we are in the case $\gamma < 1$ and the eigenvalue $\hat{E}_1$ is absent. If $l \to 0$ we have $\hat{E}_0 \to -2m\alpha^2/\hbar^2$, which is the eigenvalue in the case of one point interaction with strength $2\alpha < 0$.

The Hamiltonian $H_{\alpha,l}$ for $\alpha < 0$ and $\gamma > 1$ can be considered as a simplified model to describe a particle in a double well potential (see, e.g., [1], Sect. 50, Problem 3, [8] Sect. 8.5). In particular, it is interesting to characterize the behavior of the eigenvalues and the eigenvectors in the semiclassical regime which, in this case, can be characterized by the condition $\gamma \gg 1$. Note that the condition can also be rewritten as

$$\gamma = \frac{\sqrt{2m|E_0|}\,l}{\hbar} \gg 1. \tag{8.104}$$

Exercise 8.12 Verify that

$$\xi_0 = \gamma + \gamma e^{-\gamma} + R_0(\gamma), \qquad \xi_1 = \gamma - \gamma e^{-\gamma} + R_1(\gamma), \tag{8.105}$$

where $\gamma^{-1}e^{\gamma} R_i(\gamma) \to 0$ for $\gamma \to \infty$ and $i = 1, 2$.

Using the expansion (8.105), we find that the energy splitting $\Delta E = \hat{E}_1 - \hat{E}_0$ is exponentially small in the semiclassical parameter, i.e.,

$$\Delta E \simeq \frac{2|\alpha|}{l}\gamma e^{-\gamma}. \tag{8.106}$$

Moreover, for the two eigenvectors we have

$$\Phi_0(x) \simeq \frac{1}{\sqrt{2}}\big(\varXi_0(x) + \varXi_0(x - l)\big), \qquad \Phi_1(x) \simeq \frac{1}{\sqrt{2}}\big(\varXi_0(x) - \varXi_0(x - l)\big),$$
(8.107)

where $\varXi_0$ is the eigenvector in the case of one point interaction with strength $\alpha < 0$ (see (8.17)).

Exercise 8.13 Compute the time evolution corresponding to the initial states

$$\Phi_L = \frac{1}{\sqrt{2}}(\Phi_0 + \Phi_1), \qquad \Phi_R = \frac{1}{\sqrt{2}}(\Phi_0 - \Phi_1),$$
(8.108)

i.e., for a particle initially localized on the left, around $x = 0$, or on the right, around $x = l$. Chosen the initial state Φ_L, show that for $\gamma \gg 1$

$$\langle x \rangle(t) \simeq \frac{l}{2}\left(1 - \cos\frac{\Delta E}{\hbar}t\right),$$
(8.109)

where $\langle x \rangle(t)$ is the mean value of the position at time t. Therefore, in the semiclassical regime $\langle x \rangle(t)$ exhibits a periodic motion between $x = 0$ and $x = l$ with the very long period $T = 2\pi\hbar/\Delta E$ (beating motion).

8.6 Self-adjoint Extensions of A_0

Here, we prove Proposition 8.1 by constructing the self-adjoint extensions H_α, $D(H_\alpha)$ of the operator A_0, $D(A_0)$ defined in (8.2).
It is convenient to work in the Fourier space. Therefore, we consider the operator $B_0 = \mathscr{F} A_0 \mathscr{F}^{-1}$ explicitly given by

$$(B_0 u)(p) = \frac{\hbar^2 p^2}{2m} u(p), \qquad D(B_0) = \left\{u \in D(\widetilde{H}_0) \mid \int dp\, u(p) = 0\right\}, \quad (8.110)$$

where $D(\widetilde{H}_0)$ has been defined in (6.7). The first step is to find the orthogonal complement of $Ran\,(B_0 + \lambda)$, with $\lambda > 0$.

Proposition 8.6

$$Ran\,(B_0 + \lambda)^\perp = \left\{v \in L^2(\mathbb{R}) \mid v = q\,\tilde{G}^\lambda,\ q \in \mathbb{C}\right\} \quad (8.111)$$

where $\tilde{G}^\lambda$ is the Fourier transform of G^λ, defined in (8.4).

Proof We look for $v \in L^2(\mathbb{R})$ satisfying the equation

$$\int dp\, \overline{v}(p)\Big(\frac{\hbar^2 p^2}{2m} + \lambda\Big)u(p) = 0, \qquad \text{for any } u \in D(B_0) \tag{8.112}$$

A solution is $v = q\, \tilde{G}^\lambda$, for any $q \in \mathbb{C}$. It remains to show that the solution is unique. Defined $g(p) := \overline{v}(p)(\frac{\hbar^2 p^2}{2m} + \lambda)$, Eq. (8.112) is rewritten as

$$\int dp\, g(p)\, u(p) = 0 \qquad \text{for any } u \in D(B_0). \tag{8.113}$$

Let us prove that the constant function is the only solution in the class of the step functions. Let $g(p) = \sum_{i=1}^{n} c_i \chi_{\Omega_i}(p)$, where c_i are constants and χ_{Ω_i} denotes the characteristic function of the interval $\Omega_i \subset \mathbb{R}$, with $\Omega_i \cap \Omega_j = \emptyset$ if $i \neq j$, and $\cup_{i=1}^{n}\Omega_i = \mathbb{R}$. We proceed by contradiction and, without loss of generality, we assume that $c_1 \neq c_2$. Then the equation reads

$$c_1 \int_{\Omega_1} dp\, u(p) + c_2 \int_{\Omega_2} dp\, u(p) + \sum_{j \neq 1,2} c_j \int_{\Omega_j} dp\, u(p) = 0. \tag{8.114}$$

Let us consider $u_0 = \chi_{(-R,R)}(b_1 \chi_{\Omega_1} + b_2 \chi_{\Omega_2})$, where $b_1, b_2 \in \mathbb{R} \setminus \{0\}$, $R > 0$ is sufficiently large so that $m_i := |(-R, R) \cap \Omega_i| > 0$ for $i = 1, 2$ and $b_1 m_1 + b_2 m_2 = 0$. Then $u_0 \in D(B_0)$ and Eq. (8.114) reduces to

$$0 = c_1 \int_{\Omega_1} dp\, u_0(p) + c_2 \int_{\Omega_2} dp\, u_0(p) = c_1 b_1 m_1 + c_2 b_2 m_2 = (c_1 - c_2)b_1 m_1. \tag{8.115}$$

Hence, $c_1 = c_2$ and we conclude that the constant function is the only solution of (8.113) in the class of the step functions. By a density argument, one shows that the constant function is the only solution also in the class $L^2_{loc}(\mathbb{R})$ and then the proof is complete.

$\square$

The second step is the characterization of the adjoint of B_0.

Proposition 8.7

$$D(B_0^*) = \Big\{v \in L^2(\mathbb{R}) \mid v = w + q\, \tilde{G}^\lambda,\ w \in D(\widetilde{H_0}),\ q \in \mathbb{C}\Big\}, \tag{8.116}$$

$$(B_0^* + \lambda)v = (\widetilde{H_0} + \lambda)w, \tag{8.117}$$

where $\widetilde{H_0}$ is the free Hamiltonian in the Fourier space (see (6.7)).

Proof Let us observe that

$$(\tilde{G}^\lambda, (B_0 + \lambda)u) = 0 \tag{8.118}$$

for any $u \in D(B_0)$. Then, $\tilde{G}^\lambda \in D(B_0^*)$ and $(B_0^* + \lambda)\tilde{G}^\lambda = 0$. Moreover, for $w \in D(\widetilde{H_0})$ we have

$$(w, (B_0 + \lambda)u) = ((\widetilde{H_0} + \lambda)w, u) \tag{8.119}$$

for any $u \in D(B_0)$. Then, $w \in D(B_0^*)$ and $(B_0^* + \lambda)w = (\widetilde{H_0} + \lambda)w$. Thus, we have shown that if $v = w + q\,\tilde{G}^\lambda$, with $w \in D(\widetilde{H_0})$ and $q \in \mathbb{C}$, then $v \in D(B_0^*)$ and Eq. (8.117) holds. Conversely, let us assume that $v \in D(B_0^*)$. Since $Ran\,(\widetilde{H_0} + \lambda) = L^2(\mathbb{R})$, there exists $w \in D(\widetilde{H_0})$ such that

$$(B_0^* + \lambda)v = (\widetilde{H_0} + \lambda)w\,. \tag{8.120}$$

Moreover, for any $u \in D(B_0)$ we have $((B_0 + \lambda)u, w) = (u, (\widetilde{H_0} + \lambda)w) = (u, (B_0^* + \lambda)v)$, which implies $w \in D(B_0^*)$ and $(B_0^* + \lambda)v = (B_0^* + \lambda)w$. Therefore, for any $u \in D(B_0)$ we can write $((B_0 + \lambda)u, v - w) = (u, (B_0^* + \lambda)(v - w)) = 0$, i.e., $v - w \in Ran\,(B_0 + \lambda)^\perp$. Using the previous proposition, we conclude the proof. $\square$

At this point, one could use a general theorem saying that any self-adjoint extension of B_0 is given by a suitable restriction of the adjoint B_0^* (see, e.g., [9, 10]). In this way, one obtains all the self-adjoint extensions H_α, $D(H_\alpha)$ defined in (8.5), (8.6).

Here, we prefer to construct such self-adjoint extensions following a more direct procedure. We first note that B_0^* is not symmetric. Indeed, for $v_1, v_2 \in D(B_0^*)$ one has

$$(v_2, (B_0^* + \lambda)v_1) = (w_2 + q_2\,\tilde{G}^\lambda, (\widetilde{H_0} + \lambda)w_1) = ((\widetilde{H_0} + \lambda)w_2, w_1) + \overline{q}_2 \int dp\, w_1(p)$$

$$= ((\widetilde{H_0} + \lambda)w_2, v_1) - q_1 \int dp\, \overline{w_2(p)} + \overline{q}_2 \int dp\, w_1(p)$$

$$= ((B_0^* + \lambda)v_2, v_1) + \overline{q}_2 \int dp\, v_1(p) - q_1 \int dp\, \overline{v_2(p)}\,. \tag{8.121}$$

We know that a self-adjoint extension of B_0 is a restriction of B_0^*. A good candidate is the symmetric restriction of B_0^*, obtained requiring that the ratio

$$\frac{q}{\int dp\, v(p)} \tag{8.122}$$

is a real constant. Thus, for any $\alpha \in \mathbb{R} \setminus \{0\}$, we define the symmetric operator

$$D(B_\alpha) = \left\{ u \in D(B_0^*) \mid q = -\frac{\alpha}{\sqrt{2\pi}} \int dp\, u(p) \right\}, \qquad B_\alpha = B_0^*|_{D(B_\alpha)}. \tag{8.123}$$

In the next proposition, we show that such operator is self-adjoint and we compute its resolvent.

Proposition 8.8 $B_\alpha, D(B_\alpha)$ *is self-adjoint and*

$$(B_\alpha - z)^{-1} = (\widetilde{H}_0 - z)^{-1} - \frac{\alpha}{1 + \alpha\, G^{-z}(0)} \left(\overline{\tilde{G}^{-z}}, \cdot \right) \tilde{G}^{-z} \qquad (8.124)$$

for $z \in \mathbb{C}$, with $\Im z \neq 0$.

Proof We show that $B_\alpha, D(B_\alpha)$ is self-adjoint by using the self-adjointness criterion (Proposition 4.5). Then, for any $z \in \mathbb{C}$, with $\Im z \neq 0$, and $f \in L^2(\mathbb{R})$, we look for the solution of the problem

$$(B_\alpha - z)\, u = f, \qquad u \in D(B_\alpha). \qquad (8.125)$$

Recall that $(B_\alpha - z)\, u = (B_\alpha + \lambda)\, u - (\lambda + z)u = (\widetilde{H}_0 - z)w^\lambda - (\lambda + z)q\, \tilde{G}^\lambda$, where $q = -\frac{\alpha}{\sqrt{2\pi}} \int\! dp\, u(p)$. Then, the equation can be rewritten as

$$(\widetilde{H}_0 - z)w^\lambda - (\lambda + z)q\, \tilde{G}^\lambda = f. \qquad (8.126)$$

Denoted $(\widetilde{H}_0 - z)^{-1} = \sqrt{2\pi}\, \tilde{G}^{-z}$ and using the identity $\tilde{G}^{-z} - \tilde{G}^\lambda = \sqrt{2\pi}\, (z + \lambda)\tilde{G}^{-z}\, \tilde{G}^\lambda$, we obtain the equation

$$u + \frac{\alpha}{\sqrt{2\pi}} \left(\int\! dp\, u(p) \right) \tilde{G}^{-z} = \sqrt{2\pi}\, \tilde{G}^{-z} f. \qquad (8.127)$$

Integrating on $\mathbb{R}$, we find

$$\int\! dp\, u(p) = \frac{\sqrt{2\pi}}{1 + \alpha\, G^{-z}(0)} \int\! dp\, \tilde{G}^{-z}(p) f(p). \qquad (8.128)$$

Inserting (8.128) into (8.127), we have

$$u = (\widetilde{H}_0 - z)^{-1} f - \frac{\alpha}{1 + \alpha\, G^{-z}(0)} \left(\int\! dp\, \tilde{G}^{-z}(p) f(p) \right) \tilde{G}^{-z} \qquad (8.129)$$

which is the required solution of problem (8.125). This implies that $B_\alpha, D(B_\alpha)$ is self-adjoint and the r.h.s. of (8.129) defines the corresponding resolvent operator. $\qquad \Box$

It is now easy to verify that $B_\alpha = \mathscr{F} H_\alpha \mathscr{F}^{-1}$, where H_α is the Hamiltonian defined in (8.5), (8.6), and that formula (8.7) holds.

8.7   Proof of the Eigenfunction Expansion Theorem

We give the proof of Proposition 8.3 for $\alpha > 0$, while the case $\alpha < 0$ is left as an exercise. In particular, we shall prove that the operator W_+ is well defined in $L^2(\mathbb{R})$, it is unitary and it diagonalizes the Hamiltonian H_α. The proof for W_- is analogous. We shall obtain the result following the line of the standard proof of Plancherel theorem for the Fourier transform. As a first step, we prove the inversion formula for functions in $C_0^\infty(\mathbb{R})$. In Proposition 8.10 we construct the isometric operator W_+ by means of a limiting procedure. In Proposition 8.11, we show that also W_+^* is isometric and, therefore, W_+ is unitary. In the last step, it is easily seen that W_+ diagonalizes H_α.

Proposition 8.9 *For any $f \in C_0^\infty(\mathbb{R})$, we have*

$$f(x) = \frac{1}{\sqrt{2\pi}} \int dk \, \hat{f}_+(k) \phi_+(x, k) \,. \tag{8.130}$$

Proof By definition of the generalized transform and taking into account that f is smooth and has compact support, we write

$$\hat{f}_+(k) = \tilde{f}(k) + \frac{\overline{\mathscr{R}(k)}}{\sqrt{2\pi}} \int dx \, f(x) \, e^{i|k||x|} \,. \tag{8.131}$$

Since f is integrable, we have that $\hat{f}_+$ is bounded and continuous. Moreover, integrating by parts

$$\int dx \, f(x) \, e^{i|k||x|} = \int_0^\infty dx \, (f(x) + f(-x)) \, e^{i|k|x}$$
$$= 2i \frac{f(0)}{|k|} + \frac{i}{|k|} \int_0^\infty dx \, \frac{d}{dx} (f(x) + f(-x)) \, e^{i|k|x} \,, \tag{8.132}$$

we obtain the estimate $|\hat{f}_+(k)| \leq c|k|^{-2}$ and then $\hat{f}_+ \in L^1(\mathbb{R})$. It follows that

$$J(x) := \frac{1}{\sqrt{2\pi}} \int dk \, \hat{f}_+(k) \phi_+(x, k) \tag{8.133}$$

is a bounded and continuous function. We have to show that $J(x) = f(x)$ for any $x \in \mathbb{R}$. Let us fix $x \geq 0$ and let us write

$$J(x) = \lim_{N \to \infty} J_N(x) \,, \tag{8.134}$$

where

$$J_N(x) = \frac{1}{\sqrt{2\pi}} \int_{|k|<N} dk \, \hat{f}_+(k) \phi_+(x, k) = \frac{1}{2\pi} \int dy \, f(y) \int_{|k|<N} dk \, \overline{\phi_+(y, k)} \phi_+(x, k) \,. \tag{8.135}$$

Using the explicit expression of ϕ_+ in (8.135), we obtain

$$
\begin{aligned}
J_N(x) = \frac{1}{2\pi} &\int dy\, f(y) \int_{|k|<N} dk\, e^{ik(x-y)} \\
+ \frac{1}{2\pi} &\int_0^\infty dy\, (f(y) + f(-y)) \Bigg[\int_0^N dk\, \mathscr{R}(k) \left(e^{-iky-ikx} + e^{iky-ikx} \right) \\
&+ \int_0^N dk\, \overline{\mathscr{R}(k)} \left(e^{iky+ikx} + e^{iky-ikx} \right) + 2 \int_0^N dk\, |\mathscr{R}(k)|^2 e^{ik(y-x)} \Bigg].
\end{aligned}
\tag{8.136}
$$

By the inversion formula for the Fourier transform, the first term in the right-hand side of (8.136) converges to $f(x)$ for $N \to \infty$. It remains to verify that the remaining terms converge to zero for $N \to \infty$. The expression in square brackets in (8.136) can be written as

$$
\int_0^N dk\, \left(\mathscr{R}(k) + \overline{\mathscr{R}(k)} + 2|\mathscr{R}(k)|^2 \right) e^{ik(y-x)} + 2\Re \int_0^N dk\, \mathscr{R}(k) e^{-ik(x+y)}.
\tag{8.137}
$$

The first integral in (8.137) is zero due to the identity $\mathscr{R}(k) + \overline{\mathscr{R}(k)} + 2|\mathscr{R}(k)|^2 = 0$. For the second integral, an explicit computation yields

$$
2\Re \int_0^N dk\, \mathscr{R}(k) e^{-ik(x+y)} = 2\alpha_0 \int_0^N dk\, \frac{2k \sin k(x+y) - \alpha_0 \cos k(x+y)}{\alpha_0^2 + 4k^2}.
\tag{8.138}
$$

We note that

$$
\int_0^\infty dz\, \frac{\cos az}{b^2 + z^2} = \frac{\pi}{2b} e^{-ab}, \qquad \int_0^\infty dz\, \frac{z \sin az}{b^2 + z^2} = \frac{\pi}{2} e^{-ab}, \qquad a,b > 0.
\tag{8.139}
$$

Then, we obtain

$$
\lim_{N\to\infty} \int_0^\infty dy\, (f(y) + f(-y)) \int_0^N dk\, \frac{2k \sin k(x+y) - \alpha_0 \cos k(x+y)}{\alpha_0^2 + 4k^2} = 0
\tag{8.140}
$$

and we conclude $J(x) = \lim_{N\to\infty} J_N(x) = f(x)$. The case $x < 0$ is analogous and then the proposition is proved. $\qquad\square$

By the inversion formula (8.130), for $f \in C_0^\infty(\mathbb{R})$, we find

$$
\begin{aligned}
\int dx\, |f(x)|^2 &= \frac{1}{\sqrt{2\pi}} \int dx\, \overline{f(x)} \int dk\, \hat{f}_+(k) \phi_+(x,k) \\
&= \frac{1}{\sqrt{2\pi}} \int dk\, \hat{f}_+(k) \int dx\, \overline{f(x)} \phi_+(x,k) = \int dk\, |\hat{f}_+(k)|^2,
\end{aligned}
\tag{8.141}
$$

where the exchange of the order of integration is allowed since $\hat{f}_+ \bar{f}$ is integrable in $\mathbb{R}^2$. In the next proposition, we construct the operator W_+ in $L^2(\mathbb{R})$.

Proposition 8.10 *Let $f \in L^2(\mathbb{R})$, $N > 0$ and define*

$$\hat{f}_{+,N}(k) = \frac{1}{\sqrt{2\pi}} \int_{|x|<N} dx\, f(x)\, \overline{\phi_+(x,k)}\,. \tag{8.142}$$

Then we have

(i) *$\hat{f}_{+,N} \in L^2(\mathbb{R})$,*
(ii) *there exists $\hat{f}_+ \in L^2(\mathbb{R})$ such that $\|\hat{f}_{+,N} - \hat{f}_+\| \to 0$ for $N \to \infty$,*
(iii) *the linear operator $W_+ : f \to \hat{f}_+$ is isometric.*

Proof Let us consider $f_N = f\chi_N$, where χ_N is the characteristic function of the interval $(-N, N)$. We have $f_N \in L^1(\mathbb{R}) \cap L^2(\mathbb{R})$ and $\|f - f_N\| \to 0$ for $N \to \infty$. Using the definition of ϕ_+, we write

$$\hat{f}_{+,N}(k) = \tilde{f}_N(k) + \frac{\overline{\mathscr{R}(k)}}{\sqrt{2\pi}} \int dx\, f_N(x)\, e^{i|k||x|} \tag{8.143}$$

and we conclude that $\hat{f}_{+,N} \in L^2(\mathbb{R})$. Let us prove *ii*). For any N, let $f_N^m \in C_0^\infty((-N, N))$ be a sequence such that $\|f_N - f_N^m\| \to 0$ for $m \to \infty$. Then

$$\hat{f}_{+,N}(k) - \hat{f}_{+,N}^m(k) = \tilde{f}_N(k) - \tilde{f}_N^m(k) + \frac{\overline{\mathscr{R}(k)}}{\sqrt{2\pi}} \int dx\, (f_N(x) - f_N^m(x))\, e^{i|k||x|}\,. \tag{8.144}$$

Note that $\|\tilde{f}_N - \tilde{f}_N^m\| \to 0$ and $\|f_N - f_N^m\|_{L^1(\mathbb{R})} \to 0$ for $m \to \infty$. So, we also have $\|\hat{f}_{+,N} - \hat{f}_{+,N}^m\| \to 0$ for $m \to \infty$. Moreover, using (8.141), we obtain

$$\|f_N\| = \lim_{m\to\infty} \|f_N^m\| = \lim_{m\to\infty} \|\hat{f}_{+,N}^m\| = \|\hat{f}_{+,N}\|\,. \tag{8.145}$$

Since f_N is a Cauchy sequence in $L^2(\mathbb{R})$, by (8.145), the same is true for $\hat{f}_{+,N}$. Thus, there exists $\hat{f}_+ \in L^2(\mathbb{R})$ such that $\|\hat{f}_{+,N} - \hat{f}_+\| \to 0$ for $N \to \infty$. For the proof of (iii), it is sufficient to observe that

$$\|f\| = \lim_{N\to\infty} \|f_N\| = \lim_{N\to\infty} \|\hat{f}_{+,N}\| = \|\hat{f}_+\|\,. \tag{8.146}$$

$\square$

In order to prove that W_+ is unitary, it is sufficient to show that W_+^* is isometric.

Proposition 8.11 *For any $h \in L^2(\mathbb{R})$, we have*

$$(W_+^* h)(x) = \frac{1}{\sqrt{2\pi}} \lim_{N \to \infty} \int_{|k|<N} dk \, h(k) \, \phi_+(x, k) . \tag{8.147}$$

Moreover, W_+^* is isometric.

Proof Let $g \in C_0^\infty(\mathbb{R})$, $h \in L^2(\mathbb{R})$ and $h_N = h\chi_N$. Then

$$(g, W_+^* h_N) = (W_+ g, h_N) = \frac{1}{\sqrt{2\pi}} \int dk \, h_N(k) \int dx \, \overline{g(x)} \phi_+(x, k)$$

$$= \frac{1}{\sqrt{2\pi}} \int dx \, \overline{g(x)} \int dk \, h_N(k) \phi_+(x, k) . \tag{8.148}$$

Since W_+^* is a bounded operator, $W_+^* h_N \to W_+^* h$ for $N \to \infty$ and we find (8.147).

Let us prove that the inversion formula

$$v(k) = \frac{1}{\sqrt{2\pi}} \int dx \, (W_+^* v)(x) \, \overline{\phi_+(x, k)} \tag{8.149}$$

holds for any $v \in C_0^\infty(\mathbb{R} \setminus \{0\})$. Indeed, it is easy to see that $W_+^* v$ is a bounded and integrable function, so that the right-hand side of (8.149) is well defined. Moreover,

$$\frac{1}{\sqrt{2\pi}} \int dx \, (W_+^* v)(x) \, \overline{\phi_+(x, k)} = \lim_{N \to \infty} \frac{1}{2\pi} \int dp \, v(p) \int_{|x|<N} dx \, \phi_+(x, p) \overline{\phi_+(x, k)}$$

$$= \lim_{N \to \infty} \frac{1}{2\pi} \int dp \, v(p) \int_{|x|<N} dx \, e^{-i(k-p)x}$$

$$+ \lim_{N \to \infty} \frac{1}{2\pi} \int dp \, v(p) \int_{|x|<N} dx \left(\phi_+(x, p) \overline{\phi_+(x, k)} - e^{-i(k-p)x} \right)$$

$$:= \lim_{N \to \infty} I_N^{(1)}(k) + \lim_{N \to \infty} I_N^{(2)}(k) . \tag{8.150}$$

By the inversion formula for the Fourier transform, we have $\lim_{N \to \infty} I_N^{(1)}(k) = v(k)$. On the other hand, one can verify that $\lim_{N \to \infty} I_N^{(2)}(k) = 0$ and then (8.149) is proved.

Using (8.149), we obtain $\|W_+^* v\| = \|v\|$ for any $v \in C_0^\infty(\mathbb{R} \setminus \{0\})$. By a density argument, we also have $\|W_+^* h\| = \|h\|$ for any $h \in L^2(\mathbb{R})$, i.e., W_+^* is isometric. $\square$

The last step is to prove that W_+ diagonalizes the Hamiltonian H_α. For $u \in D(H_\alpha)$ we have

$$(W_+ H_\alpha u)(k) = \frac{1}{\sqrt{2\pi}} \int dy \, \overline{\phi_+(y, k)} (H_\alpha u)(y) = \frac{1}{\sqrt{2\pi}} \lim_{\varepsilon \to 0} \int_{|y|>\varepsilon} dy \, \overline{\phi_+(y, k)} (H_0 u)(y) . \tag{8.151}$$

Integrating by parts and using (8.29), (8.30), (8.9) we obtain

$$
\begin{aligned}
\left(W_+ H_\alpha u\right)(k) &= \frac{1}{\sqrt{2\pi}} \lim_{\varepsilon \to 0} \int_{|y|>\varepsilon} dy \, \overline{(H_0 \phi_+)(y,k)} \, u(y) \\
&+ \frac{\hbar^2}{2m\sqrt{2\pi}} \lim_{\varepsilon \to 0} \left(\overline{\phi'_+(-\varepsilon,k)} u(-\varepsilon) - \overline{\phi_+(-\varepsilon,k)} u'(-\varepsilon) - \overline{\phi'_+(\varepsilon,k)} u(\varepsilon) + \overline{\phi_+(\varepsilon,k)} u'(\varepsilon) \right) \\
&= \frac{\hbar^2 k^2}{2m} \frac{1}{\sqrt{2\pi}} \int dy \, \overline{\phi_+(y,k)} u(y) = \frac{\hbar^2 k^2}{2m} \hat{u}_+(k) .
\end{aligned}
\tag{8.152}
$$

This concludes the proof of Proposition 8.3 for $\alpha > 0$.

References

1. Landau, L.D., Lifshitz, E.M.: Quantum Mechanics, 3th edn. Pergamon Press, Oxford (1977)
2. Albeverio, S., Gesztesy, F., Hoegh-Krohn, R., Holden, H.: Solvable Models in Quantum Mechanics, 2th edn. with an Appendix by P. Exner. AMS Chelsea Publ, Providence (2004)
3. Agmon, S.: Spectral Properties of Schrödinger Operators and Scattering Theory. Ann. Scuola Norm. Sup. Pisa Cl. Sci. 4(2), 151–218 (1975)
4. Reed, M., Simon, B.: Methods of Modern Mathematical Physics, III: Scattering Theory. Academic Press, New York (1979)
5. Simon, B.: Quantum Mechanics for Hamiltonians Defined as Quadratic Forms. Princeton University Press, Princeton (1971)
6. Thaller, B.: Visual Quantum Mechanics. Springer, New York (2000)
7. Gottfried, K., Yan, T.: Quantum Mechanics: Fundamentals, 2th edn. Springer, New York (2004)
8. Merzbacher, E.: Quantum Mechanics, 3th edn. Wiley, New York (1998)
9. Alonso, A., Simon, B.: The Birman-Krein-Vishik theory of selfadjoint extensions of semi-bounded operators. J. Oper. Theory **4**, 251–270 (1980)
10. Flamand, G.: Mathematical theory of non-relativistic two- and three-particle systems with point interactions. In Cargese Lectures in Theoretical Physics: Application of Mathematics to Problems in Theoretical Physics, pp. 247–287. Gordon and Breach Science Publ., New York (1967)

Chapter 9
Hydrogen Atom

9.1 Self-adjointness

The hydrogen atom is a system made of a proton and an electron subject to their mutual interaction. We study the system under a number of approximations. We consider the nonrelativistic regime and assume that proton and electron are point-like particles. Since the mass of the proton is much larger than the mass of the electron, we consider the proton as a fixed center of force placed at the origin of the reference frame. Moreover, we assume that the interaction between the electron, with mass μ and negative charge $-e$, and the proton, with positive charge e, is described by the attractive electrostatic (or Coulomb) force. Therefore, the electron at the position x is only subject to the Coulomb potential

$$V_e(x) = -\frac{e^2}{|x|}. \tag{9.1}$$

Under these assumptions, the quantum Hamiltonian H_e in $L^2(\mathbb{R}^3)$ describing the motion of the electron is

$$H_e = H_0 + V_e, \qquad H_0 = -\frac{\hbar^2}{2\mu}\Delta, \tag{9.2}$$

In the following, we study the various properties of the Hamiltonian (9.2) (for other treatments of the hydrogen atom which take care of the mathematical details involved we refer to [1–4]). In the first five sections of the chapter, the analysis will be carried out introducing and exploiting some general mathematical techniques which are useful to study also other Hamiltonians of the type $H_0 + V$ in $L^2(\mathbb{R}^3)$, with $V(x) \to 0$ for $|x| \to \infty$. Therefore, this part can be considered as a first introduction to the methods of qualitative analysis of such Hamiltonians.

The first step is to give a meaning to (9.2) as a self-adjoint operator in the Hilbert space of the system $L^2(\mathbb{R}^3)$. To this aim, in the next proposition, we show that the

© Springer International Publishing AG, part of Springer Nature 2018
A. Teta, *A Mathematical Primer on Quantum Mechanics*,
UNITEXT for Physics, https://doi.org/10.1007/978-3-319-77893-8_9

Coulomb potential, as a multiplication operator, is a small perturbation with respect to the free Hamiltonian (recall Definition 4.12).

Proposition 9.1 *There exist* a, $0 < a < 1$, *and* $b > 0$ *such that*

$$\| V_e \phi \| \le a \, \| H_0 \phi \| + b \, \| \phi \| \tag{9.3}$$

for any $\phi \in D(H_0)$.

Proof Denoted by χ_R the characteristic function of $B_R = \{x \in \mathbb{R}^3 \mid |x| < R\}$, we write

$$V_1(x) = V_e(x) \chi_R(x) \,, \qquad V_2(x) = V_e(x) \left(1 - \chi_R(x)\right) , \tag{9.4}$$

so that $V_e = V_1 + V_2$. For any $\phi \in D(H_0)$ we have

$$\| V_e \phi \| \le \| V_1 \phi \| + \| V_2 \phi \| \le \| V_1 \| \sup_{x \in \mathbb{R}^3} |\phi(x)| + \sup_{x \in \mathbb{R}^3} |V_2(x)| \, \| \phi \| , \tag{9.5}$$

where

$$\| V_1 \| = e^2 \, (4\pi R)^{1/2} \,, \qquad \sup_{x \in \mathbb{R}^3} |V_2(x)| = e^2 \, R^{-1} \,. \tag{9.6}$$

It remains to estimate $\sup_{x \in \mathbb{R}^3} |\phi(x)|$ for $\phi \in D(H_0)$. The estimate is obtained using the definition of $D(H_0)$ (see 6.9) and Schwarz inequality. Indeed, fixed an arbitrary $\eta > 0$, we have

$$
\begin{aligned}
|\phi(x)| &\le \frac{1}{(2\pi)^{3/2}} \int dk \, |\tilde{\phi}(k)| = \frac{1}{(2\pi)^{3/2}} \int dk \, \frac{1}{k^2 + \eta^2} \, (k^2 + \eta^2) \, |\tilde{\phi}(k)| \\
&\le \frac{1}{(2\pi)^{3/2}} \left(\int dk \, \frac{1}{(k^2 + \eta^2)^2} \right)^{1/2} \left(\int dk \, (k^2 + \eta^2)^2 \, |\tilde{\phi}(k)|^2 \right)^{1/2} \\
&= \frac{1}{(2\pi)^{3/2}} (4\pi)^{1/2} \eta^{-1/2} \left(\int_0^\infty dy \, \frac{y^2}{(y^2 + 1)^2} \right)^{1/2} \left\| \frac{2\mu}{\hbar^2} H_0 \phi + \eta^2 \phi \right\| \\
&\le \frac{\mu}{\sqrt{2\pi} \, \hbar^2} \, \eta^{-1/2} \, \| H_0 \phi \| + \frac{1}{2\sqrt{2\pi}} \, \eta^{3/2} \, \| \phi \| \,.
\end{aligned}
\tag{9.7}
$$

By (9.7) and (9.5), we find

$$\| V_e \phi \| \le \frac{\mu \, \| V_1 \|}{\sqrt{2\pi} \, \hbar^2} \, \eta^{-1/2} \, \| H_0 \phi \| + \left(\frac{\| V_1 \|}{2\sqrt{2\pi}} \, \eta^{3/2} + \sup_{x \in \mathbb{R}^3} |V_2(x)| \right) \| \phi \| \,. \tag{9.8}$$

Taking η sufficiently large, we obtain (9.3). $\qquad\qquad\qquad\qquad\qquad\qquad\qquad\square$

By Kato–Rellich Theorem 4.8 and Proposition 9.1, we conclude that the operator H_e, with $D(H_e) = D(H_0)$, is self-adjoint in $L^2(\mathbb{R}^3)$.

Remark 9.1 The fact that H_e, $D(H_0)$ is self-adjoint implies that the corresponding unitary group $e^{-i\frac{t}{\hbar}H_e}$ (and then the solution of the Schrödinger equation for any initial datum) is well defined for any t. We recall that the situation in the classical case is different. Indeed, in the case of zero angular momentum, one can choose initial conditions such that the corresponding solutions of Newton's equations exist only for a finite time interval (fall in the center). In this sense, quantum dynamics is more regular than the classical one.

Remark 9.2 One can check that any $\phi \in D(H_0) = H^2(\mathbb{R}^3)$ is a Hölder continuous function of order α, with $\alpha < 1/2$. Indeed, for any $\phi \in H^2(\mathbb{R}^3)$ one has

$$(2\pi)^{3/2}|\phi(x) - \phi(y)| \leq \int dk \, |1 - e^{ik\cdot(x-y)}||\tilde{\phi}(k)| \leq \int dk \, \max\{2, |k||x-y|\}|\tilde{\phi}(k)|$$

$$= \int dk \, \frac{\max\{2, |k||x-y|\}}{k^2+1}|(k^2+1)\tilde{\phi}(k)| . \qquad (9.9)$$

Then, note that $\max\{2, |k||x-y|\} \leq 2^{1-\alpha}|k|^\alpha |x-y|^\alpha$ and use Schwarz inequality to obtain $|\phi(x) - \phi(y)| \leq C_\alpha |x-y|^\alpha \|(-\Delta+1)\phi\|$, where C_α is a positive constant depending on α.

Using the same kind of arguments of Kato–Rellich theorem, we obtain a first spectral information on the Hamiltonian H_e.

Proposition 9.2 *There exists $\lambda_0 > 0$ such that* $\inf \sigma(H_e) \geq -\lambda_0$.

Proof Let $R_0(-\lambda) = (H_0 + \lambda)^{-1}$, with $\lambda > 0$, and let $\psi \in L^2(\mathbb{R}^3)$. Since $R_0(-\lambda)\psi \in D(H_0)$, we have

$$\|V_e \, R_0(-\lambda)\psi\| \leq a \, \|H_0 R_0(-\lambda)\psi\| + b \, \|R_0(-\lambda)\psi\| \leq \left(a + \frac{b}{\lambda}\right)\|\psi\| . \qquad (9.10)$$

Therefore, we can find $\lambda_0 > 0$ such that for $\lambda > \lambda_0$ the operator $V_e \, R_0(-\lambda)$ has a norm less than one. This implies that the corresponding Neumann series

$$\sum_{n=0}^{\infty} (-1)^n [V_e \, R_0(-\lambda)]^n \qquad (9.11)$$

converges in norm to the bounded operator $(I + V_e \, R_0(-\lambda))^{-1}$. It follows that, for $\lambda > \lambda_0$, the bounded operator

$$R_0(-\lambda) \, (I + V_e \, R_0(-\lambda))^{-1} \qquad (9.12)$$

coincides with the resolvent

$$R_e(-\lambda) = (H_e + \lambda)^{-1} . \qquad (9.13)$$

As a consequence, $-\lambda$ belongs to the resolvent set for $\lambda > \lambda_0$ and this concludes the proof.   □

Remark 9.3 It is easy to see that Propositions 9.1 and 9.2 still hold if the Coulomb potential is replaced by a generic potential V written as the sum $V = V_1 + V_2$, with $V_1 \in L^2(\mathbb{R}^3)$ and $V_2 \in L^\infty(\mathbb{R}^3)$. Therefore, we conclude that any Hamiltonian $H_0 + V$, with $V = V_1 + V_2$, where $V_1 \in L^2(\mathbb{R}^3)$ and $V_2 \in L^\infty(\mathbb{R}^3)$, is self-adjoint on $D(H_0)$ and its spectrum is bounded from below. In particular, the result holds for the Hamiltonian

$$H = H_0 - \frac{k}{|x|^\alpha} \tag{9.14}$$

if $\alpha \in (0, 3/2)$ and $k \in \mathbb{R}$. In fact, a more accurate analysis [5] shows that (9.14) is self-adjoint and bounded from below also for $\alpha \in [3/2, 2)$.

9.2 Essential Spectrum

We want to characterize the essential spectrum of the Hamiltonian H_e making use of Weyl's Theorem 4.30. Let us consider the resolvents of H_e and H_0 evaluated for $z = -\lambda$, $-\lambda < \inf \sigma(H_e)$ and let us look for a representation of $R_e(-\lambda)$ in terms of $R_0(-\lambda)$. The resolvent identity (4.81) yields $R_e(-\lambda) = R_0(-\lambda) - R_e(-\lambda)V_eR_0(-\lambda)$. Using the fact that R_e is given by (9.12), we find the representation

$$R_e(-\lambda) - R_0(-\lambda) = -R_0(-\lambda)(I + V_eR_0(-\lambda))^{-1}V_eR_0(-\lambda). \tag{9.15}$$

In particular, (9.15) shows that it is sufficient to prove that $V_eR_0(-\lambda)$ is compact to obtain that the difference $R_e(-\lambda) - R_0(-\lambda)$ is compact and then to apply Weyl's Theorem 4.30.

Proposition 9.3 *The difference $R_e(-\lambda) - R_0(-\lambda)$ is a compact operator and then*

$$\sigma_{ess}(H_e) = [0, \infty). \tag{9.16}$$

Proof We decompose the Coulomb potential as in the proof of Proposition 9.1

$$V_e = V_n + W_n, \qquad V_n = V_e\chi_n, \qquad W_n = V_e(1-\chi_n), \tag{9.17}$$

where we recall that χ_n is the characteristic function of the sphere of radius n and center in the origin. Let us observe that $V_n \in L^2(\mathbb{R}^3)$, $W_n \in L^\infty(\mathbb{R}^3)$ and $\lim_n \|W_n\|_{L^\infty(\mathbb{R}^3)} = 0$. The operator $V_nR_0(-\lambda)$ is an integral operator with kernel

$$V_n(x)\frac{2\mu}{\hbar^2}\frac{e^{-\sqrt{\lambda}\frac{\sqrt{2\mu}}{\hbar}|x-y|}}{4\pi|x-y|}, \tag{9.18}$$

where we have used the explicit form of the kernel of the free Hamiltonian (see 6.36). A straightforward computation for $c > 0$

$$\int dxdy\, |V_n(x)|^2 \frac{e^{-c|x-y|}}{|x-y|^2} = \int dx\, |V_n(x)|^2 \int dy\, \frac{e^{-c|x-y|}}{|x-y|^2}$$

$$= 4\pi \int dx\, |V_n(x)|^2 \int_0^\infty dr\, e^{-cr} < \infty \qquad (9.19)$$

shows that $V_n R_0(-\lambda)$ is a Hilbert–Schmidt, and then compact, operator. Moreover, we observe that

$$\| V_e R_0(-\lambda) - V_n R_0(-\lambda) \| = \| W_n R_0(-\lambda) \| \leq \| W_n \|_{L^\infty(\mathbb{R}^3)} \| R_0(-\lambda) \| \to 0 \quad (9.20)$$

for $n \to \infty$. It follows that the operator $V_e R_0(-\lambda)$ is the limit in norm of the sequence of compact operators $V_n R_0(-\lambda)$ and, therefore, it is compact. As a consequence, the same is true for the right-hand side of (9.15) and thus we obtain that $R_e(-\lambda) - R_0(-\lambda)$ is a compact operator. By Weyl's theorem, we conclude that $\sigma_{ess}(H_e) = \sigma_{ess}(H_0) = [0, \infty)$. $\qquad\qquad\qquad \Box$

Remark 9.4 It is easy to see that proposition 9.3 still holds if the Coulomb potential is replaced by a generic potential V written as the sum $V = V_n + W_n$, with $V_n \in L^2(\mathbb{R}^3)$, $W_n \in L^\infty(\mathbb{R}^3)$ and $\lim_n \| W_n \|_{L^\infty(\mathbb{R}^3)} = 0$. Therefore, for any V of this type, we have $\sigma_{ess}(H_0 + V) = [0, \infty)$.

In particular, the result hold for the Hamiltonian (9.14) with $\alpha \in (0, 3/2)$. With a different method [6], it also extends to the case $\alpha \in [3/2, 2)$.

9.3 Absence of Positive Eigenvalues

We shall prove that the eigenvalues of H_e are negative and then $\sigma_{ess}(H_e) = \sigma_c(H_e)$. The proof is based on the properties of the family of operators in $L^2(\mathbb{R}^3)$ defined by

$$(D_s f)(x) = e^{3s/2} f(e^s x), \qquad s \in \mathbb{R}. \qquad (9.21)$$

It is immediately seen that (9.21) is a strongly continuous one-parameter group of unitary operators, called dilation group.

Exercise 9.1 Verify that

$$\lim_{s \to 0} \frac{D_s f - f}{s} = x \cdot \nabla f + \frac{3}{2} f, \qquad f \in \mathscr{S}(\mathbb{R}^3). \qquad (9.22)$$

For $f \in D(H_0)$, we have $D_s f \in D(H_0)$ and

$$(H_0\, D_{-s}\, f)(x) = -\frac{\hbar^2}{2m}\Delta\, e^{-3s/2} f(e^{-s}x) = e^{-3s/2}e^{-2s}(H_0 f)(e^{-s}x)\,. \quad (9.23)$$

Then

$$(D_s\, H_0\, D_{-s}\, f)(x) = e^{-2s}(H_0 f)(x)\,. \qquad (9.24)$$

Moreover,

$$(D_s\, V_e\, D_{-s}\, f)(x) = e^{-s}(V_e f)(x)\,. \qquad (9.25)$$

Formulas (9.24), (9.25) are the main ingredients of the proof of the following proposition.

Proposition 9.4 (virial theorem)
Let E be an eigenvalue of H_e, $D(H_0)$ and let $\psi \in D(H_0)$, $\|\psi\| = 1$, be a correspondent eigenvector. Then

$$E = -(\psi,\, H_0\, \psi) = \frac{1}{2}(\psi,\, V_e\, \psi)\,. \qquad (9.26)$$

In particular, it follows that $E < 0$.

Proof Using $(H_e - E)\psi = 0$, Eqs. (9.24), (9.25) and unitarity of D_s, we have

$$
\begin{aligned}
0 = \big(D_{-s}\psi,\, (H_e - E)\psi\big) &= \big(\psi,\, D_s(H_e - E)\psi\big) = \big(\psi,\, (D_s H_e D_{-s} - E)D_s\psi\big) \\
&= \big(\psi,\, (e^{-2s}H_0 + e^{-s}V_e - E)\, D_s\,\psi\big) = \big((e^{-2s}H_0 + e^{-s}V_e - E)\,\psi,\, D_s\psi\big)\,.
\end{aligned}
$$
$$(9.27)$$

Taking also into account that $\big((H_e - E)\psi,\, D_s\psi\big) = 0$, we write

$$
\begin{aligned}
0 &= \big((H_e - E)\psi,\, D_s\psi\big) - \big((e^{-2s}H_0 + e^{-s}V_e - E)\,\psi,\, D_s\psi\big) \\
&= \big((1 - e^{-2s})\, H_0\psi + (1 - e^{-s})\, V_e\psi,\, D_s\psi\big)\,.
\end{aligned}
$$
$$(9.28)$$

Multiplying by s^{-1} and taking the limit $s \to 0$, we obtain

$$
\begin{aligned}
0 &= \lim_{s\to 0}\left(\frac{1 - e^{-2s}}{s}\, H_0\psi + \frac{1 - e^{-s}}{s}\, V_e\psi,\, D_s\,\psi\right) = (2H_0\psi + V_e\psi,\, \psi) \\
&= 2(\psi,\, H_0\psi) + (\psi,\, V_e\psi)\,.
\end{aligned}
$$
$$(9.29)$$

Equation (9.29) implies

$$-(\psi,\, H_0\psi) = (\psi,\, H_0\psi) + (\psi,\, V_e\psi) = (\psi,\, H_e\psi) = E \qquad (9.30)$$

Replacing (9.30) in (9.29), we also have $(\psi,\, V_e\psi) = 2E$ and this concludes the proof. $\square$

Remark 9.5 Virial theorem also implies that the Hamiltonian with repulsive Coulomb potential $-V_e(x)$ has no eigenvalues.

Exercise 9.2 Prove the absence of positive eigenvalues for

$$H = H_0 - \frac{k}{|x|^\alpha}, \qquad \alpha \in (0, 2). \tag{9.31}$$

Remark 9.6 Virial theorem is based on properties (9.24), (9.25) and then it cannot be generalized to arbitrary potentials. However, following different, and nontrivial, methods, the absence of positive eigenvalues can be proved for a large class of Hamiltonians $H_0 + V$, with $\lim_{|x| \to \infty} V(x) = 0$ (see, e.g., [6], Chap. XIII, Sect. 13).

9.4 Existence of Negative Eigenvalues

From the analysis developed in the previous sections, it follows that the negative eigenvalues of H_e, if they exist, are contained in $[\gamma_\sigma, 0)$, where $\gamma_\sigma := \inf \sigma(H_e)$. Moreover, from Weyl's theorem we know that $[\gamma_\sigma, 0)$ can only contain the discrete spectrum and this means that the eigenvalues are isolated and have finite multiplicity. Here, we prove that there are infinite negative eigenvalues which accumulate to zero.

Proposition 9.5 *The point spectrum of H_e consists of an infinite sequence $\{E_n\}$ of negative eigenvalues with finite multiplicity, where*

$$E_n < E_{n+1}, \qquad E_1 = \inf \sigma(H_e), \qquad \lim_n E_n = 0. \tag{9.32}$$

Proof We first observe that for $\psi \in D(H_0)$ and $s > 0$ sufficiently large one has

$$(D_{-s}\psi, H_e D_{-s}\psi) = e^{-2s}(\psi, H_0\psi) + e^{-s}(\psi, V_e\psi) < 0. \tag{9.33}$$

By Proposition 4.22, this implies that $\gamma_\sigma < 0$ and then there exists at least one negative eigenvalue.

Now we fix $f \in C_0^\infty(\mathbb{R}^3)$ with $\|f\| = 1$, supp $f \subset \{x \in \mathbb{R}^3 \mid 1 \le |x| \le 2\}$ and consider

$$f_n(x) = (D_{-s_n} f)(x) = 3^{-\frac{3}{2}n} f(3^{-n}x), \qquad s_n = n \log 3. \tag{9.34}$$

Since supp $f_n \subset \{x \in \mathbb{R}^3 \mid 3^n \le |x| \le 2 \cdot 3^n\}$, we obtain

$$\text{supp } f_n \cap \text{supp } f_m = \emptyset, \qquad n \ne m \tag{9.35}$$

and then the sequence $\{f_n\}$ is orthonormal. Moreover, there exists an integer n_0 such that for $n, m > n_0$ one has

$$(f_n, H_e f_m) = 0 \quad \text{if } n \ne m, \qquad (f_n, H_e f_m) < 0 \quad \text{if } n = m. \tag{9.36}$$

Making use of the sequence $\{f_{n_0+n}\}$, which we still denote $\{f_n\}$, we want to prove that the dimension of $Ran\, E_{H_e}((-\infty, 0)) = Ran\, E_{H_e}([\gamma_\sigma, 0))$ is infinite.

Let X be the subspace generated by $\{f_n\}$, i.e., $g \in X$ iff $g = \sum_{j=1}^{\infty} a_j f_j$, with $a_j = (f_j, g)$. We note that the following implication holds:

$$g \in X, \qquad E_{H_e}((-\infty, 0))g = 0 \qquad \Rightarrow \qquad g = 0. \tag{9.37}$$

Indeed, assuming that $g \in X$ and $E_{H_e}((-\infty, 0))g = 0$, we have

$$(g, H_e g) = \int \lambda\, d(E_{H_e}(\lambda)g, g) = \int_{[0,\infty)} \lambda\, d(E_{H_e}(\lambda)g, g) \geq 0 \tag{9.38}$$

and

$$(g, H_e g) = \sum_{j,l=1}^{\infty} \overline{a_j} a_l (f_j, H_e f_l) = \sum_{j=1}^{\infty} |a_j|^2 (f_j, H_e f_j) \leq 0. \tag{9.39}$$

The two inequalities imply $a_j = 0$ for any j and then $g = 0$.

Let us consider $\varphi_j = E_{H_e}((-\infty, 0))f_j \in Ran\, E_{H_e}((-\infty, 0))$. The vectors φ_j are linearly independent. In fact, if

$$0 = \sum_{j=1}^{\infty} c_j \varphi_j = \sum_{j=1}^{\infty} c_j E_{H_e}((-\infty, 0))f_j = E_{H_e}((-\infty, 0)) \sum_{j=1}^{\infty} c_j f_j \tag{9.40}$$

then, by (9.37), we have $\sum_{j=1}^{\infty} c_j f_j = 0$ and this implies $c_j = 0$ for any j. Thus, the subspace $Ran\, E_{H_e}((-\infty, 0)) = Ran\, E_{H_e}([\gamma_\sigma, 0))$ is infinite dimensional since it contains the infinite set of linearly independent vectors $\{\varphi_1, \ldots \varphi_j, \ldots\}$. Let us now recall that the bounded interval $[\gamma_\sigma, 0)$ can only contain isolated eigenvalues with finite multiplicity. Then, using Proposition 4.27, we conclude that there is an infinite sequence of negative eigenvalues which accumulate to zero and that the minimum eigenvalue coincides with γ_σ.   □

Exercise 9.3 Study the existence of negative eigenvalues for

$$H = H_0 - \frac{k}{|x|^\alpha}, \qquad \alpha \in (0, 2), \qquad k > 0. \tag{9.41}$$

Remark 9.7 The existence of negative eigenvalues, and then the characterization of the discrete spectrum, can be studied for generic Hamiltonians $H = H_0 + V$, with $V(x) \to 0$ for $|x| \to \infty$, using the min-max principle (see, e.g., [2, 6]). It turns out that the analysis strongly depends on the behavior of $V(x)$ for $|x|$ large. More precisely, it is possible to prove that $\sigma_d(H)$ is infinite if there exist $R_0 > 0$, $a > 0$, $\varepsilon > 0$ such that $V(x) \leq -a|x|^{-2+\varepsilon}$ for $|x| > R_0$. On the other hand, $\sigma_d(H)$ is finite if there exist $R_0 > 0$, $b < 1$ such that $V(x) \geq -\frac{\hbar^2 b}{8m}|x|^{-2}$ for $|x| > R_0$. Moreover, when $\sigma_d(H)$ is finite, various different bounds on the number of negative eigenvalues can be proved (for details we refer to [6], Chap. XIII, Sect. 3).

To summarize, by the analysis developed so far, we have obtained

$$\sigma_c(H_e) = [0, \infty), \qquad \sigma_p(H_e) = \sigma_d(H_e) = \{E_n\}, \qquad (9.42)$$

with E_n satisfying (9.32).

Remark 9.8 It is also possible to prove that $\sigma_{sc}(H_e) = \emptyset$ (see, e.g., [4], Proposition 4.1.16), so that $\sigma_c(H_e) = \sigma_{ac}(H_e)$. For general conditions that guarantee $\sigma_{sc}(H) = \emptyset$ for a Hamiltonian $H = H_0 + V$ we refer to [6], Chap. XIII, Sect. 6.7.

9.5   Estimate of the Minimum Eigenvalue

In this section, we describe an elementary estimate of the minimum eigenvalue (or ground state energy). For the estimate from below, it is useful the following inequality.

Proposition 9.6   (Hardy inequality)
For any $\psi \in H^1(\mathbb{R}^3)$, we have

$$\int dx \, |\nabla \psi(x)|^2 \geq \int dx \, \frac{|\psi(x)|^2}{4|x|^2} . \qquad (9.43)$$

Proof For any $a \in \mathbb{R}$, we write

$$0 \leq \int dx \left| \nabla \psi(x) + a \frac{x}{|x|^2} \psi(x) \right|^2$$

$$= \int dx \, |\nabla \psi(x)|^2 + a^2 \int dx \, \frac{|\psi(x)|^2}{|x|^2} + 2a \, \Re \int dx \, \psi(x) \frac{x}{|x|^2} \cdot \nabla \bar{\psi}(x) . (9.44)$$

Taking into account that $\nabla |\psi(x)|^2 = 2 \Re(\psi(x) \nabla \bar{\psi}(x))$ and integrating by parts in the last integral, we find

$$0 \leq \int dx \, |\nabla \psi(x)|^2 + a^2 \int dx \, \frac{|\psi(x)|^2}{|x|^2} - a \int dx \left(\nabla \cdot \frac{x}{|x|^2} \right) |\psi(x)|^2 . \qquad (9.45)$$

Since $\nabla \cdot (x \, |x|^{-2}) = |x|^{-2}$, we obtain

$$0 \leq \int dx \, |\nabla \psi(x)|^2 + (a^2 - a) \int dx \, \frac{|\psi(x)|^2}{|x|^2} . \qquad (9.46)$$

Taking $a = 1/2$ we conclude the proof.   □

Hardy inequality provides an estimate from below of the mean value of the kinetic energy (for $\hbar = 2\mu = 1$) in terms of the mean value of the "potential" $(4|x|^2)^{-1}$. This allows a simple estimate from below of the minimum eigenvalue. Indeed, for any $\psi \in D(H_0)$, $\|\psi\| = 1$, we have

$$
\begin{aligned}
(\psi, H_e\psi) &= \frac{\hbar^2}{2\mu} \int dx\, |\nabla\psi(x)|^2 - e^2 \int dx\, \frac{|\psi(x)|^2}{|x|} \\
&\geq \int dx\, \left(\frac{\hbar^2}{8\mu|x|^2} - \frac{e^2}{|x|} \right) |\psi(x)|^2 \geq \inf_{r\geq 0} \left(\frac{\hbar^2}{8\mu r^2} - \frac{e^2}{r} \right) \\
&= -\frac{2\mu e^4}{\hbar^2}.
\end{aligned}
\tag{9.47}
$$

It follows the estimate

$$
E_1 = \inf_{\psi \in D(H_0), \|\psi\|=1} (\psi, H_e\psi) \geq -\frac{2\mu e^4}{\hbar^2}.
\tag{9.48}
$$

A simple estimate from above of the minimum eigenvalue is obtained using trial functions. More precisely, given a function $\phi_{\alpha,\beta,\ldots} \in D(H_0)$, $\|\phi_{\alpha,\beta,\ldots}\| = 1$, depending on some parameters $\alpha, \beta, \ldots$, one has

$$
E_1 \leq \inf_{\alpha,\beta,\ldots} (\phi_{\alpha,\beta,\ldots}, H_e\, \phi_{\alpha,\beta,\ldots}).
\tag{9.49}
$$

Exercise 9.4 Given the trial function $\phi_\alpha(x) = A\, e^{-\alpha|x|}$, verify that

$$
E_1 \leq -\frac{\mu e^4}{2\hbar^2}.
\tag{9.50}
$$

We shall see later that the above estimate is optimal, i.e., the right-hand side coincides with the exact value of the minimum eigenvalue.

Exercise 9.5 Estimate from below and from above the minimum eigenvalue of the Hamiltonian (9.41).

9.6 Eigenvalue Problem in Spherical Coordinates

In the previous sections, we have developed the qualitative analysis of the Hamiltonian H_e, obtaining various information concerning the structure of the spectrum. From now on, we concentrate on the explicit solution of the eigenvalue problem, i.e., on the computation of the eigenvalues, or energy levels, and of the eigenvectors, or stationary states. More precisely, we approach the solution of the following problem:

$$H_e u = E u, \quad u \in D(H_0), \quad \|u\| = 1, \quad E < 0. \tag{9.51}$$

Remark 9.9 We know that a solution of problem (9.51) is a Hölder continuous function of order α, with $\alpha < 1/2$, in $\mathbb{R}^3$ (see Remark 9.2). Using the fact that the only singularity of the Coulomb potential is at the origin, by a bootstrap argument one can show that $u \in C^\infty(\mathbb{R}^3 \backslash \{0\})$ (see, e.g., [7], Sect. 11.10).

We note that the invariance under rotations of the Hamiltonian H_e implies, via Noether's theorem, the conservation law of the angular momentum (see Sect. 5.2 and exercise below). We shall use this fact to solve problem (9.51), in complete analogy with the study of the Kepler problem in Classical Mechanics.

Exercise 9.6 Consider the family of operators $R^{(3)}(\lambda)$, $\lambda \in \mathbb{R}$, in $L^2(\mathbb{R}^3)$

$$(R^{(3)}(\lambda)f)(x_1, x_2, x_3) := f(x_1 \cos \lambda + x_2 \sin \lambda, -x_1 \sin \lambda + x_2 \cos \lambda, x_3). \tag{9.52}$$

For each λ, (9.52) defines the rotation around the x_3-axis of the angle λ. Verify that $R^{(3)}(\lambda)$ is a symmetry for H_e. Moreover, for any f differentiable prove that

$$\frac{d}{d\lambda}(R^{(3)}(\lambda)f)(x_1, x_2, x_3)\Big|_{\lambda=0} = -i \hbar^{-1}(L_3 f)(x_1, x_2, x_3), \tag{9.53}$$

where L_3 is the third component of the angular momentum (see Exercise 3.3). This means that $R^{(3)}(\lambda) = e^{-i\lambda\hbar^{-1}L_3}$ and the generator $\hbar^{-1}L_3$ is a constant of motion. The same is obviously true for the other components of the angular momentum.

The solution of problem (9.51) will be obtained through various steps that we summarize for the convenience of the reader.

- In the rest of this section, we shall find the expression of the Hamiltonian and of the angular momentum in spherical coordinates and we shall reformulate problem (9.51) in such coordinates. Noting that H_e, L_3 and the square L^2 of the angular momentum commute, we look for a system of common eigenvectors of the three operators.
- In Sect. 9.7, we determine the eigenvalues and the common eigenvectors (the spherical harmonics) of L_3 and L^2.
- In Sect. 9.8, we show that the search of the eigenvectors of H_e that are also eigenvectors of L_3 and L^2 reduces to the study of the eigenvalue problem for an ordinary differential equation in the radial variable.
- Finally, in Sect. 9.9 we solve such eigenvalue problem and thus determine eigenvalues and eigenvectors of H_e.

Given $x = (x_1, x_2, x_3)$, we introduce the change of coordinates

$$x_1 = r \sin \theta \cos \phi, \quad x_2 = r \sin \theta \sin \phi, \quad x_3 = r \cos \theta, \tag{9.54}$$

with inverse

$$r = \sqrt{x_1^2 + x_2^2 + x_3^2}, \qquad \theta = \cos^{-1} \frac{x_3}{\sqrt{x_1^2 + x_2^2 + x_3^2}}, \qquad \phi = \tan^{-1} \frac{x_2}{x_1}. \quad (9.55)$$

Moreover, denoted by S^2 the unit sphere in $\mathbb{R}^3$ equipped with the measure $d\Omega = \sin\theta d\theta d\phi$, we consider the Hilbert space

$$\mathcal{H}_S = L^2((0, \infty), \ r^2 dr) \otimes L^2(S^2, d\Omega), \qquad (9.56)$$

with norm

$$\|f\|_{\mathcal{H}_S}^2 = \int_0^\infty dr\, r^2 \int_{S^2} d\Omega\, |f(r, \theta, \phi)|^2, \qquad (9.57)$$

the unitary operator

$$U_S : L^2(\mathbb{R}^3) \to \mathcal{H}_S, \qquad (9.58)$$
$$u(x_1, x_2, x_3) \to (U_S u)(r, \theta, \phi) = u(r \sin\theta \cos\phi, r \sin\theta \sin\phi, r \cos\theta) \quad (9.59)$$

and the Hamiltonian in spherical coordinates

$$\hat{H}_e := U_S H_e U_S^{-1}. \qquad (9.60)$$

In order to find the explicit expression of $\hat{H}_e$, we observe that

$$\left(\frac{\partial}{\partial x_j} U_S^{-1} f \right) = \frac{\partial f}{\partial r} \frac{\partial r}{\partial x_j} + \frac{\partial f}{\partial \theta} \frac{\partial \theta}{\partial x_j} + \frac{\partial f}{\partial \phi} \frac{\partial \phi}{\partial x_j} \qquad (9.61)$$

and

$$\begin{aligned}
\left(\frac{\partial^2}{\partial x_j^2} U_S^{-1} f \right) &= \frac{\partial}{\partial x_j} \left(\frac{\partial f}{\partial r} \frac{\partial r}{\partial x_j} + \frac{\partial f}{\partial \theta} \frac{\partial \theta}{\partial x_j} + \frac{\partial f}{\partial \phi} \frac{\partial \phi}{\partial x_j} \right) \\
&= \left(\frac{\partial^2 f}{\partial r^2} \frac{\partial r}{\partial x_j} + \frac{\partial^2 f}{\partial \theta \partial r} \frac{\partial \theta}{\partial x_j} + \frac{\partial^2 f}{\partial \phi \partial r} \frac{\partial \phi}{\partial x_j} \right) \frac{\partial r}{\partial x_j} + \frac{\partial f}{\partial r} \frac{\partial^2 r}{\partial x_j^2} \\
&\quad + \left(\frac{\partial^2 f}{\partial r \partial \theta} \frac{\partial r}{\partial x_j} + \frac{\partial^2 f}{\partial \theta^2} \frac{\partial \theta}{\partial x_j} + \frac{\partial^2 f}{\partial \phi \partial \theta} \frac{\partial \phi}{\partial x_j} \right) \frac{\partial \theta}{\partial x_j} + \frac{\partial f}{\partial \theta} \frac{\partial^2 \theta}{\partial x_j^2} \\
&\quad + \left(\frac{\partial^2 f}{\partial r \partial \phi} \frac{\partial r}{\partial x_j} + \frac{\partial^2 f}{\partial \theta \partial \phi} \frac{\partial \theta}{\partial x_j} + \frac{\partial^2 f}{\partial \phi^2} \frac{\partial \phi}{\partial x_j} \right) \frac{\partial \phi}{\partial x_j} + \frac{\partial f}{\partial \phi} \frac{\partial^2 \phi}{\partial x_j^2}. \quad (9.62)
\end{aligned}$$

Making use of (9.55) and (9.62), we find

$$\begin{aligned}
\hat{H}_0 f &:= U_S H_0 U_S^{-1} f \\
&= -\frac{\hbar^2}{2\mu r^2} \frac{\partial}{\partial r} \left(r^2 \frac{\partial f}{\partial r} \right) - \frac{\hbar^2}{2\mu r^2} \left[\frac{1}{\sin\theta} \frac{\partial}{\partial \theta} \left(\sin\theta \frac{\partial f}{\partial \theta} \right) + \frac{1}{\sin^2\theta} \frac{\partial^2 f}{\partial \phi^2} \right]. \quad (9.63)
\end{aligned}$$

Adding the Coulomb potential, we obtain

$$\hat{H}_e f = -\frac{\hbar^2}{2\mu r^2}\frac{\partial}{\partial r}\left(r^2\frac{\partial f}{\partial r}\right) - \frac{\hbar^2}{2\mu r^2}\left[\frac{1}{\sin\theta}\frac{\partial}{\partial\theta}\left(\sin\theta\frac{\partial f}{\partial\theta}\right) + \frac{1}{\sin^2\theta}\frac{\partial^2 f}{\partial\phi^2}\right] - \frac{e^2}{r}f$$

(9.64)

and the eigenvalue problem (9.51) is equivalently reformulated as

$$\hat{H}_e\psi = E\psi, \quad \psi \in \mathscr{H}_S, \quad \psi = U_S u, \quad u \in D(H_0), \quad \|\psi\|_{\mathscr{H}_S} = 1, \quad E < 0.$$

(9.65)

For the angular momentum, we have

$$L_1 f = U_S\frac{\hbar}{i}\left(x_2\frac{\partial}{\partial x_3} - x_3\frac{\partial}{\partial x_2}\right)U_S^{-1}f = i\hbar\left(\sin\phi\frac{\partial f}{\partial\theta} + \frac{\cos\phi}{\tan\theta}\frac{\partial f}{\partial\phi}\right), \quad (9.66)$$

$$L_2 f = U_S\frac{\hbar}{i}\left(x_3\frac{\partial}{\partial x_1} - x_1\frac{\partial}{\partial x_3}\right)U_S^{-1}f = i\hbar\left(-\cos\phi\frac{\partial f}{\partial\theta} + \frac{\sin\phi}{\tan\theta}\frac{\partial f}{\partial\phi}\right), \quad (9.67)$$

$$L_3 f = U_S\frac{\hbar}{i}\left(x_1\frac{\partial}{\partial x_2} - x_2\frac{\partial}{\partial x_1}\right)U_S^{-1}f = \frac{\hbar}{i}\frac{\partial f}{\partial\phi} \quad (9.68)$$

and

$$L^2 f = L_1^2 f + L_2^2 f + L_3^2 f = -\frac{\hbar^2}{\sin\theta}\frac{\partial}{\partial\theta}\left(\sin\theta\frac{\partial f}{\partial\theta}\right) + \frac{1}{\sin^2\theta}L_3^2 f . (9.69)$$

The operators L_1, L_2, L_3, L^2 depend only on the angular variables, i.e., they act nontrivially only in $L^2(S_r^2, d\Omega)$. We define them on the domain of smooth functions

$$D_L = \left\{f \mid f \in C^2([0,\pi] \times [0,2\pi]), f(\theta,0) = f(\theta,2\pi),\right.$$
$$\left.\frac{\partial f}{\partial\phi}(\theta,0) = \frac{\partial f}{\partial\phi}(\theta,2\pi) \quad \forall\theta \in [0,\pi]\right\}.$$

(9.70)

Exercise 9.7 Verify that L_1, L_2, L_3, L^2 defined on D_L are symmetric in $L^2(S_r^2, d\Omega)$. Moreover, show that L^2 is positive and $L^2 f = 0$ if and only if f is constant.

Note that $L^2 = -\hbar^2\Delta_\Omega$, where Δ_Ω is the Laplace–Beltrami operator on S^2.

Remark 9.10 The three components L_1, L_2, L_3 do not commute. Indeed, for $f \in D_L$, one has

$$[L_1, L_2]f = i\hbar L_3 f, \quad [L_2, L_3]f = i\hbar L_1 f, \quad [L_3, L_1]f = i\hbar L_2 f . \quad (9.71)$$

On the other hand, it is easy to see that each component commutes with L^2. Then it is possible to determine a system of common eigenvectors of L^2 and one of the component. In the next section, we shall determine the common eigenvectors of L_3 and L^2.

Comparing the expression of L^2 with (9.64), we can write

$$\hat{H}_e f = -\frac{\hbar^2}{2\mu r^2} \frac{\partial}{\partial r} \left(r^2 \frac{\partial f}{\partial r} \right) + \frac{1}{2\mu r^2} L^2 f - \frac{e^2}{r} f . \tag{9.72}$$

From formulas (9.68), (9.69), (9.72), it follows that L_3, L^2, $\hat{H}_e$ commute. This implies that L_3 and L^2 are constants of motion and, as we shall see in Sects. 9.8 and 9.9, we can determine a system eigenvectors of $\hat{H}_e$ that are also eigenvectors of L_3 and L^2.

9.7   Eigenvalue Problem for the Angular Momentum

Let us consider the operators L_3 and L^2 in the Hilbert space $L^2(S_r^2, d\Omega)$, where norm and scalar product are denoted (only in this section) by $\| \cdot \|$, $(\cdot, \cdot)$. We want to find $\nu, \lambda \in \mathbb{R}$ and $F_{\lambda\nu}$, such that

$$L_3 F_{\lambda\nu} = \hbar \nu F_{\lambda\nu} , \tag{9.73}$$

$$L^2 F_{\lambda\nu} = \hbar^2 \lambda F_{\lambda\nu} , \tag{9.74}$$

$$F_{\lambda\nu} \in D_L , \qquad \| F_{\lambda\nu} \| = 1 . \tag{9.75}$$

Taking into account that L_3 acts only on the variable ϕ, it is convenient to first solve equation (9.73). It is easily seen that problem (9.73), (9.75) has solution if and only if $\nu = m$, with $m \in \mathbb{Z}$, and the solution is

$$F_{\lambda m}(\theta, \phi) = \Theta_{\lambda m}(\theta) \Phi_m(\phi) , \qquad \Phi_m(\phi) = \frac{e^{im\phi}}{\sqrt{2\pi}} \tag{9.76}$$

for any function $\Theta_{\lambda m}$ satisfying the normalization condition

$$\int_0^\pi d\theta \, \sin\theta \, |\Theta_{\lambda m}(\theta)|^2 = 1 . \tag{9.77}$$

Then, the problem is reduced to find $\Theta_{\lambda m}$ in such a way that $F_{\lambda m}$ is also solution of (9.74), (9.75). We shall follow an algebraic method similar to that used for the harmonic oscillator and based on the factorization property of L^2. More precisely, we define

$$L_\pm = L_1 \pm i L_2 = \hbar e^{\pm i\phi} \left(\pm \frac{\partial}{\partial\theta} + i \cot\theta \frac{\partial}{\partial\phi} \right) \tag{9.78}$$

and we observe that the following properties hold for $f, g \in D_L$

$$(g, L_\pm f) = (L_\mp g, f),\tag{9.79}$$

$$[L_+, L_-]f = 2\hbar L_3 f, \quad [L_3, L_\pm]f = \pm\hbar L_\pm f, \quad [L^2, L_\pm]f = 0,\tag{9.80}$$

$$L^2 f = L_+ L_- f + L_3^2 f - \hbar L_3 f,\tag{9.81}$$

$$L^2 f = L_- L_+ f + L_3^2 f + \hbar L_3 f.\tag{9.82}$$

The operators $L_\pm$ behave as creation and annihilation operators with respect to problem (9.73), (9.75).

Proposition 9.7 *Let $F_{\lambda m}$ be solution of problem (9.73), (9.74), (9.75).*

(i) If $\lambda > m(m+1)$ then

$$\frac{L_+ F_{\lambda m}}{\hbar\sqrt{\lambda - m(m+1)}}\tag{9.83}$$

is eigenvector of L_3 with eigenvalue $\hbar(m+1)$ and of L^2 with the same eigenvalue $\hbar^2\lambda$.

(ii) If $\lambda > m(m-1)$ then

$$\frac{L_- F_{\lambda m}}{\hbar\sqrt{\lambda - m(m-1)}}\tag{9.84}$$

is eigenvector of L_3 with eigenvalue $\hbar(m-1)$ and of L^2 with the same eigenvalue $\hbar^2\lambda$.

Proof Let us consider the vector $L_+ F_{\lambda m}$. Using the equation $[L_3, L_+] = \hbar L_+$, we have

$$L_3 L_+ F_{\lambda m} = L_+ L_3 F_{\lambda m} + \hbar L_+ F_{\lambda m} = \hbar m L_+ F_{\lambda m} + \hbar L_+ F_{\lambda m} = \hbar(m+1)L_+ F_{\lambda m}.\tag{9.85}$$

Moreover, using (9.79), (9.82), we obtain

$$\|L_+ F_{\lambda m}\|^2 = \hbar^2(\lambda - m(m+1)) > 0.\tag{9.86}$$

Thus,

$$F_{\lambda, m+1} := \frac{L_+ F_{\lambda m}}{\|L_+ F_{\lambda m}\|}\tag{9.87}$$

is eigenvector of L_3 with eigenvalue $\hbar(m+1)$. On the other hand, the equation $[L^2, L_\pm] = 0$ implies that $F_{\lambda, m+1}$ is also eigenvector of L^2 with the same eigenvalue $\hbar^2\lambda$. Point (ii) is proved in a similar way. $\square$

Using the properties of $L_\pm$, we characterize the possible eigenvalues of L^2 and L_3.

Proposition 9.8 *Let $F_{\lambda m}$ be solution of problem (9.73), (9.74), (9.75). Then $\lambda = l(l+1)$ and $-l \leq m \leq l$, with l nonnegative integer.*

Proof Making use of (9.82), (9.81), (9.79), we have

$$\hbar^2\lambda = (F_{\lambda m}, (L_{\mp}L_{\pm} + L_3^2 \pm \hbar L_3)F_{\lambda m}) = \|L_{\pm}F_{\lambda m}\|^2 + \hbar^2 m(m \pm 1). \quad (9.88)$$

The above equation implies

$$m(m \pm 1) \le \lambda. \quad (9.89)$$

We observe that the two inequalities (9.89) are equivalent to the condition

$$-l \le m \le l, \quad (9.90)$$

where

$$l = \frac{-1 + \sqrt{1 + 4\lambda}}{2} \ge 0. \quad (9.91)$$

We now prove that l is an integer. By absurd, let us assume that l is not integer and let $F_{\lambda m'}$ be solution of the problem with m' satisfying (9.89) or, equivalently, (9.90). Then, there exists k, nonnegative integer, such that

$$m' + k < l < m' + k + 1. \quad (9.92)$$

By the previous proposition, we know that the vector

$$F_{\lambda, m'+1} := \frac{L_+ F_{\lambda m'}}{\|L_+ F_{\lambda m'}\|} \quad (9.93)$$

is eigenvector of L_3 with eigenvalue $\hbar(m'+1)$ (and eigenvector of L^2 with eigenvalue $\hbar^2\lambda$). Iterating k times the procedure, we find that

$$F_{\lambda, m'+k} := \frac{L_+ F_{\lambda, m'+k-1}}{\|L_+ F_{\lambda, m'+k-1}\|} \quad (9.94)$$

is eigenvector of L_3 with eigenvalue $m'+k$. By a further application of L_+ to $F_{\lambda, m'+k}$, we find an eigenvector of L_3 with eigenvalue $m' + k + 1 > l$, and this contradicts (9.90). Therefore, it is proved that l is a nonnegative integer. Solving (9.91) with respect to λ, we obtain $\lambda = l(l + 1)$ and then the proposition is proved. $\square$

In Proposition 9.8, we have shown that problem (9.73), (9.74), (9.75) can be solved only if $\lambda = l(l + 1)$, with l nonnegative integer, $\nu = m \in \mathbb{Z}$ and $-l \le m \le l$. In the next proposition, we prove that if such conditions are satisfied then the problem admits a unique solution. Such a solution, denoted by Y_{lm}, is called spherical harmonics of order l, m and it can be explicitly computed.

Proposition 9.9 *Let us consider problem (9.73), (9.74), (9.75) with $\lambda = l(l + 1)$, l nonnegative integer, $\nu = m \in \mathbb{Z}$ and $-l \le m \le l$. Then*

(i) the problem admits a unique solution, given by

$$Y_{lm}(\theta, \phi) = \frac{1}{\hbar^{l+m}} \sqrt{\frac{(l-m)!}{(l+m)!(2l)!}} \, (L_+^{l+m} Y_{l,-l})(\theta, \phi), \qquad (9.95)$$

$$Y_{l,-l}(\theta, \phi) = \frac{1}{2^l l!} \sqrt{\frac{(2l+1)!}{2}} \, \sin^l \theta \, \frac{e^{-il\phi}}{\sqrt{2\pi}}, \qquad (9.96)$$

(ii) the degeneracy of the eigenvalue $\hbar^2 l(l+1)$ of L^2 is $2l+1$,
(iii) the system (9.95), (9.96) of common eigenvectors of L^2 and L_3 is orthonormal and complete in $L^2(S^2, d\Omega)$ and, therefore, the two operators L^2 and L_3 are essentially self-adjoint.

Proof We shall prove (i) by successive steps. We first fix the integer l and we consider $m = -l$. By (9.81) and (9.76), we have that $Y_{l,-l}$ is solution of the problem if and only if

$$L_- Y_{l,-l} = 0. \qquad (9.97)$$

Using (9.78) and the fact that

$$Y_{l,-l}(\theta, \phi) = \Theta_{l,-l}(\theta) \, \frac{e^{-il\phi}}{\sqrt{2\pi}}, \qquad (9.98)$$

equation (9.97) reduces to the following equation for $\Theta_{l,-l}$

$$\frac{1}{\Theta_{l,-l}} \frac{d\Theta_{l,-l}}{d\theta} = l \cot \theta. \qquad (9.99)$$

Solving by separation of variables, we find

$$\Theta_{l,-l}(\theta) = c_l \sin^l \theta, \qquad (9.100)$$

where the constant c_l is fixed by the normalization condition

$$\int_{S^2} d\Omega \, |Y_{l,-l}(\theta, \phi)|^2 = c_l^2 \int_0^\pi d\theta \, \sin\theta \, \sin^{2l}\theta = 1. \qquad (9.101)$$

Introducing the change of the integration variable $x = \cos\theta$, for the integral in (9.101), we have

$$I_l = \int_0^\pi d\theta \, \sin\theta \, \sin^{2l}\theta = \int_{-1}^1 dx \, (1-x^2)^l = I_{l-1} - \int_{-1}^1 dx \, x^2(1-x^2)^{l-1}$$

$$= I_{l-1} + \frac{1}{2l} \int_{-1}^1 dx \, x \frac{d}{dx}(1-x^2)^l = I_{l-1} - \frac{1}{2l} \int_{-1}^1 dx \, (1-x^2)^l = I_{l-1} - \frac{I_l}{2l}$$

$$(9.102)$$

and we obtain the recursive formula $I_l = \frac{2l}{2l+1} I_{l-1}$, with $I_0 = 2$. Hence,

$$I_l = \frac{2l\, 2(l-1)}{(2l+1)(2l-1)} I_{l-2} = \frac{2l\, 2(l-1)\, 2(l-2) \cdots 2}{(2l+1)(2l-1)(2l-3) \cdots 3} I_0$$

$$= \frac{2^l\, l!}{(2l+1)(2l-1)(2l-3) \cdots 3} I_0 = \frac{2^l\, l!\, 2l\, 2(l-1)\, 2(l-2) \cdots 2}{(2l+1)!} I_0$$

$$= \frac{2^{2l+1}(l!)^2}{(2l+1)!}. \tag{9.103}$$

Using (9.103) in (9.101), we find c_l and then

$$\Theta_{l,-l}(\theta) = \frac{1}{2^l\, l!} \sqrt{\frac{(2l+1)!}{2}} \sin^l \theta. \tag{9.104}$$

Replacing such expression in (9.98), we conclude that the unique solution $Y_{l,-l}$ of the problem for $m = -l$ coincides with the r.h.s. of (9.96).

As second step, we consider the case $m = -l + 1$. By Proposition 9.7, we know that

$$Y_{l,-l+1} = \frac{1}{\hbar\sqrt{2l}} L_+ Y_{l,-l} \tag{9.105}$$

is solution of the problem with $m = -l + 1$. Let us show that such a solution is unique. Let $Z_{l,-l+1}$ be a solution for $m = -l + 1$. Then, by Proposition 9.7, we have that $\frac{1}{\hbar\sqrt{2l}} L_- Z_{l,-l+1}$ is solution for $m = -l$ and, therefore,

$$\frac{1}{\hbar\sqrt{2l}} L_- Z_{l,-l+1} = Y_{l,-l}. \tag{9.106}$$

Applying L_+ and using (9.81), we find

$$\frac{1}{\hbar\sqrt{2l}} L_+ L_- Z_{l,-l+1} = \frac{1}{\hbar\sqrt{2l}} (L^2 - L_3^2 + \hbar L_3) Z_{l,-l+1} = \hbar\sqrt{2l}\, Z_{l,-l+1} = L_+ Y_{l,-l}. \tag{9.107}$$

Thus, by (9.105), we have that $Z_{l,-l+1}$ coincides with $Y_{l,-l+1}$ and this proves uniqueness in the case $m = -l + 1$. In similar way, one shows that

$$Y_{l,-l+2} = \frac{1}{\hbar\sqrt{2(2l-1)}} L_+ Y_{l,-l+1} = \frac{1}{\hbar^2 \sqrt{2 \cdot 2l(2l-1)}} L_+^2 Y_{l,-l} \tag{9.108}$$

is the unique solution of the problem for $m = -l + 2$ and so on.

After $l + m$ steps we find that

$$Y_{lm} = \frac{1}{\hbar^{l+m} \sqrt{(l+m)!\, 2l(2l-1)(2l-2) \cdots (2l-(l+m)+1)}} L_+^{l+m} Y_{l,-l} \tag{9.109}$$

is the unique solution of the problem for a generic m, $-l \leq m \leq l$, and it coincides with the r.h.s. of (9.95), concluding the proof of (i).

The proof of (ii) is an immediate consequence of (i).

Concerning (iii), we observe that the family of vectors $\{Y_{lm}\}$ is an orthonormal system in $L^2(S^2, d\Omega)$ since they are eigenvectors of the symmetric operators L^2 and L_3. We refer to Sect. 9.10 for the proof of the completeness. $\qquad \square$

The spherical harmonics Y_{lm} can be written in terms of the so-called associated Legendre functions (see, e.g., [8]). Indeed, we recall that the Legendre polynomial of order l is defined by

$$P_l(x) := \frac{(-1)^l}{2^l l!} \frac{d^l}{dx^l} (1 - x^2)^l \qquad (9.110)$$

for $x \in [-1, 1]$. Note that $P_l(-x) = (-1)^l P_l(x)$. The associated Legendre function of order l, m, with $0 \leq m \leq l$, is given by

$$P_l^m(x) := \frac{(-1)^{l+m}}{2^l l!} (1 - x^2)^{\frac{m}{2}} \frac{d^{l+m}}{dx^{l+m}} (1 - x^2)^l = (-1)^m (1 - x^2)^{\frac{m}{2}} \frac{d^m}{dx^m} P_l(x) \qquad (9.111)$$

for $x \in [-1, 1]$. We have

Proposition 9.10

$$Y_{lm}(\theta, \phi) = (-1)^{\frac{m+|m|}{2}} \sqrt{\frac{(2l+1)(l - |m|)!}{2(l + |m|)!}} \, P_l^{|m|}(\cos\theta) \, \frac{e^{im\phi}}{\sqrt{2\pi}} . \qquad (9.112)$$

Proof Let us compute the action of L_+ on $Y_{l,-l}$

$$
\begin{aligned}
(L_+ Y_{l,-l})(\theta, \phi) &= \frac{\hbar}{2^l l!} \sqrt{\frac{(2l+1)!}{2}} \, \frac{e^{-i(l-1)\phi}}{\sqrt{2\pi}} \left(\frac{d}{d\theta} + l\cot\theta \right) \sin^l \theta \\
&= \frac{\hbar}{2^l l!} \sqrt{\frac{(2l+1)!}{2}} \, \frac{e^{-i(l-1)\phi}}{\sqrt{2\pi}} \, 2l \sin^{l-1}\theta \cos\theta \\
&= -\frac{\hbar}{2^l l!} \sqrt{\frac{(2l+1)!}{2}} \, \frac{e^{-i(l-1)\phi}}{\sqrt{2\pi}} \, \frac{1}{\sin^{l-1}\theta} \frac{d}{dx} (1 - x^2)^l \Big|_{x=\cos\theta} \quad (9.113)
\end{aligned}
$$

Analogously

$$(L_+^2 Y_{l,-l})(\theta, \phi) = \frac{\hbar^2}{2^l l!} \sqrt{\frac{(2l+1)!}{2}} \, \frac{e^{-i(l-2)\phi}}{\sqrt{2\pi}} \, \frac{1}{\sin^{l-2}\theta} \frac{d^2}{dx^2} (1 - x^2)^l \Big|_{x=\cos\theta} . \qquad (9.114)$$

Iterating the computation, after $l + m$, with $m \geq 0$, we find

$$(L_+^{l+m} Y_{l,-l})(\theta, \phi) = \frac{(-\hbar)^{l+m}}{2^l l!} \sqrt{\frac{(2l+1)!}{2}} \frac{e^{im\phi}}{\sqrt{2\pi}} (1-x^2)^{\frac{m}{2}} \frac{d^{l+m}}{dx^{l+m}} (1-x^2)^l \Big|_{x=\cos\theta}.$$

$$(9.115)$$

Replacing the above expression in (9.95), we obtain (9.112) for $m \geq 0$. The computation for $m < 0$ is analogous.   □

We note that the condition of orthonormality of the spherical harmonics in $L^2(S^2, d\Omega)$ implies the orthogonality of Legendre polynomials and of the associated Legendre functions in $L^2([-1, 1])$. Indeed, by $(Y_{l0}, Y_{l'0}) = \delta_{ll'}$ we have

$$\int_{-1}^{1} dx\, P_l(x) P_{l'}(x) = \frac{2}{2l+1} \delta_{ll'}$$

$$(9.116)$$

and by $(Y_{lm}, Y_{l'm}) = \delta_{ll'}$, $m \geq 0$, we have

$$\int_{-1}^{1} dx\, P_l^m(x) P_{l'}^m(x) = \frac{2(l+m)!}{(2l+1)(l-m)!} \delta_{ll'}.$$

$$(9.117)$$

We also observe that, using the system $\{Y_{lm}\}$, we can represent an arbitrary solution of the Laplace equation in spherical coordinates (see exercise below).

Exercise 9.8 Verify that a harmonic function, i.e., a solution of the Laplace equation, in spherical coordinates can be written as

$$u(r, \theta, \phi) = \sum_{l=0}^{\infty} \sum_{m=-l}^{l} (a_{lm} r^l + b_{lm} r^{-l-1}) Y_{lm}(\theta, \phi),$$

$$(9.118)$$

where a_{lm}, b_{lm} are arbitrary constants.
(Hint: if u is a solution of the Laplace equation then

$$u(r, \theta, \phi) = \sum_{l=0}^{\infty} \sum_{m=-l}^{l} f_{lm}(r) Y_{lm}(\theta, \phi)$$

$$(9.119)$$

with f_{lm} solution of

$$-\frac{d}{dr}\left(r^2 \frac{df_{lm}}{dr}\right) + l(l+1) f_{lm} = 0.$$

$$(9.120)$$

Defining $g_{lm}(r) = r f_{lm}(r)$, we have

$$-\frac{d^2 g_{lm}}{dr^2} + \frac{l(l+1)}{r^2} g_{lm} = 0.$$

$$(9.121)$$

Equation (9.121) can be written in factorized form

$$\left(-\frac{d}{dr} + \frac{l}{r}\right)\left(\frac{d}{dr} + \frac{l}{r}\right) g_{lm} = 0 . \tag{9.122}$$

If we set $h_{lm} = (d/dr + l\,r^{-1})\, g_{lm}$, we find $h_{lm}(r) = c_{lm}\, r^l$, with c_{lm} constant. From h_{lm}, we obtain g_{lm} and then f_{lm}).

Remark 9.11 In particular, the solution (9.118) allows to prove the following useful formula:

$$\frac{1}{|x - x'|} = \sum_{l=0}^{\infty} \frac{r_<^l}{r_>^{l+1}} P_l(\cos \gamma) , \tag{9.123}$$

where $r_< = \min\{|x|, |x'|\}$, $r_> = \max\{|x|, |x'|\}$ and γ is the angle between x and x'.

Indeed, let us fix x' along the z-axis. The function $|x - x'|^{-1}$ is harmonic for $|x| > |x'|$, it is invariant under rotation around the z-axis and it goes to zero for $|x| \to \infty$. Then

$$\frac{1}{|x - x'|} = \sum_{l=0}^{\infty} \frac{b_l}{|x|^{l+1}} Y_{l0}(\theta) , \qquad |x| > |x'| . \tag{9.124}$$

In order to compute b_l, it is sufficient to evaluate the above equation for $\theta = 0$. For $|x| < |x'|$, the proof is analogous.

Exercise 9.9 Let us consider a sphere of radius R made of dielectric material, with dielectric constant ε, in presence of a point charge e placed at a distance $R_1 > R$ from the center of the sphere. Compute the electrostatic potential.

9.8 Reduction to the Radial Problem

As a consequence of the results of the previous section, an element ψ of the Hilbert space $\mathcal{H}_S = L^2(\mathbb{R}^+, r^2 dr) \otimes L^2(S_1^2, d\Omega)$ can be written as

$$\psi(r, \theta, \phi) = \sum_{l=0}^{\infty} \sum_{m=-l}^{l} f_{lm}(r) Y_{lm}(\theta, \phi) , \tag{9.125}$$

where $f_{lm} \in L^2(\mathbb{R}^+, r^2 dr)$. Here, we shall verify that in order to determine the eigenvectors of $\hat{H}_e$ which are also eigenvectors of L_3 and L^2 it is sufficient to solve an ordinary differential equation in the radial variable r. Let us define

$$D_{h_l} := \left\{ f \mid f \in L^2(\mathbb{R}^+, r^2 dr), \int_0^\infty dr\, r^2 \left| \frac{-\hbar^2}{2\mu r^2} \frac{d}{dr} \left(r^2 \frac{df(r)}{dr} \right) + \frac{\hbar^2 l(l+1)}{2\mu r^2} f(r) \right|^2 < \infty \right\} . \tag{9.126}$$

Then

Proposition 9.11 *The vector ψ is eigenvector of $\hat{H}_e$ (see 9.65) with eigenvalue $E < 0$ and, moreover, eigenvector of L_3 and L^2 with eigenvalues $\hbar m$, $\hbar^2 l(l + 1)$ respectively if and only if*

$$\psi(r, \theta, \phi) = R_{El}(r) Y_{lm}(\theta, \phi) \tag{9.127}$$

with R_{El} solution of

$$h_l R_{El} := -\frac{\hbar^2}{2\mu r^2} \frac{d}{dr}\left(r^2 \frac{dR_{El}}{dr}\right) + \frac{\hbar^2 l(l+1)}{2\mu r^2} R_{El} - \frac{e^2}{r} R_{El} = E\, R_{El}, \tag{9.128}$$

$$R_{El} \in D_{h_l}, \quad \int_0^\infty dr\, r^2 |R_{El}(r)|^2 = 1, \quad E < 0. \tag{9.129}$$

Proof If ψ is eigenvector of L_3 and L^2 then it has the form (9.127). Moreover, the condition $\|\psi\|_{\mathscr{H}_s} = 1$ implies

$$\int_0^\infty dr\, r^2 |R_{El}(r)|^2 = 1. \tag{9.130}$$

Moreover, $u := U_S^{-1}\psi \in D(H_0)$ is equivalent to $\hat{H}_0\psi := U_S H_0 U_S^{-1}\psi \in \mathscr{H}_S$ and, taking into account of the explicit expression of $\hat{H}_0$ (see 9.63), one obtains

$$\int_0^\infty dr\, r^2 \left| \frac{-\hbar^2}{2\mu r^2} \frac{d}{dr}\left(r^2 \frac{dR_{El}(r)}{dr}\right) + \frac{\hbar^2 l(l+1)}{2\mu r^2} R_{El}(r) \right|^2 < \infty. \tag{9.131}$$

Finally, it is easy to verify that if $\hat{H}_e\psi = E\psi$ and $\psi = R_{El} Y_{lm}$ then $h_l R_{El} = E R_{El}$.

The sufficient condition is also evident. If ψ has the form (9.127) then it is eigenvector of L^2 and L_3. Moreover, if R_{El} is solution of the problem (9.129) then ψ has norm one, $U_S^{-1}\psi \in D(H_0)$ and $\hat{H}_e\psi = E\psi$. $\qquad\square$

A further simplification is obtained if we introduce the dimensionless variable

$$x = \frac{r}{a}, \tag{9.132}$$

where $a = \frac{\hbar^2}{\mu e^2}$ is the Bohr radius. More precisely, it is easy to verify that if R_{El} is solution of the problem (9.128), (9.129) then

$$G_{\eta l}(x) := a^{3/2} x\, R_{\frac{\mu e^4}{\hbar^2}\eta, l}(ax), \qquad \eta := \frac{\hbar^2}{\mu e^4} E \tag{9.133}$$

is solution of the problem

$$\hat{h}_l G_{\eta l} := -\frac{1}{2}\frac{d^2 G_{\eta l}}{dx^2} + \frac{l(l+1)}{2x^2}G_{\eta l} - \frac{1}{x}G_{\eta l} = \eta G_{\eta l}, \qquad (9.134)$$

$$G_{\eta l} \in D_{\hat{h}_l}, \qquad \int_0^\infty dx\, |G_{\eta l}(x)|^2 = 1, \qquad \eta < 0, \qquad (9.135)$$

where

$$D_{\hat{h}_l} := \left\{ f \mid f \in L^2(\mathbb{R}^+),\ f(0) = 0,\ \int_0^\infty dx\, \left| -\frac{1}{2}\frac{d^2 f(x)}{dx^2} + \frac{l(l+1)}{2x^2}f(x) \right|^2 < \infty \right\}.$$
$$(9.136)$$

Conversely, if $G_{\eta l}$ is solution of the problem (9.134), (9.135) then

$$R_{El}(r) = \frac{1}{\sqrt{a}\, r}\, G_{\frac{\hbar^2}{\mu e^4}E,l}(a^{-1}r) \qquad (9.137)$$

is solution of the problem (9.128), (9.129).

Exercise 9.10 Let us consider the Hamiltonian

$$H = -\frac{\hbar^2}{2\mu}\Delta + V(r), \qquad (9.138)$$

where $V(r) = -V_0 < 0$ for $r < a$ and $V(r) = 0$ otherwise. Study the eigenvalue problem in the subspace with zero angular momentum.

9.9 Energy Levels and Stationary States

We are now in position to determine eigenvalues and eigenvectors of $\hat{H}_e$ by solving problem (9.134), (9.135). Once again, we shall use an algebraic method based on the factorization property of the operator $\hat{h}_l$. In this section, we shall denote norm and scalar product in $L^2(\mathbb{R}^+)$ by $\|\cdot\|$, $(\cdot,\cdot)$.

Let us consider the following two operators defined on $D_{\hat{h}_l}$:

$$A_l^\pm := \frac{1}{\sqrt{2}}\left(\mp\frac{d}{dx} + \frac{l+1}{x} - \frac{1}{l+1} \right). \qquad (9.139)$$

By a direct computation, one verifies that

$$(g, A_l^+ f) = (A_l^- g, f), \qquad f, g \in D_{\hat{h}_l} \cap C^1([0,\infty)), \qquad (9.140)$$

$$\hat{h}_l = A_l^- A_l^+ - \frac{1}{2(l+1)^2}, \qquad (9.141)$$

$$\hat{h}_{l+1} = A_l^+ A_l^- - \frac{1}{2(l+1)^2}. \qquad (9.142)$$

As an immediate consequence, we have that if $G_{\eta l}$ is solution of the problem (9.134), (9.135) then

$$\eta + \frac{1}{2(l+1)^2} \geq 0 \,. \tag{9.143}$$

Indeed, using (9.140) and (9.141), one finds

$$\eta = (G_{\eta l}, \hat{h}_l G_{\eta l}) = \left(G_{\eta l}, \left(A_l^- A_l^+ - \frac{1}{2(l+1)^2}\right) G_{\eta l}\right) = \|A_l^+ G_{\eta l}\|^2 - \frac{1}{2(l+1)^2} \,. \tag{9.144}$$

Moreover, the operators $A_l^\pm$ act as creation and annihilation operators with respect to the quantum number l.

Proposition 9.12 *Let $G_{\eta l}$ be solution of the problem (9.134), (9.135).*

(i) If $\eta + 2^{-1}(l+1)^{-2} > 0$ then

$$\frac{A_l^+ G_{\eta l}}{\sqrt{\eta + \frac{1}{2(l+1)^2}}} \tag{9.145}$$

is eigenvector of $\hat{h}_{l+1}$ with eigenvalue η;
(ii) if $l > 0$ and $\eta + 2^{-1} l^{-2} > 0$ then

$$\frac{A_{l-1}^- G_{\eta l}}{\sqrt{\eta + \frac{1}{2l^2}}} \tag{9.146}$$

is eigenvector of $\hat{h}_{l-1}$ with eigenvalue η.

Proof Concerning (i), we have

$$\hat{h}_{l+1} A_l^+ G_{\eta l} = \left(A_l^+ A_l^- - \frac{1}{2(l+1)^2}\right) A_l^+ G_{\eta l} = A_l^+ \left(A_l^- A_l^+ - \frac{1}{2(l+1)^2}\right) G_{\eta l}$$
$$= A_l^+ \hat{h}_l G_{\eta l} = \eta A_l^+ G_{\eta l} \tag{9.147}$$

Moreover, by (9.144), we have $\|A_l^+ G_{\eta l}\|^2 = \eta + \frac{1}{2(l+1)^2}$. The proof of (ii) is analogous. $\square$

In the next proposition, we characterize the possible eigenvalues of $\hat{h}_l$.

Proposition 9.13 *Let $G_{\eta l}$ be solution of the problem (9.134), (9.135). Then there exists a positive integer n, with $n \geq l + 1$, such that*

$$\eta = -\frac{1}{2n^2} \,. \tag{9.148}$$

Proof We know that if $G_{\eta l}$ is solution of (9.134), (9.135) then

$$\eta \geq -\frac{1}{2(l+1)^2} . \tag{9.149}$$

If the equality sign holds in (9.149) then the thesis is proved with $n = l + 1$. Let us consider the case

$$\eta > -\frac{1}{2(l+1)^2} . \tag{9.150}$$

If we write

$$\eta = -\frac{1}{2\gamma^2} \tag{9.151}$$

with $\gamma > l + 1$, the problem is reduced to prove that γ is a positive integer. Let us assume that γ is not a positive integer and let $G_{\eta l}$ be solution of (9.134), (9.135) with $\eta > -2^{-1}(l+1)^{-2}$. Since $\gamma > l + 1$, there exists a positive integer k such that

$$l + k < \gamma < l + k + 1 . \tag{9.152}$$

By Proposition 9.12, we know that

$$G_{\eta,l+1} := \frac{A_l^+ G_{\eta l}}{\|A_l^+ G_{\eta l}\|} \tag{9.153}$$

is eigenvector of $\hat{h}_{l+1}$ with eigenvalue η. Iterating k times, we find that

$$G_{\eta,l+k} := \frac{A_{l+k-1}^+ G_{\eta,l+k-1}}{\|A_{l+k-1}^+ G_{\eta l+k-1}\|} \tag{9.154}$$

is eigenvector of $\hat{h}_{l+k}$ with eigenvalue η. Proceeding as in (9.144), we have

$$\|A_{l+k}^+ G_{\eta,l+k}\|^2 = \eta + \frac{1}{2(l+k+1)^2} = -\frac{1}{2\gamma^2} + \frac{1}{2(l+k+1)^2} \geq 0 \quad (9.155)$$

which implies $\gamma \geq l + k + 1$. This last inequality contradicts (9.152). Then γ must be integer and the proposition is proved. $\qquad \square$

We have shown that if the problem (9.134), (9.135) has a solution then η has the form (9.148), with $n \geq l + 1$. It remains to prove that if η is given by (9.148), with $n \geq l + 1$, then the problem admits a unique solution. With an abuse of notation, the solution will be denoted by G_{nl}.

Proposition 9.14 *Let us consider the problem (9.134), (9.135) with η given by (9.148) and $n \geq l + 1$. Then the problem has the unique solution*

$$G_{nl} = \frac{(\sqrt{2}\,n)^{n-1-l}(n-1)!\sqrt{(n+l)!}}{l!\sqrt{(n-1-l)!}\sqrt{(2n-1)!}} \, A_l^- \cdots A_{n-3}^- A_{n-2}^- \, G_{n,n-1} \,, \quad (9.156)$$

$$G_{n,n-1}(x) = \left(\frac{2}{n}\right)^{n+1/2} \frac{1}{\sqrt{(2n)!}} \, x^n \, e^{-\frac{x}{n}} \,. \quad (9.157)$$

Moreover, for $n \neq n'$, the vectors G_{nl} and $G_{n'l}$ are orthogonal in $L^2(\mathbb{R}^+)$.

Proof Given the positive integer n, let us consider $l = n - 1$. Taking into account of (9.141), we have that $G_{n,n-1}$ is solution of the problem if and only if

$$A_{n-1}^+ G_{n,n-1} = 0 \,. \quad (9.158)$$

Using the explicit expression of A_{n-1}^+, the above equation reads

$$\frac{1}{G_{n,n-1}} \frac{dG_{n,n-1}}{dx} = \frac{n}{x} - \frac{1}{n} \,. \quad (9.159)$$

The solution is

$$G_{n,n-1}(x) = c_n \, x^n \, e^{-\frac{x}{n}} \,, \quad (9.160)$$

where c_n is an arbitrary constant. The normalization condition implies

$$(c_n)^{-2} = \int_0^\infty dx \, x^{2n} e^{-\frac{2x}{n}} = \left(\frac{n}{2}\right)^{2n+1} \int_0^\infty dy \, y^{2n} e^{-y} = \left(\frac{n}{2}\right)^{2n+1} \int_0^\infty dy \, y^{2n} \frac{d^{2n}}{dy^{2n}} e^{-y}$$

$$= \left(\frac{n}{2}\right)^{2n+1} \int_0^\infty dy \, \frac{d^{2n}}{dy^{2n}} y^{2n} e^{-y} = \left(\frac{n}{2}\right)^{2n+1} (2n)! \quad (9.161)$$

By (9.160), (9.161), we find that the unique solution of the problem for $l = n - 1$ is given by (9.157).

Let us consider the case $l = n - 2$. By Proposition 9.12, point (ii), we know that

$$G_{n,n-2} = \frac{\sqrt{2}\,n(n-1)}{\sqrt{2n-1}} \, A_{n-2}^- G_{n,n-1} \quad (9.162)$$

is solution of the problem for $l = n - 2$. Let us see that the solution is unique. Let $W_{n,n-2}$ be solution for $l = n - 2$. Then, by Proposition 9.12, point (i), we know that $\frac{\sqrt{2}n(n-1)}{\sqrt{2n-1}} A_{n-2}^+ W_{n,n-2}$ is solution for $l = n - 1$ and therefore

$$\frac{\sqrt{2}\,n(n-1)}{\sqrt{2n-1}} \, A_{n-2}^+ W_{n,n-2} = G_{n,n-1} \,. \quad (9.163)$$

Applying A_{n-2}^- and using (9.141), we have

$$A_{n-2}^- G_{n,n-1} = \frac{\sqrt{2}n(n-1)}{\sqrt{2n-1}} A_{n-2}^- A_{n-2}^+ W_{n,n-2}$$

$$= \frac{\sqrt{2}n(n-1)}{\sqrt{2n-1}} \left(-\frac{1}{2n^2} + \frac{1}{2(n-1)^2} \right) W_{n,n-2} = \frac{\sqrt{2n-1}}{\sqrt{2}n(n-1)} W_{n,n-2}.$$

$$(9.164)$$

Comparing with (9.162), we conclude that $W_{n,n-2}$ coincides with $G_{n,n-2}$ and this prove uniqueness in the case $l = n - 2$. Iterating the procedure, after $k = n - 1 - l$ steps we obtain that

$$G_{nl} = \frac{(\sqrt{2}n)^{n-1-l}(n-1)(n-2)\cdots(n-(n-1-l))}{\sqrt{1\cdot2\cdots(n-1-l)}\sqrt{(2n-1)(2n-2)\cdots(2n-(n-1-l))}} A_l^- \cdots A_{n-2}^- G_{n,n-1}$$

$$= \frac{(\sqrt{2}n)^{n-1-l}(n-1)!\sqrt{(n+l)!}}{l!\sqrt{(n-1-l)!}\sqrt{(2n-1)!}} A_l^- \cdots A_{n-2}^- G_{n,n-1}$$

$$(9.165)$$

is the unique solution of the problem for a generic l, $0 \le l \le n - 1$, and it coincides with the r.h.s. of (9.156).

Finally, the orthogonality of G_{nl} and $G_{n'l}$ for $n \ne n'$ is a consequence of the symmetry of the operator $\hat{h}_l$. □

Remark 9.12 The function G_{nl} can be written in terms of the so-called generalized Laguerre polynomials. We recall that the Laguerre polynomial of order k is

$$L_k(x) := e^x \frac{d^k}{dx^k}(e^{-x}x^k) \qquad x \in [0, \infty). \qquad (9.166)$$

The generalized Laguerre polynomial of order k, j, $0 \le j \le k$, is

$$L_k^j(x) := (-1)^j \frac{k!}{(k-j)!} e^x x^{-j} \frac{d^{k-j}}{dx^{k-j}}(e^{-x}x^k) \qquad x \in [0, \infty) \qquad (9.167)$$

and $L_k^0(x) = L_k(x)$. As an exercise, one can prove that

$$G_{nl}(x) = -\frac{1}{n}\sqrt{\frac{(n-l-1)!}{[(n+l)!]^3}} e^{-\frac{x}{n}} \left(\frac{2x}{n}\right)^{l+1} L_{n+l}^{2l+1}\left(\frac{2x}{n}\right). \qquad (9.168)$$

We have thus solved problem (9.134), (9.135), formulated in the dimensionless variable x. Then, we have solved problem (9.128), (9.129), formulated in the radial variable r and, finally, by Proposition 9.11 we have also solved the eigenvalue problem for $\hat{H}_e$.

We summarize the result in the following proposition.

Proposition 9.15

(i) *The negative eigenvalues of $\hat{H}_e$ are*

$$E_n = -\frac{\mu\, e^4}{2\hbar^2 n^2}\,, \qquad n = 1, 2, \ldots . \qquad (9.169)$$

(ii) *The minimum eigenvalue, or ground state energy, E_1 is nondegenerate and the corresponding eigenfunction, or ground state, is*

$$\psi_{E_1}(r) := \psi_{100}(r) = \frac{1}{\sqrt{4\pi}\, ar}\, G_{10}\left(\frac{r}{a}\right) = -\frac{1}{\sqrt{\pi}\, a^{3/2}}\, e^{-\frac{r}{a}} L_1^1\left(\frac{2r}{a}\right) = \frac{1}{\sqrt{\pi}\, a^{3/2}}\, e^{-\frac{r}{a}}. \qquad (9.170)$$

(iii) *The eigenvalue E_n, $n > 1$, is degenerate with multiplicity*

$$\sum_{l=0}^{n-1} \sum_{m=-l}^{l} 1 = \sum_{l=0}^{n-1} (2l + 1) = n^2 \qquad (9.171)$$

and an associated eigenfunction, or excited state, is written as the linear combination

$$\psi_{E_n}(r, \theta, \phi) = \sum_{l=0}^{n-1} \sum_{m=-l}^{l} c_{lm}\, R_{nl}(r)\, Y_{lm}(\theta, \phi) \qquad (9.172)$$

where c_{lm} arbitrary constants such that $\sum_{l=0}^{n-1} \sum_{m=-l}^{l} |c_{lm}|^2 = 1$ and $R_{nl}(r)$ is

$$R_{nl}(r) = \frac{1}{\sqrt{ar}} G_{nl}\left(\frac{r}{a}\right) = -\frac{2}{n^2} \sqrt{\frac{(n-l-1)!}{[(n+l)!]^3 a^3}}\, e^{-\frac{r}{na}} \left(\frac{2r}{na}\right)^l L_{n+l}^{2l+1}\left(\frac{2r}{na}\right). \qquad (9.173)$$

(iv) *The degeneracy is removed if one requires that an eigenfunction of $\hat{H}_e$ is also eigenfunction of L^2 and L_3. More precisely, the only common eigenfunction of $\hat{H}_e$, L^2, L_3, associated to the eigenvalues E_n, $\hbar^2 l(l+1)$, $\hbar m$ respectively, is*

$$\psi_{nlm}(r, \theta, \phi) = R_{nl}(r)\, Y_{lm}(\theta, \phi), \qquad (9.174)$$

where $n = 1, 2, \ldots,\ \ l = 0, \ldots, n-1,\ \ m = -l, \ldots, l$.

Let us comment on the above result.

- The negative eigenvalues of $\hat{H}_e$ coincide with the energy levels found by Bohr in his first model of hydrogen atom.
- The ground state is the only spherically symmetric eigenfunction of $\hat{H}_e$. Moreover, the probability to find the electron at a distance from the nucleus between r_1 and r_2 is

$$\int_{r_1}^{r_2} dr\, r^2 \int_{S^2} d\Omega\, |\psi_{100}(r)|^2 = \int_{r_1}^{r_2} dr\, D(r) , \qquad D(r) := \frac{4r^2}{a^3}\, e^{-\frac{2r}{a}} \qquad (9.175)$$

It is easy to verify that the radial probability density $D(r)$ has a maximum for r equal to the Bohr's radius a and that the mean value of the distance from the nucleus is $\langle r \rangle = (3/2)a$.

- The negative eigenvalues of $\hat{H}_e$ depend only on the quantum number n (and not on l or m). Such degeneracy is typical of the Coulomb potential and it is a consequence of a special symmetry property of the Hamiltonian (see, e.g., [4]). On the other hand, in the case of a generic central potential, the eigenvalues of the Hamiltonian depend also on the quantum number l, as it is natural to expect from Eq. (9.128).

- Finally, for the convenience of the reader, we explicitly write the first common eigenfunctions of $\hat{H}_e$, L^2 and L_3

$$\psi_{100}(r,\theta,\phi) = R_{10}(r)Y_{00}(\theta,\phi) = \frac{1}{\sqrt{\pi}\, a^{3/2}}\, e^{-\frac{r}{a}} ,$$

$$\psi_{200}(r,\theta,\phi) = R_{20}(r)Y_{00}(\theta,\phi) = \frac{1}{4\sqrt{2\pi}\, a^{3/2}}\left(2 - \frac{r}{a}\right) e^{-\frac{r}{2a}} ,$$

$$\psi_{211}(r,\theta,\phi) = R_{21}(r)Y_{11}(\theta,\phi) = -\frac{1}{8\sqrt{\pi}\, a^{3/2}}\left(\frac{r}{a}\right) e^{-\frac{r}{2a}} \sin\theta\, e^{i\phi} ,$$

$$\psi_{210}(r,\theta,\phi) = R_{11}(r)Y_{10}(\theta,\phi) = \frac{1}{4\sqrt{2\pi}\, a^{3/2}}\left(\frac{r}{a}\right) e^{-\frac{r}{2a}} \cos\theta ,$$

$$\psi_{21-1}(r,\theta,\phi) = R_{21}(r)Y_{1-1}(\theta,\phi) = \frac{1}{8\sqrt{\pi}\, a^{3/2}}\left(\frac{r}{a}\right) e^{-\frac{r}{2a}} \sin\theta\, e^{-i\phi} .$$

$$(9.176)$$

Exercise 9.11 Compute the probability distribution for the momentum of the electron in the state ψ_{100} and in the state ψ_{210}.

Exercise 9.12 Determine the electrostatic potential produced by the hydrogen atom (i.e., by the nucleus and the electronic cloud) when the electron is in the state ψ_{100} and in the state ψ_{200}.

Exercise 9.13 Let us consider a hydrogen atom in a constant magnetic field B with strength $|B| > 0$ and directed along the z-axis. In the case of weak magnetic field (i.e., neglecting terms of order $|B|^2$), verify that the Hamiltonian is

$$H = H_e + \frac{e|B|}{2\mu c} L_3 . \qquad (9.177)$$

Determine the eigenvalues of H and study their degeneracy (Zeeman effect).

Exercise 9.14 Let us consider the Hamiltonian in $L^2(\mathbb{R}^6)$

$$H = -\Delta_{x_1} - \Delta_{x_2} - \frac{2}{|x_1|} - \frac{2}{|x_2|} + \frac{1}{|x_1 - x_2|} , \qquad (9.178)$$

which describes the helium atom in natural units and neglecting the antisymmetry property of the wave function. Verify that H is self-adjoint and bounded from below on the domain $H^2(\mathbb{R}^6)$. Estimate from above the infimum of the spectrum using the trial function

$$\phi_\alpha(x_1)\phi_\alpha(x_2)\,, \qquad \phi_\alpha(x) = \frac{\alpha^{3/2}}{\sqrt{8\pi}}\, e^{-\frac{\alpha}{2}|x|}\,, \qquad \alpha > 0\,. \qquad (9.179)$$

9.10 Completeness of the Spherical Harmonics

In Sect. 9.7, we have seen that the system of the spherical harmonics

$$Y_{lm}(\theta, \phi) = c_{lm}\, P_l^{|m|}(\cos\theta)\, \frac{e^{im\phi}}{\sqrt{2\pi}}\,, \qquad (9.180)$$

where the constant c_{lm} are defined in (9.112), is orthonormal in $L^2(S^2)$. Here, we show that the system is also complete following the line of [9], Chap. 7. In the proof, we shall use Weierstrass theorem on the approximation of a continuous functions by polynomials, which can be precisely formulated as follows:

let I be a closed and bounded interval and let $f : I \to \mathbb{C}$ be a continuous function. Then for any $\varepsilon > 0$ there exists a polynomial P with complex coefficients such that $\sup_{x \in I} |f(x) - P(x)| < \varepsilon$.

As a preliminary result, we prove completeness of the Legendre polynomials (9.110) and of the associated Legendre functions (9.111).

Proposition 9.16 *The two systems of vectors* $\sqrt{\frac{2l+1}{2}}\, P_l$, *for* $l = 0, 1, 2\ldots$, *and* $\sqrt{\frac{(2l+1)(l-m)!}{2(l+m)!}}\, P_l^m$, *for* $l \geq m$ *and* $m > 0$ *fixed, are orthonormal and complete in* $L^2([-1, 1])$.

Proof Orthonormality has been already proved (see 9.116, 9.117). By Weierstrass theorem, the subspace $D \subset L^2([-1, 1])$ made of finite linear combinations of the polynomials $p_n(x) := x^n$, $n = 0, 1, 2\ldots$, is dense in $L^2([-1, 1])$. On the other hand, each p_n can be written as a linear combination of the Legendre polynomials $P_0, \ldots, P_n$. Hence, D coincides with the subspace made of finite linear combinations of the Legendre polynomials and therefore the system of Legendre polynomials is complete.

Concerning the associate Legendre functions, we fix $f \in L^2([-1, 1])$, $m > 0$ and we assume that

$$(P_l^m, f) = \int_{-1}^{1} dx\, f(x)(1 - x^2)^{\frac{m}{2}} \frac{d^m}{dx^m} P_l(x) = 0 \qquad (9.181)$$

for any $l \geq m$. Note that $h(x) := f(x)(1 - x^2)^{\frac{m}{2}}$ belongs to $L^2([-1, 1])$. Moreover, $\frac{d^m}{dx^m} P_l$ is a polynomial of order $l - m$ and each p_n can be written as a linear combination of the polynomials $\frac{d^m}{dx^m} P_m, \frac{d^m}{dx^m} P_{m+1}, \ldots, \frac{d^m}{dx^m} P_{m+n}$. As in the previous case, this means that the subspace made of finite linear combinations of such polynomials is dense. It follows that $h(x) = 0$ a.e. and then we also have $f(x) = 0$ a.e., concluding the proof of the proposition. $\qquad\square$

From the above proposition, we deduce that for each $m \geq 0$ the system $c_{lm} P_l^m(\cos\theta)$, $l \geq m$, is orthonormal and complete in $L^2([0, \pi], \sin\theta d\theta)$. Taking into account that $(2\pi)^{-1/2} e^{im\phi}$, $m \in \mathbb{Z}$, is orthonormal and complete in $L^2([0, 2\pi])$, we conclude that the system of the spherical harmonics (9.180) is complete in $L^2([0, \pi], \sin\theta d\theta) \otimes L^2([0, 2\pi])$ which is isomorphic to $L^2(S^2)$.

References

1. Faddeev, L.D., Yakubovskii, O.A.: Lectures on Quantum Mechanics for Mathematics Students. AMS (2009)
2. Teschl, G.: Mathematical Methods in Quantum Mechanics. American Mathematical Society, Providence (2009)
3. Thaller, B.: Visual Quantum Mechanics. Springer, New York (2000)
4. Thirring, W.: Quantum Mechanics of Atoms and Molecules. Springer, New York (1981)
5. Reed, M., Simon, B.: Methods of Modern Mathematical Physics, II: Fourier Analysis, Self-Adjointness. Academic Press, New York (1975)
6. Reed, M., Simon, B.: Methods of Modern Mathematical Physics, IV: Analysis of Operators. Academic Press, New York (1978)
7. Lieb, E.H., Loss, M.: Analysis, 2nd edn. American Mathematical Society, Providence, Rhode Island (2001)
8. Lebedev, N.N.: Special Functions and Their Applications. Dover Publication, New York (1972)
9. Prugovecki, E.: Quantum Mechanics in Hilbert Space. Academic Press, New York (1981)

Appendix A
Semiclassical Evolution

A.1 Introduction

In this appendix, we give an introduction to the problem of the classical limit for a quantum system made of a single particle with mass m and subject to a potential $V(x)$. For simplicity, in the first three sections, we consider the one dimensional case.

As we already mentioned in Sect. 3.6, by classical limit we mean that, under certain conditions, the solution of the Schrödinger equation for a particle approximately reproduces the classical evolution of the particle itself.

Roughly speaking, the conditions to be satisfied are: $\hbar$ is much smaller than the typical action of the system and the initial datum depends on $\hbar$ in a suitable way.

From the mathematical point of view, the first condition is realized by taking the limit $\hbar \to 0$. For the second condition, we choose an initial datum having the form of the Gaussian state ψ_0 defined in (7.56) (another possible choice is briefly discussed in Sect. A.4). We rewrite the initial datum for the convenience of the reader

$$\psi_0(x) = \frac{1}{(\pi \hbar)^{1/4} \sqrt{a_0}} \, e^{-b_0 a_0^{-1} \frac{(x-q_0)^2}{2\hbar} + i \frac{p_0}{\hbar}(x-q_0)}, \tag{A.1}$$

where $q_0, p_0 \in \mathbb{R}$ and a_0, b_0 are two complex numbers such that $\Re(\bar{a}_0 b_0) = 1$. We also have

$$\Re(b_0 a_0^{-1}) = \frac{1}{|a_0|^2}, \qquad \Re(a_0 b_0^{-1}) = \frac{1}{|b_0|^2} \tag{A.2}$$

and

$$\langle x \rangle(0) = q_0, \qquad \Delta x(0) = \sqrt{\frac{\hbar}{2}} |a_0|, \qquad \langle p \rangle(0) = p_0, \qquad \Delta p(0) = \sqrt{\frac{\hbar}{2}} |b_0|. \tag{A.3}$$

Observe that for $\hbar \to 0$ the state ψ_0 is well concentrated in position around q_0 and in momentum around p_0. Roughly speaking, this means that the quantum state ψ_0

© Springer International Publishing AG, part of Springer Nature 2018
A. Teta, *A Mathematical Primer on Quantum Mechanics*,
UNITEXT for Physics, https://doi.org/10.1007/978-3-319-77893-8

is a quasi-classical state in the sense that it is "close" (as far as it is permitted by the Heisenberg uncertainty principle) to the classical state (q_0, p_0) for $\hbar \to 0$.

Then, we consider the quantum evolution of the particle described by the solution $\psi(t)$ of the Cauchy problem for the one-dimensional Schrödinger equation

$$i\hbar\frac{\partial \psi(t)}{\partial t} = -\frac{\hbar^2}{2m}\Delta\psi(t) + V(x)\psi(t),\qquad\qquad\text{(A.4)}$$

$$\psi(0) = \psi_0.\qquad\qquad\text{(A.5)}$$

The corresponding classical evolution of the particle is described by the path $(q(t), p(t))$, $t \in [0, T]$, $T > 0$, in the phase space $\mathbb{R}^2$ given by the solution of the Cauchy problem for Hamilton's equations

$$\dot{q} = \frac{\partial H_{cl}}{\partial q},\qquad \dot{p} = -\frac{\partial H_{cl}}{\partial q},\qquad q(0) = q_0,\quad p(0) = p_0,\qquad\text{(A.6)}$$

$$H_{cl}(q, p) = \frac{p^2}{2m} + V(q).\qquad\qquad\text{(A.7)}$$

The aim is to show that, for $\hbar \to 0$ and for a time interval "not too long", the solution $\psi(t)$ of the quantum evolution problem (A.4), (A.5) is approximated by a Gaussian wave function $\phi(t)$ well concentrated in position around $q(t)$ and in momentum around $p(t)$, where $(q(t), p(t))$ is the solution of the classical evolution problem (A.6), (A.7). In this sense, one can affirm that the approximate wave function $\phi(t)$ reproduces the classical motion of the particle.

A.2   Approximate Wave Function

The approximate wave function $\phi(t)$ can be explicitly constructed following the approach discussed for the harmonic oscillator in Sect. 7.3. The starting point is to observe that, for $\hbar \to 0$, the approximate solution of problem (A.6), (A.7) must be substantially different from zero only for $x \simeq q(t)$. Therefore, the effect of the interaction potential $V(x)$ must be relevant only for x close to $q(t)$. This observation suggests to approximate the potential $V(x)$ using Taylor's formula with initial point $q(t)$

$$V(x) = V(q(t)) + V'(q(t))(x - q(t)) + \frac{1}{2}V''(q(t))(x - q(t))^2 + R_2(x),\quad\text{(A.8)}$$

$$R_2(x) = \frac{1}{6}V'''(z)(x - q(t))^3\qquad\qquad\text{(A.9)}$$

for some z between x and $q(t)$ and assuming $V \in C^3(\mathbb{R})$. The fact that a second-order approximation is required will be clear during the proof in the next section. Using such second-order approximation of $V(x)$, we define the following approximate evolution problem:

$$i\hbar\frac{\partial\phi(t)}{\partial t} = H_2(t)\phi(t)\,, \tag{A.10}$$

$$\phi(0) = \psi_0\,, \tag{A.11}$$

where $H_2(t)$ is the quadratic and time-dependent Hamiltonian

$$H_2(t) = -\frac{\hbar^2}{2m}\Delta + V(q(t)) + V'(q(t))(x - q(t)) + \frac{1}{2}V''(q(t))(x - q(t))^2\,. \tag{A.12}$$

As a matter of fact, the solution of problem (A.10), (A.11) can be explicitly found following exactly the same procedure discussed for the harmonic oscillator in Sect. 7.3. Indeed, a direct computation shows that the solution is

$$\phi(x,t) = e^{\frac{i}{\hbar}S(t)}\frac{1}{(\pi\hbar)^{1/4}\sqrt{a(t)}}e^{-b(t)a(t)^{-1}\frac{(x-q(t))^2}{2\hbar} + i\frac{p(t)}{\hbar}(x-q(t))}\,, \tag{A.13}$$

where $S(t)$, $a(t)$, $b(t)$, $q(t)$, $p(t)$ are functions satisfying the following equations:

$$\dot{q}(t) = \frac{p(t)}{m}\,, \qquad \dot{p}(t) = -V'(q(t))\,, \tag{A.14}$$

$$\dot{a}(t) = \frac{i}{m}b(t)\,, \qquad \dot{b}(t) = i\,V''(q(t))a(t)\,, \tag{A.15}$$

$$\dot{S}(t) = \frac{p^2(t)}{2m} - V(q(t)) := L(q(t), p(t))\,, \tag{A.16}$$

with initial conditions $q(0) = q_0$, $p(0) = p_0$, $a(0) = a_0$, $b(0) = b_0$, $S(0) = 0$. We note that Eqs. (A.14) coincide with Hamilton's equations (A.6), (A.7) and $S(t)$ is the corresponding classical action. Moreover, if we set $u = a$, $v = ib$, Eqs. (A.15) are rewritten as

$$\dot{u}(t) = \frac{v(t)}{m}\,, \qquad \dot{v}(t) = -V''(q(t))u(t)\,. \tag{A.17}$$

The solutions of Eqs. (A.17) define the linearized flow around the solution $(q(t), p(t))$ (see Remark 1.7).

We also observe that the solution $\phi(t)$ has the same form of the initial state ψ_0 and therefore it has the required property to be well concentrated in position around $q(t)$ and in momentum around $p(t)$ for $\hbar \to 0$. In other words, the quantum state $\phi(t)$ is close to the classical state $(q(t), p(t))$ for $\hbar \to 0$. In the next section, we shall prove that $\phi(t)$ is a good approximation of the solution $\psi(t)$ of problem (A.6), (A.7) for $\hbar \to 0$. We shall follow the line of [3] (see also [1]) but, in order to simplify the proof, we introduce stronger assumptions on $V(x)$.

A.3   Estimate for $\hbar \to 0$

Here we give an estimate of the L^2-norm of the difference $\psi(t) - \phi(t)$. As a preliminary step, we prove the following technical lemma.

Lemma A.1 *Let us assume $V \in L^\infty(\mathbb{R})$ and consider the self-adjoint Hamiltonian in $L^2(\mathbb{R})$*

$$H = -\frac{\hbar^2}{2m}\Delta + V(x), \qquad D(H) = D(H_0). \tag{A.18}$$

Suppose that $\xi(t)$ belongs to $D(H_0)$, is continuously differentiable in t and satisfies

$$i\hbar\frac{\partial \xi(t)}{\partial t} = H\xi(t) + \zeta(t), \tag{A.19}$$

where $\zeta(t) \in L^2(\mathbb{R})$. Then,

$$\left\| e^{-i\frac{t}{\hbar}H}\xi(0) - \xi(t) \right\| \le \frac{1}{\hbar}\int_0^t ds\, \|\zeta(s)\|. \tag{A.20}$$

Proof Using the unitarity of the propagator, the fundamental theorem of calculus and Eq. (A.19), we have

$$\left\| e^{-i\frac{t}{\hbar}H}\xi(0) - \xi(t) \right\| = \left\| \xi(0) - e^{i\frac{t}{\hbar}H}\xi(t) \right\| = \left\| \int_0^t ds\, \frac{\partial}{\partial s}\left(e^{i\frac{s}{\hbar}H}\xi(s) \right) \right\|$$

$$= \left\| \int_0^t ds\, \left(\frac{i}{\hbar}e^{i\frac{s}{\hbar}H}H\xi(s) + e^{i\frac{s}{\hbar}H}\frac{\partial \xi(s)}{\partial s} \right) \right\|$$

$$= \left\| \int_0^t ds\, \left(\frac{1}{i\hbar}e^{i\frac{s}{\hbar}H}\zeta(s) \right) \right\|$$

$$\le \frac{1}{\hbar}\int_0^t ds\, \|\zeta(s)\|. \tag{A.21}$$

$\square$

We are now in position to formulate and prove our result on the semiclassical evolution of a Gaussian state in dimension one.

Proposition A.1 *Let us assume $V \in C^3(\mathbb{R})$ and $\|V\|_{L^\infty} + \|V'''\|_{L^\infty} < \infty$. Let $S(t)$, $a(t)$, $b(t)$, $q(t)$, $p(t)$ be solutions for $t \in [0, T]$, $T > 0$, of (A.14), (A.15), (A.16) with initial conditions $q(0) = q_0$, $p(0) = p_0$, $a(0) = a_0$, $b(0) = b_0$, $S(0) = 0$.*
Then there exists $C(T) > 0$ such that

$$\sup_{t\in[0,T]} \|\psi(t) - \phi(t)\| \le C(T)\sqrt{\hbar}, \tag{A.22}$$

where $\psi(t)$ is the solution of the Cauchy problem (A.4), (A.5) *and $\phi(t)$ is given by* (A.13).

Proof Let us observe that the difference $\psi(t) - \phi(t)$ satisfies the equation

$$i\hbar\frac{\partial}{\partial t}(\psi(t) - \phi(t)) = H(\psi(t) - \phi(t)) + (H - H_2(t))\phi(t). \tag{A.23}$$

Using Lemma A.1 and taking into account that $\psi(0) - \phi(0) = 0$, we find

$$\|\psi(t) - \phi(t)\| \leq \frac{1}{\hbar}\int_0^t ds\, \|(H - H_2(s))\phi(s)\|$$

$$= \frac{1}{\hbar}\int_0^t ds\left[\int dx\,\left|\frac{1}{6}V'''(z)(x - q(s))^3\,\phi(x,s)\right|^2\right]^{1/2}$$

$$\leq \frac{1}{6\hbar}\|V'''\|_{L^\infty}\int_0^t ds\left[\int dx\,|x - q(s)|^6\,|\phi(x,s)|^2\right]^{1/2}. \tag{A.24}$$

Using (A.13) and the change of the integration variable $y = \frac{x - q(s)}{\sqrt{\hbar}|a(s)|}$, we obtain

$$\|\psi(t) - \phi(t)\| \leq \frac{\sqrt{\hbar}}{6\pi^{1/4}}\|V'''\|_{L^\infty}\left(\int dy\, y^6\, e^{-y^2}\right)^{1/2}\int_0^T ds\,|a(s)|^3 \tag{A.25}$$

and, therefore, the proposition is proved.   □

Remark A.1 The above result can be extended to other space dimensions and the assumptions on $V(x)$ can be weakened [3]. We also stress that the constant $C(T)$ in (A.22) increases with time T and this is an unavoidable, typical feature of the Schrödinger equation.

It is worth mentioning that the type of result given in Proposition A.1 cannot be considered a satisfactory solution of the problem of the classical limit of Quantum Mechanics. The reason is that the result strongly depends on the choice of the initial datum of the type (A.1), i.e., a quasi-classical state. On the other hand, there are many important physical situations where a quantum system is characterized by a completely different type of initial state, e.g., a superposition state, and it nevertheless exhibits a classical behavior. This fact cannot be explained analyzing the quantum evolution of the isolated system. A possible explanation can be obtained only if one takes into account the role of the environment, i.e., if one studies the quantum evolution of "system + environment". This approach is the subject of the so called decoherence theory, which goes beyond the scope of these notes. The interested reader is referred to [2, 4, 5] and references therein.

A.4 WKB Method

In this section, we briefly mention the classical limit according to the WKB method (by Wentzel, Kramers and Brillouin). The starting point is to fix an initial state $\psi_0 \in L^2(\mathbb{R}^3)$ of the form

$$\psi_0(x) = \sqrt{\rho_0(x)}\, e^{\frac{i}{\hbar} S_0(x)}\,, \tag{A.26}$$

i.e., a state written as an amplitude independent of $\hbar$ and a phase that is rapidly oscillating for $\hbar \to 0$. Then, one looks for an approximate solution of the Schrödinger equation written in the same form and describing the evolution of a "fluid" of particles moving along classical trajectories. We emphasize that in the WKB method the localization in position of the approximate wave function is not small for $\hbar \to 0$. The reader should note the analogy with the short wavelength limit of Wave Optics discussed in Sect. 1.10.

We illustrate the method without going into technical details (for a complete treatment see [6]). Let us consider the solution $\psi(t)$ of the Schrödinger equation

$$i\hbar \frac{\partial \psi(t)}{\partial t} = -\frac{\hbar^2}{2m} \Delta \psi(t) + V(x)\psi(t) \tag{A.27}$$

with initial datum (A.26). Writing the solution as

$$\psi(x, t) = \sqrt{\hat{\rho}(x, t)}\, e^{\frac{i}{\hbar} \hat{S}(x,t)}\,, \tag{A.28}$$

we find

$$\sqrt{\hat{\rho}}\left(\frac{\partial \hat{S}}{\partial t} + \frac{|\nabla \hat{S}|^2}{2m} + V\right) - \frac{i\hbar}{2\sqrt{\hat{\rho}}}\left(\frac{\partial \hat{\rho}}{\partial t} + \nabla \hat{\rho} \cdot \frac{\nabla \hat{S}}{m} + \hat{\rho}\frac{\Delta \hat{S}}{m}\right) - \frac{\hbar^2}{2m} \Delta \sqrt{\hat{\rho}} = 0\,. \tag{A.29}$$

Separating real and imaginary parts, we obtain the two coupled equations

$$\frac{\partial \hat{S}}{\partial t} + \frac{|\nabla \hat{S}|^2}{2m} + V - \frac{\hbar^2}{2m}\frac{\Delta \sqrt{\hat{\rho}}}{\sqrt{\hat{\rho}}} = 0\,, \tag{A.30}$$

$$\frac{\partial \hat{\rho}}{\partial t} + \nabla \cdot \left(\frac{\nabla \hat{S}}{m}\hat{\rho}\right) = 0 \tag{A.31}$$

which are obviously equivalent to the Schrödinger equation (A.27). Note that (A.30) is a Hamilton–Jacobi equation, where the interaction term is the sum of the potential V acting on the particle plus the "quantum potential"

$$V_Q := -\frac{\hbar^2}{2m}\frac{\Delta \sqrt{\hat{\rho}}}{\sqrt{\hat{\rho}}}\,. \tag{A.32}$$

Moreover, (A.31) is a transport equation with a velocity field given by $m^{-1}\nabla S$. We emphasize that the solutions $\hat{S}(x,t)$, $\hat{\rho}(x,t)$ depend on $\hbar$.

The idea of the WKB method is that, for $\hbar \to 0$, the quantum potential becomes negligible and therefore an approximate solution $\eta(t)$ of the Schrödinger equation can be written again in the form

$$\eta(x,t) = \sqrt{\rho(x,t)}\, e^{\frac{i}{\hbar} S(x,t)}, \qquad (A.33)$$

where $\rho(x,t)$ and $S(x,t)$ are solutions of

$$\frac{\partial S}{\partial t} + \frac{|\nabla S|^2}{2m} + V = 0, \qquad (A.34)$$

$$\frac{\partial \rho}{\partial t} + \nabla \cdot \left(\frac{\nabla S}{m} \rho \right) = 0. \qquad (A.35)$$

Equations (A.34), (A.35) are two classical evolution equations (they do not depend on $\hbar$). In particular, (A.34) is the Hamilton–Jacobi equation for a classical particle of mass m, subject to the potential V and (A.35) is the corresponding transport equation. We assume that there exists $t_0 > 0$ such that for $t \in [0, t_0]$ Eqs. (A.34), (A.35) admit unique (smooth) solutions $S(x,t)$ and $\rho(x,t)$, with given (smooth) initial conditions $S_0(x)$, $\rho_0(x)$ (the existence of global in time solutions is a nontrivial mathematical problem).

The proof that $\eta(t)$, $t \in [0, t_0]$, is a good approximation of the solution $\psi(t)$ of the Schrödinger equation (A.27) for $\hbar \to 0$ is not difficult and it is left as an exercise.

Exercise A.1 Verify that there exists $c(t_0) > 0$ such that

$$\sup_{t \in [0,t_0]} \| \psi(t) - \eta(t) \| \leq c(t_0)\, \hbar. \qquad (A.36)$$

(Hint: using Eqs. (A.34), (A.35), show that $\eta(t)$ satisfies $i\hbar \frac{\partial \eta(t)}{\partial t} = H\eta(t) + \mathcal{R}(t)$, where $H = -\frac{\hbar^2}{2m}\Delta + V$ and $\mathcal{R}(x,t) = \frac{\hbar^2}{2m} e^{\frac{i}{\hbar} S(x,t)} \Delta \sqrt{\rho(x,t)}$. Taking into account Lemma A.1 and the fact that $\eta(0) = \psi_0$, one obtains the result).

We conclude by explaining the meaning of the approximate solution $\eta(t)$. Let us assume that $\rho_0(x)$ is significantly different from zero only for $x \simeq x_0$. As we have seen in Sect. 1.7, for $t \in [0, t_0]$ the solution $\rho(x,t)$ is significantly different from zero only for $x \simeq z(t, x_0)$, where $z(t, x_0)$ is the solution of

$$\frac{dz}{dt} = \frac{1}{m}\nabla S(z,t), \qquad z(0) = x_0. \qquad (A.37)$$

Computing the derivative of (A.37) with respect to t and using (A.34), one has

$$m\frac{d^2z_i}{dt^2} = \frac{d}{dt}\frac{\partial S}{\partial z_i} = \sum_{j=1}^{3}\frac{\partial^2 S}{\partial z_j \partial z_i}\frac{dz_j}{dt} + \frac{\partial^2 S}{\partial t \partial z_i} = \frac{1}{m}\sum_{j=1}^{3}\frac{\partial^2 S}{\partial z_j \partial z_i}\frac{\partial S}{\partial z_j} + \frac{\partial^2 S}{\partial t \partial z_i}$$

$$= \frac{\partial}{\partial z_i}\left(\frac{|\nabla S|^2}{2m} + \frac{\partial S}{\partial t}\right) = -\frac{\partial V}{\partial z_i}, \tag{A.38}$$

i.e., $z(t, x_0)$ is the classical motion with initial conditions $(x_0, m^{-1}\nabla S_0(x_0))$.

Therefore, we find that the probability density of the position $\rho(x, t) = |\eta(x, t)|^2$ is localized around the classical trajectory of the particle. In this sense, one can say that the approximate solution $\eta(t)$ reproduces the classical motion for $\hbar \to 0$.

References

1. Combescure, M., Robert, D.: Coherent States and Applications in Mathematical Physics. Springer (2012)
2. Figari, R., Teta, A.: Quantum Dynamics of a Particle in a Tracking Chamber. Springer Briefs in Physics, Springer (2014)
3. Hagedorn, G.: Raising and lowering operators for semiclassical wave packets. Ann. Phys. **269**, 77–104 (1998)
4. Hornberger, K.: Introduction to decoherence theory. In: Buchleitner, A., Viviescas, C., Tiersch, M. (eds.) Entanglement and Decoherence. Foundations and Modern Trends. Lecture Notes Physics, vol. 768, 221–276. Springer (2009)
5. Joos, E., Zeh, H.D., Kiefer, C., Giulini, D., Kupsch, J., Stamatescu, I.O.: Decoherence and the Appearance of a Classical World in Quantum Theory. Springer (2003)
6. Maslov, V.P., Fedoriuk, M.V.: Semi-Classical Approximation in Quantum Mechanics. Springer (1981)

Appendix B
Basic Concepts of Scattering Theory

B.1 Formulation of the Problem

In Sect. 8.4, we described a scattering problem in the particular case of a particle subject to a point interaction in dimension one. Here, we give an introduction to the basic notions of scattering theory which can be considered as a prerequisite for the approach to a general scattering problem in Quantum Mechanics (for a detailed treatment we refer to [1–4]).

We start with an elementary description of a scattering experiment in a laboratory.

Let us consider a particle sent from a large distance toward a target placed in a given position of the laboratory. Initially, the particle does not "feel" the presence (and the action) of the target and then its motion is essentially a free motion fixed by the experimenter. As the particle approaches the target, it begins to feel the action of the target and this means that its motion is no longer free. This fact remains true for all the time that the particle spends in the vicinity of the target. As time goes by and assuming that the target does not capture the particle, the particle moves away from the target and then its motion becomes again a free motion, in general different from the initial one. Basically, the scattering experiment consists in the following: given the initial free motion, i.e., the free motion prepared by the experimenter much time before the interaction with the target, determine (some parameters of) the final free motion, i.e., the free motion emerging much time after the interaction with the target. We shall limit to consider the case in which the kinetic energy of the initial and the final free motions are the same (elastic scattering).

We are interested in formulating a mathematical model able to describe such elastic scattering experiment for a quantum particle and to provide theoretical predictions of the experimental results. The basic assumptions are

(i) the motion of the particle is described by the Schrödinger equation;
(ii) the action of the target on the particle is described by a potential $V(x)$ decaying to zero sufficiently fast for $|x| \to \infty$;
(iii) the internal dynamics of the target is neglected.

A. Teta, *A Mathematical Primer on Quantum Mechanics*,
UNITEXT for Physics, https://doi.org/10.1007/978-3-319-77893-8

From these assumptions, it follows that we are studying a one-body problem for the quantum particle, whose state at time t is

$$\psi_t = e^{-i\frac{t}{\hbar}H}\psi, \qquad H = H_0 + V(x), \qquad H_0 = -\frac{\hbar^2}{2m}\Delta, \qquad (B.1)$$

where $\psi \in L^2(\mathbb{R}^d)$, $d = 1, 2, 3$, is the initial state.

According to the description of the scattering experiment, in our mathematical model, we are concerned with the comparison of the interacting $e^{-i\frac{t}{\hbar}H}$ and the free $e^{-i\frac{t}{\hbar}H_0}$ unitary groups for $t \to -\infty$, i.e., much time before the interaction with the target, and for $t \to +\infty$, i.e., much time after the interaction with the target.

In the next two sections, we introduce the main concepts (wave and scattering operators) required to arrive at a mathematical description of the scattering experiment.

B.2 Wave Operators

The first step is to characterize the solutions of the Schrödinger equation having the property to resemble a free evolution for t large.

Definition B.1 The state $\psi \in L^2(\mathbb{R}^d)$ is called asymptotically free for $t \to +\infty$ if there exists $f \in L^2(\mathbb{R}^d)$ such that

$$\lim_{t \to +\infty} \left\| e^{-i\frac{t}{\hbar}H}\psi - e^{-i\frac{t}{\hbar}H_0}f \right\| = 0. \qquad (B.2)$$

Analogously, one defines an asymptotically free state for $t \to -\infty$. The sets of the asymptotically free states for $t \to -\infty$ and for $t \to +\infty$ are denoted by $\mathscr{H}_{in}$ and $\mathscr{H}_{out}$ respectively.

It is immediately seen that not all states are asymptotically free. For example, a stationary state, i.e., an eigenvector of the Hamiltonian H, is not asymptotically free. Therefore, the first problem is to show the existence of asymptotically free states. In mathematical terms, the problem is formulated as follows: given $f \in L^2(\mathbb{R}^d)$, prove that there exists $\psi \in L^2(\mathbb{R}^d)$ such that (B.2) holds. Roughly speaking, we are dealing with a Cauchy problem for the Schrödinger equation with an initial datum assigned for "$t = +\infty$" (or for "$t = -\infty$").

Using the unitarity of the evolution operator $e^{-i\frac{t}{\hbar}H}$, Eq. (B.2) is equivalent to

$$\lim_{t \to +\infty} \| \psi - e^{i\frac{t}{\hbar}H}e^{-i\frac{t}{\hbar}H_0}f \| = 0 \qquad (B.3)$$

and then the existence problem can be reformulated as: for any $f \in L^2(\mathbb{R}^d)$, prove that $e^{i\frac{t}{\hbar}H}e^{-i\frac{t}{\hbar}H_0}f$ converges in $L^2(\mathbb{R}^d)$ for $t \to +\infty$. This suggests the following important definition.

Definition B.2 For any $f \in L^2(\mathbb{R}^d)$, the wave operator Ω_+ is defined by

$$\Omega_+ f := \lim_{t \to +\infty} e^{i\frac{t}{\hbar}H} e^{-i\frac{t}{\hbar}H_0} f \,. \tag{B.4}$$

Analogously, one defines the wave operator Ω_-.

Note that if $\Omega_\pm$ exist then $\mathscr{H}_{in} = Ran(\Omega_-)$, $\mathscr{H}_{out} = Ran(\Omega_+)$ and for any $f \in L^2(\mathbb{R}^d)$ we have

$$\lim_{t \to \pm\infty} \left\| e^{-i\frac{t}{\hbar}H} \Omega_\pm f - e^{-i\frac{t}{\hbar}H_0} f \right\| = 0 \,. \tag{B.5}$$

The proof of the existence of the wave operators, and then of the asymptotically free states, can be obtained under suitable assumptions on the potential V. For example, in dimension three a sufficient condition is $V \in L^2(\mathbb{R}^3)$ (see, e.g., [3], Sect. 4).

Here we list some important properties of the wave operators.

(i) The operators $\Omega_\pm$ are isometric, i.e., $\Omega_\pm^* \Omega_\pm = I$.
  In general, they are not unitary in $L^2(\mathbb{R}^d)$ since it can happen that $Ran(\Omega_\pm) \neq L^2(\mathbb{R}^d)$. By Proposition 4.16, we have that $Ran(\Omega_\pm)$ are two closed subspaces of $L^2(\mathbb{R}^d)$ and $\Omega_\pm \Omega_\pm^* = P_\pm$, where $P_\pm$ denote the orthogonal projectors on $Ran(\Omega_\pm)$. Equivalently, we have $\Omega_\pm^*|_{Ran(\Omega_\pm)} = \Omega_\pm^{-1}$ and $\Omega_\pm^*|_{Ran(\Omega_\pm)^\perp} = 0$.

(ii) $\Omega_\pm = e^{i\frac{t}{\hbar}H} \Omega_\pm e^{-i\frac{t}{\hbar}H_0}$ (interwining property).
  Indeed, for any $s, t \in \mathbb{R}$ and for any $f \in L^2(\mathbb{R}^d)$, we have $e^{i\frac{s+t}{\hbar}H} e^{-i\frac{s+t}{\hbar}H_0} f = e^{i\frac{t}{\hbar}H} e^{i\frac{s}{\hbar}H} e^{-i\frac{s}{\hbar}H_0} e^{-i\frac{t}{\hbar}H_0} f$. Taking the limit for $s \to +\infty$ we obtain the property for Ω_+. For Ω_- one proceeds in the same way.
  Moreover, as an exercise the reader can verify that if $f \in D(H_0)$ then $\Omega_\pm f \in D(H)$ and $H\Omega_\pm f = \Omega_\pm H_0 f$.

(iii) $Ran(\Omega_\pm)$ are two invariant subspaces for the action of the group $e^{-i\frac{t}{\hbar}H}$.
  The assertion follows from the interwining property, in fact if $\psi \in Ran(\Omega_+)$ then there exists $f \in L^2(\mathbb{R}^d)$ such that $\psi = \Omega_+ f$ and we have $e^{-i\frac{t}{\hbar}H} \psi = e^{-i\frac{t}{\hbar}H} \Omega_+ f = \Omega_+ e^{-i\frac{t}{\hbar}H_0} f \in Ran(\Omega_+)$. For Ω_- one proceeds in the same way.

B.3 Asymptotic Completeness and Scattering Operator

In order to describe a scattering experiment, our mathematical model should be able to give a solution to the following problem: assigned an incoming free evolution for $t \to -\infty$, find the emerging outgoing free evolution for $t \to +\infty$.

Assuming that the wave operators exist, we introduce the following crucial definition.

Definition B.3 We say that the condition of asymptotic completeness holds if

$$Ran\,(\Omega_-) = Ran\,(\Omega_+) = \mathscr{H}_p(H)^\perp. \tag{B.6}$$

Remark B.1 We recall that, according to Definition 4.27, $\mathscr{H}_p(H)$ is the (closed) subspace spanned by all the eigenvectors of H. We also note that for any reasonable interaction potential V one can prove that $\mathscr{H}_{sc}(H) = \emptyset$. In such a case, one has $\mathscr{H}_p(H)^\perp = \mathscr{H}_{ac}(H)$ and $\mathscr{H}_{ac}(H) = \mathscr{M}_\infty(H)$ (see Definitions 4.28, 5.2).

Condition (B.6) means that for any $\psi \in Ran\,(\Omega_-) = Ran\,(\Omega_+) = \mathscr{H}_p(H)^\perp$ there exist $f, g \in L^2(\mathbb{R}^d)$ such that $\psi = \Omega_- f = \Omega_+ g$ or, equivalently,

$$\lim_{t\to-\infty} \left\| e^{-i\frac{t}{\hbar}H}\psi - e^{-i\frac{t}{\hbar}H_0}f \right\| = 0\,, \qquad \lim_{t\to+\infty} \left\| e^{-i\frac{t}{\hbar}H}\psi - e^{-i\frac{t}{\hbar}H_0}g \right\| = 0\,, \tag{B.7}$$

where we have used the definition of wave operators.

The proof that the condition of asymptotic completeness holds requires a nontrivial technical work. A sufficient condition in dimension three is $V \in L^1(\mathbb{R}^3) \cap L^2(\mathbb{R}^3)$ (see, e.g., [3], Sect. 4).

From now on, we assume that the wave operators exist and asymptotic completeness holds. Under these assumptions, we can solve the main problem of scattering theory mentioned at the beginning of this section. The steps to be taken to arrive at the solution are as follows.

- For any given $f \in L^2(\mathbb{R}^d)$, let $e^{-i\frac{t}{\hbar}H_0}f$ be the incoming free evolution assigned for $t \to -\infty$.
- Using the existence of Ω_-, we construct $e^{-i\frac{t}{\hbar}H}\Omega_- f$, i.e., the solution of the Schrödinger equation with Hamiltonian H having the property

$$\lim_{t\to-\infty} \| e^{-i\frac{t}{\hbar}H}\Omega_- f - e^{-i\frac{t}{\hbar}H_0}f \| = 0\,. \tag{B.8}$$

Thus, $e^{-i\frac{t}{\hbar}H}\Omega_- f$ reduces to the assigned incoming free evolution $e^{-i\frac{t}{\hbar}H_0}f$ for $t \to -\infty$.

- It remains to characterize the behavior of $e^{-i\frac{t}{\hbar}H}\Omega_- f$ for $t \to +\infty$. Using the asymptotic completeness, we have $\Omega_- f \in Ran\,(\Omega_+)$. Therefore, there exists $g \in L^2(\mathbb{R}^d)$ such that $\Omega_- f = \Omega_+ g$ and, by definition of Ω_+, we have

$$\lim_{t\to+\infty} \| e^{-i\frac{t}{\hbar}H}\Omega_- f - e^{-i\frac{t}{\hbar}H_0}g \| = \lim_{t\to+\infty} \| e^{-i\frac{t}{\hbar}H}\Omega_+ g - e^{-i\frac{t}{\hbar}H_0}g \| = 0\,. \tag{B.9}$$

In other words, $e^{-i\frac{t}{\hbar}H}\Omega_- f$ reduces to the outgoing free evolution $e^{-i\frac{t}{\hbar}H_0}g$ for $t \to +\infty$, where g is the solution of the equation $\Omega_+ g = \Omega_- f$.

- Taking into account that Ω_+^{-1} is well defined on $Ran\,(\Omega_+) = Ran\,(\Omega_+)$ and it coincides with Ω_+^*, we obtain

$$g = \Omega_+^* \Omega_- f := Sf\,, \tag{B.10}$$

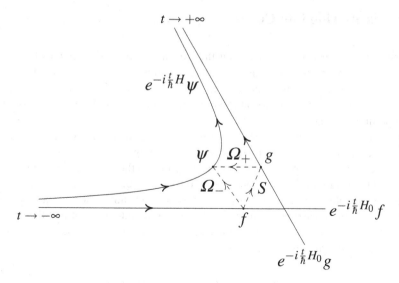

Fig. B.1 The curved line represents the interacting evolution $e^{-i\frac{t}{\hbar}H}\psi$ and the straight lines represent the incoming $e^{-i\frac{t}{\hbar}H_0}f$ and the outgoing $e^{-i\frac{t}{\hbar}H_0}g$ free evolution. The states ψ, f, g are related by $\psi = \Omega_- f$, $\psi = \Omega_+ g$, $g = Sf$

where the operator S is called scattering operator, or also S-matrix.

- To summarize, we have shown that if $e^{-i\frac{t}{\hbar}H_0}f$ is the incoming free evolution assigned for $t \to -\infty$ then $e^{-i\frac{t}{\hbar}H_0}Sf$ is the outgoing free evolution emerging for $t \to +\infty$, where the operator S is defined in (B.10) (Fig. B.1).

The scattering operator S satisfies some important properties.

(i) S is unitary.

The property follows from $SS^* = \Omega_+^* \Omega_- (\Omega_+^* \Omega_-)^* = \Omega_+^* \Omega_- \Omega_-^* \Omega_+^{**} = \Omega_+^* P_- \Omega_+^{**} = (\Omega_+^* P_- \Omega_+)^* = (\Omega_+^* P_+ \Omega_+)^* = (\Omega_+^* \Omega_+)^* = I$ and, analogously, for $S^* S = I$.

(ii) $[S, e^{-i\frac{t}{\hbar}H_0}] = 0$.

Indeed, using the interwining property of the wave operators, we have
$S e^{-i\frac{t}{\hbar}H_0} = \Omega_+^* \Omega_- e^{-i\frac{t}{\hbar}H_0} = \Omega_+^* e^{-i\frac{t}{\hbar}H}\Omega_- = (e^{i\frac{t}{\hbar}H}\Omega_+)^*\Omega_- = (\Omega_+ e^{i\frac{t}{\hbar}H_0})^*\Omega_- = e^{-i\frac{t}{\hbar}H_0}\Omega_+^*\Omega_- = e^{-i\frac{t}{\hbar}H_0}S$.
As an exercise, the reader can verify that if $f \in D(H_0)$ then $Sf \in D(H_0)$ and $[S, H_0]f = 0$.

(iii) If $f \in D(H_0)$ then $(f, H_0 f) = (Sf, H_0 Sf)$.

The assertion immediately follows from unitarity of S and (ii), in fact
$(f, H_0 f) = (Sf, SH_0 f) = (Sf, H_0 Sf)$.

This last property expresses the fact that the kinetic energies of the incoming and outgoing free evolutions coincide, i.e., the scattering is elastic.

B.4   Scattering into Cones

In the previous section we have seen that, from the theoretical point of view, all the information on the scattering process is contained in the scattering operator S, which assigns the outgoing free state to each incoming free state. Here, we show how the knowledge of S allows, at least in principle, to make theoretical predictions of experimental results.

For the sake of concreteness, we consider the three-dimensional case. Let us suppose that the experimenter prepares the incoming free evolution $e^{-i\frac{t}{\hbar}H_0} f$ for $t \to -\infty$ and then he measures the probability to find the particle in a cone $\mathscr{C}$, with vertex at the origin of the reference frame, for $t \to +\infty$.

Let us denote such a probability by $\mathscr{P}(f, \mathscr{C})$. Our aim is to give a theoretical prediction of the measured quantity $\mathscr{P}(f, \mathscr{C})$. Taking into account that the solution of the Schrödinger equation that reduces to $e^{-i\frac{t}{\hbar}H_0} f$ for $t \to -\infty$ is $e^{-i\frac{t}{\hbar}H}\Omega_- f$ and using Born's rule, we have

$$\mathscr{P}(f, \mathscr{C}) = \lim_{t \to +\infty} \int_{\mathscr{C}} dx \left| \left(e^{-i\frac{t}{\hbar}H}\Omega_- f \right)(x) \right|^2. \qquad (B.11)$$

In the next proposition, known as scattering into cones theorem, we show that $\mathscr{P}(f, \mathscr{C})$ can be computed in terms of the scattering operator S.

Proposition B.1 *For any $f \in L^2(\mathbb{R}^3)$ we have*

$$\mathscr{P}(f, \mathscr{C}) = \int_{\mathscr{C}} dk \left| \widetilde{Sf}(k) \right|^2. \qquad (B.12)$$

Proof Using the fact that $\lim_{t \to +\infty} \| e^{-i\frac{t}{\hbar}H}\Omega_- f - e^{-i\frac{t}{\hbar}H_0} Sf \| = 0$, we will show that in the r.h.s. of (B.11) we can replace $e^{-i\frac{t}{\hbar}H}\Omega_- f$ with $e^{-i\frac{t}{\hbar}H_0} Sf$. Denoted $u_t := e^{-i\frac{t}{\hbar}H}\Omega_- f$ and $v_t := e^{-i\frac{t}{\hbar}H_0} Sf$, we have

$$\left| \int_{\mathscr{C}} dx\, |u_t(x)|^2 - \int_{\mathscr{C}} dx\, |v_t(x)|^2 \right|$$

$$= \left| \int_{\mathscr{C}} dx\, \overline{u_t(x)}(u_t(x) - v_t(x)) + \int_{\mathscr{C}} dx\, (\overline{u_t(x)} - \overline{v_t(x)})v_t(x) \right|$$

$$\leq \left(\int_{\mathscr{C}} dx\, |u_t(x)|^2 \right)^{1/2} \left(\int_{\mathscr{C}} dx\, |u_t(x) - v_t(x)|^2 \right)^{1/2}$$

$$+ \left(\int_{\mathscr{C}} dx\, |u_t(x) - v_t(x)|^2 \right)^{1/2} \left(\int_{\mathscr{C}} dx\, |v_t(x)|^2 \right)^{1/2}$$

$$\leq (\|u_t\| + \|v_t\|) \|u_t - v_t\|. \qquad (B.13)$$

Since $\|u_t\| = \|v_t\| = 1$ and $\lim_{t \to +\infty} \|u_t - v_t\| = 0$, we find

$$\mathscr{P}(f,\mathscr{C}) = \lim_{t\to+\infty} \int_{\mathscr{C}} dx \left| \left(e^{-i\frac{t}{\hbar}H_0} Sf \right)(x) \right|^2. \tag{B.14}$$

Using the asymptotic expression for t large of the free evolution (Proposition 6.2) and again the estimate (B.13), we obtain

$$\mathscr{P}(f,\mathscr{C}) = \lim_{t\to+\infty} \int_{\mathscr{C}} dx \left(\frac{m}{\hbar t}\right)^3 \left| \widetilde{Sf}\left(\frac{mx}{\hbar t}\right) \right|^2 = \int_{\mathscr{C}} dk \left| \widetilde{Sf}(k) \right|^2. \tag{B.15}$$

$\square$

We stress that the l.h.s. of formula (B.12) is a quantity that, at least in principle, can be measured in the experiment while the r.h.s. can be theoretically computed within our mathematical model. Therefore, formula (B.12) provides a theoretical prediction of an experimental result.

Remark B.2 According to Born's rule for the momentum observable, the r.h.s. of formula (B.12) represents the probability that the momentum of the particle belongs to the cone $\mathscr{C}$ for $t \to +\infty$.

Remark B.3 It is easy to verify that the proposition holds also in dimension two and one. In the first case the cone reduces to an angle with vertex at the origin. In one dimension, the possible cones are the positive semiaxis $\mathbb{R}_+$ and the negative semiaxis $\mathbb{R}_-$. Therefore, the quantity $\mathscr{P}(f, \mathbb{R}_+)$ is the probability that the particle is found at the right of the origin for $t \to +\infty$ and $\mathscr{P}(f, \mathbb{R}_-)$ is the probability that the particle is found at the left of the origin for $t \to +\infty$. This means that, for a given free evolution coming from the negative semiaxis with positive momentum for $t \to -\infty$, the quantity $\mathscr{P}(f, \mathbb{R}_+)$ represents the transmission probability and $\mathscr{P}(f, \mathbb{R}_-)$ represents the reflection probability.

B.5 Scattering by a Point Interaction Revisited

In this section, we apply the notions of scattering theory introduced in the previous sections to the specific case of a particle subject to a point interaction in dimension one. This case has been already discussed in Sect. 8.4 and we will show how the results can be derived within the framework of the general theory. We also stress that the approach, based on the eigenfunction expansion of the Hamiltonian, can be extended to more general cases of scattering problems in Quantum Mechanics.

Using the results obtained in Sects. 8.2 and 8.3, we can prove the existence of the wave operators, denoted by $\Omega_{\pm,\alpha}$, and the asymptotic completeness.

Proposition B.2 *The wave operators exist and are explicitly given by*

$$\Omega_{\pm,\alpha} f = W_{\pm}^{-1} \mathscr{F} f. \tag{B.16}$$

Moreover, Ran $(\Omega_{\pm,\alpha}) = L^2(\mathbb{R})$ for $\alpha > 0$ and Ran $(\Omega_{\pm,\alpha}) = [\Xi_0]^\perp$ for $\alpha < 0$. Therefore, the condition of asymptotic completeness holds.

Proof For any $f \in L^2(\mathbb{R})$, we define $\psi := W_\pm^{-1} \mathscr{F} f$. Note that $\psi \in L^2(\mathbb{R})$ for $\alpha > 0$ and $\psi \in [\Xi_0]^\perp$ for $\alpha < 0$. Using Proposition (8.4) (and its extension to the case $\alpha < 0$), we have

$$\lim_{t \to \pm\infty} \left\| e^{i \frac{t}{\hbar} H_\alpha} e^{-i \frac{t}{\hbar} H_0} f - \psi \right\| = \lim_{t \to \pm\infty} \left\| e^{-i \frac{t}{\hbar} H_0} f - e^{-i \frac{t}{\hbar} H_\alpha} \psi \right\|$$

$$= \lim_{t \to \pm\infty} \left\| e^{-i \frac{t}{\hbar} H_0} \mathscr{F}^{-1} W_\pm \psi - e^{i \frac{t}{\hbar} H_\alpha} \psi \right\| = 0 . \quad \text{(B.17)}$$

Thus, (B.16) is proved. By Proposition 8.3, we know that $Ran (W_\pm^{-1}) = L^2(\mathbb{R})$ for $\alpha > 0$ and $Ran (W_\pm^{-1}) = [\Xi_0]^\perp$ for $\alpha < 0$. This implies the condition of asymptotic completeness. $\quad\square$

Since asymptotic completeness holds, we can define the scattering operator S_α

$$S_\alpha f := \Omega_{+,\alpha}^* \Omega_{-,\alpha} f = \mathscr{F}^{-1} W_+ W_-^{-1} \mathscr{F} f \quad \text{(B.18)}$$

or, in the Fourier space

$$\widetilde{S_\alpha f} = W_+ W_-^{-1} \tilde{f} . \quad \text{(B.19)}$$

The next step is to find an explicit representation for the scattering operator. The following technical lemma, known as abelian limit, is a useful prerequisite.

Proposition B.3 *Let $v : [0, \infty) \to \mathbb{C}$ be a continuous, bounded function such that*

$$\lim_{M \to \infty} \int_0^M ds \, v(s) = a , \qquad a \in \mathbb{R} . \quad \text{(B.20)}$$

Then

$$\lim_{\varepsilon \to 0^+} \int_0^\infty ds \, e^{-\varepsilon s} v(s) = a . \quad \text{(B.21)}$$

Proof Integrating by parts, we have

$$\int_0^\infty ds \, e^{-\varepsilon s} v(s) = \lim_{M \to \infty} \int_0^M ds \, e^{-\varepsilon s} v(s) = \lim_{M \to \infty} \int_0^M ds \, e^{-\varepsilon s} \frac{d}{ds} \int_0^s d\sigma \, v(\sigma)$$

$$= \lim_{M \to \infty} \int_0^M ds \, \varepsilon \, e^{-\varepsilon s} \int_0^s d\sigma \, v(\sigma) = \int_0^\infty ds \, \varepsilon \, e^{-\varepsilon s} \int_0^s d\sigma \, v(\sigma)$$

$$= \int_0^\infty d\tau \, e^{-\tau} \int_0^{\tau/\varepsilon} d\sigma \, v(\sigma) . \quad \text{(B.22)}$$

Taking the limit for $\varepsilon \to 0^+$ and using the dominated convergence theorem we conclude the proof. $\quad\square$

Using the explicit expression for the generalized eigenfunctions (8.27) and the previous lemma, we obtain the characterization of S_α.

Proposition B.4 *For any $f \in L^2(\mathbb{R})$ we have*

$$\left(\widehat{S_\alpha f}\right)(k) = \overline{\mathscr{T}(k)} \tilde{f}(k) + \overline{\mathscr{R}(k)} \tilde{f}(-k), \tag{B.23}$$

where $\mathscr{R}(k)$ and $\mathscr{T}(k)$ have been defined in (8.28), (8.64).

Proof It is sufficient to prove (B.23) for $f \in \mathscr{S}(\mathbb{R})$. By (8.31) and (8.33), we have

$$\left(\widehat{S_\alpha f}\right)(k) = \frac{1}{2\pi} \int dy\, \overline{\phi_+(y,k)} \int dq\, \phi_-(y,q) \tilde{f}(q) = \tilde{f}(k) + A(k) + B(k) + C(k), \tag{B.24}$$

where

$$A(k) := \frac{\overline{\mathscr{R}(k)}}{2\pi} \int dy\, e^{i|k||y|} \int dq\, \tilde{f}(q) e^{iqy}, \tag{B.25}$$

$$B(k) := \frac{\overline{\mathscr{R}(k)}}{2\pi} \int dy\, e^{i|k||y|} \int dq\, \overline{\mathscr{R}(q)} \tilde{f}(q) e^{i|q||y|}, \tag{B.26}$$

$$C(k) := \frac{1}{2\pi} \int dy\, e^{-iky} \int dq\, \overline{\mathscr{R}(q)} \tilde{f}(q) e^{i|q||y|}. \tag{B.27}$$

In order to obtain a more explicit expression for the functions A, B and C, we regularize the integral in the variable y using the previous lemma and then we interchange the order of integration of the variables y and q. We also note that A, B and C are even and then it is sufficient to consider the case $k > 0$. For A we have

$$\begin{aligned}
A(k) &= \frac{\overline{\mathscr{R}(k)}}{2\pi} \lim_{\varepsilon \to 0} \int dy\, e^{i(k+i\varepsilon)|y|} \int dq\, \tilde{f}(q) e^{iqy} \\
&= \frac{\overline{\mathscr{R}(k)}}{2\pi} \lim_{\varepsilon \to 0} \int dq\, \tilde{f}(q) \int dy\, e^{i(k+i\varepsilon)|y|+iqy} \\
&= i \frac{\overline{\mathscr{R}(k)}}{2\pi} \lim_{\varepsilon \to 0} \int dq\, \tilde{f}(q) \left(\frac{1}{k+q+i\varepsilon} + \frac{1}{k-q+i\varepsilon} \right) \\
&= i \frac{\overline{\mathscr{R}(k)}}{2\pi} \lim_{\varepsilon \to 0} \int_0^\infty dq\, (\tilde{f}(q) + \tilde{f}(-q)) \left(\frac{1}{k+q+i\varepsilon} + \frac{1}{k-q+i\varepsilon} \right) \\
&= i \frac{\overline{\mathscr{R}(k)}}{2\pi} \int_0^\infty dq\, \frac{\tilde{f}(q) + \tilde{f}(-q)}{k+q} \\
&\quad + i \frac{\overline{\mathscr{R}(k)}}{2\pi} \lim_{\varepsilon \to 0} \int_0^\infty dq\, (\tilde{f}(q) + \tilde{f}(-q)) \frac{k-q-i\varepsilon}{(k-q)^2 + \varepsilon^2}. \tag{B.28}
\end{aligned}$$

We recall that if $F : \mathbb{R} \to \mathbb{C}$ is continuous and bounded then

$$\lim_{\varepsilon \to 0} \int dx\, F(x) \frac{\varepsilon}{x^2 + \varepsilon^2} = \lim_{\varepsilon \to 0} \int dy\, F(\varepsilon y) \frac{1}{y^2 + 1} = \pi F(0). \qquad (B.29)$$

Using (B.29) we obtain

$$A(k) = \frac{\overline{\mathscr{R}(k)}}{2} \left(\tilde{f}(k) + \tilde{f}(-k)\right) + i \frac{\overline{\mathscr{R}(k)}}{2\pi} \int_0^\infty dq\, \frac{\tilde{f}(q) + \tilde{f}(-q)}{k + q}$$
$$+ i \frac{\overline{\mathscr{R}(k)}}{2\pi} \lim_{\varepsilon \to 0} \int_0^\infty dq\, \left(\tilde{f}(q) + \tilde{f}(-q)\right) \frac{k - q}{(k-q)^2 + \varepsilon^2}. \qquad (B.30)$$

We leave as an exercise the analogous computation for B

$$B(k) = i \frac{\overline{\mathscr{R}(k)}}{\pi} \int_0^\infty dq\, \left(\tilde{f}(q) + \tilde{f}(-q)\right) \frac{\overline{\mathscr{R}(q)}}{k + q}. \qquad (B.31)$$

Finally, for C we have

$$C(k) = \frac{1}{2\pi} \lim_{\varepsilon \to 0} \int dq\, \overline{\mathscr{R}(q)} \tilde{f}(q) \int_0^\infty dy\, e^{i(|q|-k+i\varepsilon)y}$$
$$+ \frac{1}{2\pi} \lim_{\varepsilon \to 0} \int dq\, \overline{\mathscr{R}(q)} \tilde{f}(q) \int_0^\infty dy\, e^{i(|q|+k+i\varepsilon)y}$$
$$= \frac{i}{2\pi} \lim_{\varepsilon \to 0} \int dq\, \overline{\mathscr{R}(q)} \tilde{f}(q) \frac{|q| - k - i\varepsilon}{(|q| - k)^2 + \varepsilon^2} + \frac{i}{2\pi} \int dq\, \overline{\mathscr{R}(q)} \tilde{f}(q) \frac{1}{|q| + k}. \qquad (B.32)$$

Let us note that

$$\frac{1}{2\pi} \lim_{\varepsilon \to 0} \int dq\, \overline{\mathscr{R}(q)} \tilde{f}(q) \frac{\varepsilon}{(|q| - k)^2 + \varepsilon^2}$$
$$= \frac{1}{2\pi} \lim_{\varepsilon \to 0} \int_0^\infty dq\, \overline{\mathscr{R}(q)} \left(\tilde{f}(q) + \tilde{f}(-q)\right) \frac{\varepsilon}{(q-k)^2 + \varepsilon^2} = \frac{\overline{\mathscr{R}(k)}}{2} \left(\tilde{f}(k) + \tilde{f}(-k)\right). \qquad (B.33)$$

Using (B.33) in (B.32), we find

$$C(k) = \frac{\overline{\mathscr{R}(k)}}{2} \left(\tilde{f}(k) + \tilde{f}(-k)\right) + \frac{i}{2\pi} \int_0^\infty dq\, \left(\tilde{f}(q) + \tilde{f}(-q)\right) \frac{\overline{\mathscr{R}(q)}}{k + q}$$
$$+ \frac{i}{2\pi} \lim_{\varepsilon \to 0} \int_0^\infty dq\, \left(\tilde{f}(q) + \tilde{f}(-q)\right) \overline{\mathscr{R}(q)} \frac{q - k}{(q - k)^2 + \varepsilon^2}. \qquad (B.34)$$

Taking into account of (B.30), (B.31) and (B.34), Eq. (B.24) reads

$$\left(\widehat{S_\alpha f}\right)(k) = \overline{\mathcal{T}(k)}\tilde{f}(k) + \overline{\mathcal{R}(k)}\tilde{f}(-k)) + \Lambda(k),$$ (B.35)

where $\mathcal{T}(k) = 1 + \mathcal{R}(k)$ and

$$
\begin{aligned}
\Lambda(k) &= \frac{i}{2\pi}\int_0^\infty dq\,(\tilde{f}(q)+\tilde{f}(-q))\frac{\overline{\mathcal{R}(k)}+2\overline{\mathcal{R}(k)}\mathcal{R}(q)+\overline{\mathcal{R}(q)}}{q+k} \\
&+ \frac{i}{2\pi}\lim_{\varepsilon\to 0}\int_0^\infty dq\,(\tilde{f}(q)+\tilde{f}(-q))\frac{(q-k)(\overline{\mathcal{R}(q)}-\overline{\mathcal{R}(k)})}{(q-k)^2+\varepsilon^2} \\
&= \frac{i}{2\pi}\int_0^\infty dq\,(\tilde{f}(q)+\tilde{f}(-q))\left(\frac{\overline{\mathcal{R}(k)}+2\overline{\mathcal{R}(k)}\mathcal{R}(q)+\overline{\mathcal{R}(q)}}{q+k}+\frac{\overline{\mathcal{R}(q)}-\overline{\mathcal{R}(k)}}{q-k}\right) \\
&= \frac{i}{\pi}\int_0^\infty dq\,(\tilde{f}(q)+\tilde{f}(-q))\frac{q\,\overline{\mathcal{R}(q)}(1+\overline{\mathcal{R}(k)})-k\,\overline{\mathcal{R}(k)}(1+\overline{\mathcal{R}(q)})}{q^2-k^2}.
\end{aligned}
$$ (B.36)

Using the identity $2q\,\mathcal{R}(q) = i\alpha(1+\mathcal{R}(q))$, we find $\Lambda = 0$ and the proposition is proved. □

The expression (B.23) for the scattering operator S_α allows a complete description of the scattering problem, in agreement with the results already found in Sect. 8.4. For simplicity, we fix the state

$$f_0(x) = \frac{1}{\sqrt{\sigma}}h\left(\frac{x}{\sigma}\right)e^{i\frac{p_0}{\hbar}x},$$ (B.37)

where $\sigma, p_0 > 0$ and $h \in \mathscr{S}(\mathbb{R})$, with h real and even. Moreover, we assume that

$$\varepsilon := \frac{\hbar}{\sigma p_0} \ll 1.$$ (B.38)

Then, we suppose that the incoming free evolution assigned for $t \to -\infty$ is

$$e^{-i\frac{t}{\hbar}H_0}f_0.$$ (B.39)

Condition (B.38) guarantees that the state $e^{-i\frac{t}{\hbar}H_0}f_0$ is, for any t, well concentrated in momentum around the mean value $p_0 > 0$ (see 6.67, 6.61, 6.62).

By Proposition 6.2, we also have

$$\left(e^{-i\frac{t}{\hbar}H_0}f_0\right)(x) \underset{t\to-\infty}{\simeq} e^{i\frac{mx^2}{2\hbar t}}\sqrt{\frac{m}{i\hbar t}}\,\tilde{h}\left(\frac{\sigma}{\hbar}\left(\frac{mx}{|t|}+p_0\right)\right).$$ (B.40)

Thus, the state (B.39) reduces, for $t \to -\infty$, to a wave packet significantly different from zero only for $x \simeq -\frac{p_0}{m}|t|$. In other words, we are considering an incoming free state $e^{-i\frac{t}{\hbar}H_0}f_0$ that, for $t \to -\infty$, is localized far on the left of the origin and, as time increases, it moves to the right with positive mean momentum p_0.

Let us characterize for $t \to +\infty$ the outgoing free evolution

$$e^{-i\frac{t}{\hbar} H_0} S_\alpha f_0 . \tag{B.41}$$

Using Propositions 6.2 and B.4, we have

$$\left(e^{-i\frac{t}{\hbar} H_0} S_\alpha f_0\right)(x) \underset{t \to +\infty}{\simeq} e^{i\frac{mx^2}{2\hbar t}} \sqrt{\frac{m}{i\hbar t}} \left[\overline{\mathscr{T}\left(\frac{mx}{\hbar t}\right)} \tilde{h}\left(\frac{\sigma}{\hbar}\left(\frac{mx}{t} - p_0\right)\right) \right.$$
$$\left. + \overline{\mathscr{R}\left(\frac{mx}{\hbar t}\right)} \tilde{h}\left(\frac{\sigma}{\hbar}\left(\frac{mx}{t} + p_0\right)\right) \right]. \tag{B.42}$$

Formula (B.42) shows that the state (B.41) reduces, for $t \to +\infty$, to the sum of two wave packets, where the first one is significantly different from zero only for $x \simeq \frac{p_0}{m} t$ and the second one only for $x \simeq -\frac{p_0}{m} t$. In other words, the outgoing free state $e^{-i\frac{t}{\hbar} H_0} S_\alpha f_0$ emerging for $t \to +\infty$ is the sum of a term localized far on the right of the origin and moving to the right with mean momentum p_0 (transmitted wave) and a term localized far on the left of the origin and moving to the left with mean momentum $-p_0$ (reflected wave).

Note that we have obtained the same type of asymptotic behavior for $t \to +\infty$ found in Sect. 8.4 (see 8.75, 8.76, 8.77).

Moreover, making use of the scattering into cones theorem, we can also explicitly compute the transmission probability $\mathscr{P}(f_0, \mathbb{R}_+)$ and the reflection probability $\mathscr{P}(f_0, \mathbb{R}_-)$.

Proposition B.5 *For any positive integer n, there exists a positive constant c_n, independent of ε, such that*

$$\mathscr{P}(f_0, \mathbb{R}_+) = \int dk \, |\mathscr{T}(k)|^2 |\tilde{f}_0(k)|^2 + \mathscr{E}_{1,n} , \tag{B.43}$$

$$\mathscr{P}(f_0, \mathbb{R}_-) = \int dk \, |\mathscr{R}(k)|^2 |\tilde{f}_0(k)|^2 + \mathscr{E}_{2,n} , \tag{B.44}$$

where

$$|\mathscr{E}_{i,n}| < c_n \varepsilon^{2n} , \qquad i = 1, 2 . \tag{B.45}$$

Proof By Propositions B.1 and B.4, we have

$$\mathscr{P}(f_0, \mathbb{R}_+) = \int_0^\infty dk \, |(\widetilde{Sf_0})(k)|^2 = \int_0^\infty dk \, |\mathscr{T}(k)|^2 |\tilde{f}_0(k)|^2 + \int_0^\infty dk \, |\mathscr{R}(k)|^2 |\tilde{f}_0(-k)|^2$$
$$+ 2\Re \int_0^\infty dk \, \overline{\mathscr{T}(k)} \mathscr{R}(k) \tilde{f}_0(k) \overline{\tilde{f}_0(-k)} . \tag{B.46}$$

Note that $\overline{\mathscr{T}(k)}\mathscr{R}(k)$ is imaginary and $\tilde{f}_0(k)$ is real. Therefore, the last term in (B.46) is zero. Moreover, for any $k \geq 0$, we have

$$
\begin{aligned}
\tilde{f}_0(-k) &= \frac{\sqrt{\sigma}}{\sqrt{2\pi}} \int dy\, h(y)\, e^{i\,y(k\sigma + \varepsilon^{-1})} = \frac{\sqrt{\sigma}}{\sqrt{2\pi}} \frac{1}{i^n(k\sigma + \varepsilon^{-1})^n} \int dy\, h(y) \frac{d^n}{dy^n} e^{i\,y(k\sigma + \varepsilon^{-1})} \\
&= \frac{\sqrt{\sigma}}{\sqrt{2\pi}} \frac{(-1)^n}{i^n(k\sigma + \varepsilon^{-1})^n} \int dy\, h^{(n)}(y) e^{i\,y(k\sigma + \varepsilon^{-1})} .
\end{aligned}
\tag{B.47}
$$

Hence

$$
\begin{aligned}
\int_0^\infty dk\, |\mathscr{R}(k)|^2 |\tilde{f}_0(-k)|^2 &\leq \frac{\sigma}{2\pi} \|h^{(n)}\|_{L^1}^2 \int dk\, |\mathscr{R}(k)|^2\, \varepsilon^{2n} \\
&= \frac{\sigma|\alpha_0|}{4} \|h^{(n)}\|_{L^1}^2\, \varepsilon^{2n} .
\end{aligned}
\tag{B.48}
$$

The rest of the proof is left as an exercise.   □

References

1. Agmon, S.: Spectral properties of Schrödinger operators and scattering theory. Ann. Sc. Norm. Super. Pisa Cl. Sci. Ser. 2 IV, 151–218 (1975)
2. Amrein, W.O.: Hilbert Space Methods in Quantum Mechanics. EPFL Press, Lausanne (2009)
3. Reed M., Simon B.: Methods of Modern Mathematical Physics, III: Scattering Theory. Academic Press, New York (1979)
4. Simon, B.: Quantum Mechanics for Hamiltonians Defined as Quadratic Forms. Princeton University Press, Princeton (1971)

Index

© Springer International Publishing AG, part of Springer Nature 2018
A. Teta, *A Mathematical Primer on Quantum Mechanics*,
UNITEXT for Physics, https://doi.org/10.1007/978-3-319-77893-8

Printed in the United States
By Bookmasters